James L. White / *Helmut Potente*

Screw Extrusion

Polymer Processing Society

Progress in Polymer Processing

Series Editor: L.A. Utracki

A.I. Isayev
Modeling of Polymer Processing

L.A. Utracki
Two-Phase Polymer Systems

A. Singh / J. Silvermann
Radiation Processing of Polymers

Series Editor: W.E. Baker

I. Manas-Zloczower / Z. Tadmor
Mixing and Compounding of Polymers

T. Kanai / G.A. Campbell
Film Processing

R.S. Davé / A.C. Loos
Processing of Composites

Series Editor: K.S. Hyun

I.M. Ward / P.D. Coates / M.M. Dumoulin
Solid Phase Processing

W.E. Baker / C.E. Scott / G.-H. Hu
Reactive Polymer Blending

J.L. White / H. Potente
Screw Extrusion

James L. White / Helmut Potente (Editors)

Screw Extrusion

Science and Technology

With Contributions from

U. Berghaus, E. Bürkle, H. Potente, H. Recker, K. Schäfer, V. Schöppner,
J.L. White, G. Wiegand, M. Würtele

HANSER

Hanser Publishers, Munich

Hanser Gardner Publications, Inc., Cincinnati

The Editors:
Prof. Dr. James L. White, Institute of Polymer Engineering, College of Polymer Science and Polymer Engineering, University of Akron, Akron, OH 44325-0301, USA
Prof. Dr.-Ing. Helmut Potente, GH-Universität Paderborn, FB 10, 33098 Paderborn, Germany

Distributed in the USA and in Canada by
Hanser Gardner Publications, Inc.
6915 Valley Avenue, Cincinnati, Ohio 45244-3029, USA
Fax: (513) 527-8950
Phone: (513) 527-8977 or 1-800-950-8977
Internet: http://www.hansergardner.com

Distributed in all other countries by
Carl Hanser Verlag
Postfach 86 04 20, 81631 München, Germany
Fax: +49 (89) 98 12 64
Internet: http://www.hanser.de

Library of Congress Cataloging-in-Publication Data
White, James Lindsay, 1938-
Screw extrusion : science and technology / James L. White, Helmut Potente ; with contributions from U. berghaus ... [et al.].
    p.cm. – (Progress in polymer processing)
Includes bibliographical references and index.
ISBN 156990-317-4 (hardcover)
1. Screws—Design and construction. 2. Plastics—Extrusion. I. Potente, H. (Helmut) II. Berghaus, U. III. Title. IV. Series.

TJ1338. W47 2001
668.4'13—dc21   2001039877

Die Deutsche Bibliothek - CIP-Einheitsaufnahme
Screw extrusion : science and technology / James L. White/Helmut Potente. With contributions from U. Berghaus .... – Munich : Hanser; Cincinnati : Hanser Gardner, 2002
(Progress in polymer processing)
ISBN 3-446-19624-2

© Carl Hanser Verlag, Munich 2003
Project Management in the UK by Martha Kürzl, Stafford
Typeset in Germany by Kösel, Kempten
Coverdesign: MCP · Susanne Kraus GbR, Holzkirchen, Germany
Printed and bound in Germany by Kösel, Kempten

PROGRESS IN POLYMER PROCESSING SERIES

Kun Sup Hyun, *Series Editor*

Editorial Advisory Board

Foreword

Since the Second World War, the industry based on polymeric materials has developed rapidly and spread widely. Polymerization of new polymeric species advanced rapidly during the sixties and the seventies, providing a wide range of properties. A plethora of specialty polymers has followed as well, including many with particularly unique characteristics. This evolution has been invigorated by the implementation of metallocene catalyst technology. The end-use of these materials has depended on the development of new techniques and methods for forming, depositing, or locating these materials in advantageous ways, which are usually quite different from those used by the metal or glass fabricating industries. The importance of this activity, "Polymer Processing", is frequently underestimated when reflecting on the growth and success of the industry.

Polymer processes such as extrusion, injection molding, thermoforming, and casting provide parts and products with specific shapes and sizes. Furthermore, they must control, beneficially, many of the unusual and complex properties of these unique materials. Because of their high molecular weights and, in many cases, a tendency to crystallize, polymer processes are called upon to control the nature and extent of orientation and crystallization, which in turn have a substantial influence on the final performance of the products made. In some cases, these processes involve synthesizing polymers within a classical polymer processing operation, such as reactive extrusion. Pultrusion and reaction injection molding both synthesize the polymer and form a finished product or part all in one step. This is an evidence of the maturing of the industry. For these reasons, successful polymer process researchers and engineers must have a broad knowledge of fundamental principles and engineering solutions.

Some polymer processes have flourished in large industrial units, as for example, synthetic fiber spinning. However the bulk of the processes are rooted in small- and medium sized entrepreneurial enterprises in both developed and new developing countries. Their energy and ingenuity have sustained growth to this point but clearly the future will belong to those who progressively adapt new scientific knowledge and engineering principles, which can be applied to the industry. Mathematical modeling, online process control and product monitoring, and characterization based on the latest scientific techniques will be important tools in keeping these organizations competitive in the future.

The Polymer Processing Society was started in Akron, Ohio in 1985 with the aim of providing a focus, on an international scale, for the development, discussion, and dissemination of new and improved polymer processing technology. The Society facilitates this by sponsoring several conferences annually and by publishing the journal International Polymer Processing, and the volume series Progress in Polymer Processing. This series of texts is dedicated to the goal of bringing together the expertise of accomplished academic and industrial professionals. The volumes have a multi-authored format, which provides a broad picture of the volume topic viewed from the perspective of contributors from around the world. To accomplish these goals, we are fortunate in having the thoughtful insight and effort of our authors and volume editors, the critical overview of our Editorial Board, and the efficient production of our Publisher.

This volume deals with the science and technology of screw extrusion. Single screw extrusion and twin screw extrusion equipment are the workhorses of the plastics industry. All resins made in reactors are converted in these equipment into useful forms for preparing polymer products. These processes have developed into what is arguably the single largest outlet for enhancing products and formulated products. Understanding the historical perspective in the development of these machines will encourage innovative engineers to develop new type of equipment. Understanding the underlying principles of operation of these equipment will help engineers to optimize the equipment and enhance production rates and the properties of synthetic polymers. They are dependent on the best achievements in screw design to provide homogeneous melts for successful post-processing such as fibre spinning, blown film, cast film and sheets. Most important in this volume are the extensive discussions on the development of process understanding governing the extrusion processes, solid conveying, melting, and metering. This volume combines numerous contributions, industrial and academic, from Europe and North America and, as such, forms a very useful contribution to the plastics industry. This volume becomes the second volume under my editorship. Thanks are due to both Profs. White and Potente and the contributing authors. We will all enjoy the book.

Livingston, New Jersey   *Kun Sup Hyun*
U.S.A.   *Series Editor*

July 2002

Contents

Contributors

Dr.-Ing. Ulrich Berghaus, Bekum Maschinenfabrik GmbH, 12107 Berlin, Germany

Dr.-Ing. E. Bürkle, Krauss-Maffei Kunststofftechnik GmbH, 80973 Munich, Germany

Prof. Dr.-Ing. Helmut Potente, GH-Universität Paderborn, FB 10, 33098 Paderborn, Germany

Dipl.-Ing. Hans Recker, Institute of Plastics Processing (IKV), 52072 Aachen, Germany

Dr. K. Schäfer, Barmag AG, 42897 Remscheid-Lennep, Germany

Dr.-Ing. Volker Schöppner, Hella KG Hueck & Co., 59552 Lippstadt, Germany

Prof. James L. White, Institute of Polymer Engineering, College of Polymer Science & Polymer Engineering, University of Akron, Akron, OH 44325-0301, USA

Dr. G. Wiegand, B.W. & Partner, 69469 Weinheim, Germany

Dipl.-Ing. Martin Würtele, Krauss-Maffei Kunststofftechnik GmbH, 80973 Munich, Germany

Preface

Screw extruders are the most important of all polymer processing machines. They are used to pump, mix, carry out chemical reactions, and devolatilize polymer systems. Screw extrusion machines used in industry have both simple rotating screws and screws which both rotate and reciprocate. Screw extrusion machines often contain multiple (usually two) screws rather than single screws, which may rotate in the same or opposite directions. A wide variety of industrial products are made using screw extruders. Some may be classified according to shape e.g. film, pipe, profiles, coated wires, tire tread sidewalls, there are also glass and mineral filled compounds, blends and polymerized and chemical modified (e.g. grafted) polymers. Screw based injection molding machines are used to produce discrete products such as automotive components.

Screw extrusion technology naturally divides into (i) the design and behavior of screw based machines (ii) post-screw processing including dies, take-up equipment and molds. This book is limited to the first of these areas. From the 1950s, there has been a realization of the need for a comprehensive book on this subject. The first major book of this type was Schenkel's "Schneckenpressen für Kunststoffe" (1959) and its second edition "Kunststoff-Extrudertechnik" (1963). Other volumes in this period generally are limited to single screw thermoplastics extruders. The second was Rauwendaal's "Polymer Extrusion" which appeared in 1985. The third major book was Knappe and Potente's "Handbuch der Kunststoff-Extrusiontechnik I Grundlagen" in 1989. All of the above books are now out of date because of the advances over the past 15 years. This book is an original volume but it is also an outgrowth of the Knappe-Potente volume mentioned in the above paragraph to which it is most similar in character with its collective authorship. This volume is being published with the encouragement of the Polymer Processing Society and is part of its Progress in Polymer Processing (PPP) series.

James L. White
Helmut Potente

1   Introduction

James L. White

1.1   Overview

Screw extrusion machines have a prominent position in the polymer industries and have an important place in many other industries where they act in many roles. In the minerals industries, screw machines are used to convey solids, such as minerals, gravel, grain, and calcite. In the petroleum and marine industries, screw machines are used to pump viscous liquids under moderate to high pressure. The polymer industry uses screw extrusion machines (1) as plasticating pumps to produce melt streams to profile dies, (2) for continuous blending and compounding, and (3) as continuous chemical reactors and (4) as devolatilizers.

A survey of industrial usage of screw extruders reveals that single screw, twin screw, and multiple screw machines are all used commercially. Twin screw machines may involve co-rotating or counter-rotating screws, which may be separated, tangential, or intermeshed (which is most common). Intermeshed screws must have the "same hand" if they are co-rotating and be of "opposite hand" if they are counter-rotating. By "hand" we refer to the directions of the helix of the screw flight along the screw axis. The same situation exists with multiple screw, intermeshed machines. Three intermeshed, co-rotating screws will all have the same hand while three linear, intermeshed, counter-rotating machines involve a central screw of one-hand interacting with two screws of opposite hand

In Section 1.2 we describe the development of screw extrusion technology from its origins to the present state of the art. In Section 1.3, we summarize the books written on screw extrusion science and technology, discussing them in chronological order from when they were written.

1.2   Historical Development

1.2.1   Early Period

The concept of screw pumping of liquids dates to ancient times [1]. It is usually associated with the famous Greek scientist, Archimedes (287–212 B.C.), who used it in the defense of the city of Syracusa (on the island of Sicily). The screw is also an important simple machine. It was one of the five simple machines (together with the wheel and axle, the lever, the pulley and the wedge) known in ancient times that could move weights through distances. The screw press was probably invented in the first century B.C. It used heavy wooden screws in applications such as squeezing olives. Screw presses are found in paintings in Pompei and were discussed by ancient authors, such as Pliny the Elder and Hero of Alexandria.

The earliest metal screws used in machines were castings finished with a file or directly produced by a triangular file. Leonardo DaVinci in the 15th century left designs of a gear-based, screw cutting lathe in his notebooks [2]. Metal screws first came into existence in the 16th century. Heavy screws of bronze and iron seem to have had wide industrial usage from about 1550 primarily in mechanical applications [1].

The first industrial extrusion processes used rams to push material through reservoirs to which were attached shaping dies. Joseph Bramah in England seems to have made lead pipe in this manner in the 1790s. Foodstuffs such as pasta and ceramic materials were similarly extruded. Patents [3, 4] were issued in the mid 19th century for the extrusion of clay through dies to produce bricks and tiles. Pumping devices other than rams were also being used. An 1866 machine of Murray and Jennings [4] used a gear pump-like device.

R.A. Brooman [5] and H. Bewley [6] in England used ram extrusion machines to produce the first continuous filaments and tubes of a thermoplastic polymeric material in the 1840s. The thermoplastic used was gutta percha (*trans*-1,4-polyisoprene), a naturally occurring crystalline resin that was imported from Malaya. During the same period, C. Hancock [7] devised a ram extrusion-based wire-coating process that placed an electrically insulating gutta percha layer on copper wires.

Processing technologies based on single screw pumping devices begin to appear in abundance in the 1870s [8–11]. In an 1871 U.S. Patent, J.D. Sturgis [8] of Chicago, Illinois, USA describes a machine consisting of a screw rotating in a tube that is used for the cooling, conveying, and mixing of soap. In an 1877 U.S. Patent, W.H. Higbie [9] of Peoria, Illinois, USA describes a screw conveying machine for the movement and drying of grain. The earliest application of screw pumping machines to the polymer industry is reported to be by the *Harburger Gummiwerke* (today the *Phoenix Gummiwerke AG* of Hamburg–Harburg) in 1873 [10]. In an 1879 patent application, Matthew Gray [11] of London, England described a screw extrusion machine for the extrusion of rubber compounds and gutta percha through extrusion shaping dies. Gray cites specific application to wire coating for the purpose of insulating electrical wires.

It is at this time that machinery companies began to commercially manufacture screw extruders. The first entrepreneurs in this business seem to have been Francis Shaw in Manchester, England and John Royle in Passaic, New Jersey, USA in 1879–1880 [12]. Paul Troester began to manufacture single screw extruders in Hannover, Germany in 1892 [13].

Hermann Berstorff subsequently began to produce screw extruders in Hannover in the same period.

The first twin screw machines occur in this period. An 1869 patent by Francois Coignet [14] of Paris, France notes intermeshing, co-rotating and counter-rotating twin screw extrusion machines and then specifically advocates an intermeshing co-rotating machine for pumping artificial stone paste. A fully intermeshing counter-rotating twin screw extruder was described in an 1874 patent by S. Lloyd Wiegand [15] of Philadelphia, Pennsylvania, USA. It is intended for pumping baking dough through profile dies. It is clear from Wiegand's patent that he fully understood that two fully intermeshed, counter-rotating screws, unlike a single screw, act as a positive displacement pump and provide a more uniform throughput. A self-wiping, co-rotating twin screw extruder was described in a 12 September, 1901 patent application [16] by Adolf Wunsche of Charlottenburg (Berlin), Germany.

1.2.2   1920 to 1945

In the 1920s, screw pumps began to be used commercially to pump viscous oils. These were used in ships, refineries, and other applications. Both single screw pumps and fully intermeshed, counter-rotating twin screw pumps were used from this period [17–20]. One of the early manufacturers was *Maschinenfabrik Paul Leistritz GmbH* of Nuremburg. Screw pumps with three or more intermeshing, counter-rotating screws were introduced [20].

The earliest efforts to analyze flow in a single screw extruder date to the 1920s when engineers, notably in England, sought to characterize single screw pumps for viscous oils [17, 18, 21]. It is in these papers that one first sees simple Newtonian fluid mechanical analyses of flow in a screw channel presuming the superposition of a forward drag flow and a backward pressure flow. R.H. Pearsall [17] seems to have been the first to model flow in a single screw pump.

Screw extruders were also being used in the food industry. A 1930 patent application by F.B. Anderson [21] of the *V.D. Anderson Company* of Cleveland, Ohio, USA, describes a pin barrel extruder for sausage stuffing. The pins were to agitate and knead the materials to remove air.

The earliest screw extruders in the polymer industry were primarily used for gutta percha and natural rubber. They were steam heated and often (especially for natural rubber) had short length/diameter ratios (3 to 5). Rubber was preheated on a mill before being added. *Paul Troester Maschinenfabrik GmbH* developed long *L/D*, cold-feed rubber extruders but they were not successful. With the coming of thermoplastics in the 1920s and 1930s, steam-heated, short *L/D* machines were seen to be inadequate. In 1935, H. Heidrich [22, 23] developed an electrically heated single screw extruder. Subsequently, H. Decker and H. Theyson of *Paul Troester Maschinenfabrik* devised the first commercial fully electrically heated single screw extruder [22, 24]. *Troester* began marketing these machines from 1940.

In the mid 1930s the *I.G. Farbenindustrie* and *Maschinenfabrik Paul Leistritz GmbH* developed intermeshing, counter-rotating twin screw kneading pumps in a joint program [10, 25–30]. These machines, which had special screw design to force the kneading of the material being processed, were intended to continuously masticate as well as pump suspensions and polymer systems. They were used by the *I.G. Farbenindustrie* on coal-oil dispersions [31]. In 1941 *Friederich Krupp* began to market an intermeshing counter-rotating twin-rotor continuous kneader with the tradename the *Knetwolf*. [10, 25–30, 32]. It was used for mastication of synthetic rubber and was marketed to the *I.G. Farbenindustrie*. *Maschinenfabrik Paul Leistritz,* in the early 1940s built, intermeshing, counter-rotating twin screw extruders for polyvinyl chloride with screw diameters of 300 mm.

The first commercial intermeshing, co-rotating twin screw extruders were developed in the same period. In 1938, R. Colombo and *Lavoriazone Materie Plastiche* [33] in Turin, Italy introduced an intermeshing, largely self-wiping, co-rotating twin screw extruder for extrusion of thermosetting resins and polyvinyl chloride. Machines were sold to the *I.G. Farbenindustrie*.

The late 1930s saw the development of a new concept in screw extrusion: a screw that is able to have an axial reciprocating motion as well as rotation. In a 1939 French patent application, H.P.M. Quillery [34] described a screw injection molding machine to produce vulcanized rubber parts. Rubber compound accumulated from the rotating screw is injected into the mold by a sudden forward axial motion of the screw.

An independent invention of the screw injection molding machine for thermoplastics was made by H. Beck [35] of the *I.G. Farbenindustrie* in a 1943 German patent application. The inventions of Quillery and Beck began a new screw injection molding technology that has become increasingly important in postwar years.

A second key invention associated with axially moving screws is the Kokneter of H. List [36]. The machine was first described in a 1945 Swiss patent application but had been conceived earlier. It contains a barrel with pins and a screw with slice in the flights. The oscillation of the screw allows the pins to continuously wipe the screw flights. This machine was marketed by *Buss AG* in Europe and initially by *Baker Perkins* in the United States as a continuous mixing machine.

1.2.3   1946 to 1959

At the end of World War II, screw extrusion and screw injection molding were seen to be two rapidly rising industries of the future. In Germany there is the story of Hans and Friederich Reifenhäuser, who from 1948 rebuilt a blacksmith shop in Troisdorf into one of the leading manufacturers of screw extruders in Europe *(Reifenhäuser KG)*. H. Heidrich helped them in the early years. Werner Battenfeld also returned to his family business in Meinerzhagen and rapidly became one of the leading producers of screw injection molding machines. In Japan, well-established machinery companies, such as *Japan Steel Works* [38] and *Ikegai Corporation* [39], began to manufacture screw extruders.

In the post-World War II period, the processing of thermoplastics in single screw extruders was thought to be of such importance that thermoplastics manufacturers around the world turned their attention to developing a scientific understanding of this processing machine. The first paper was published by A.M. Rogowsky (later Z. Rigbi) of *ICI* in the United Kingdom [40] in 1947. The subject was taken up in 1951 by W.T. Pigott of *The Goodyear Tire and Rubber Company* in the United States [41] by W. Meskat and K. Riess of *Farbenfabriken Bayer* (now *Bayer AG*) in Germany [42–44], and, most importantly, in

1953 by J.F. Carley, R.S. Mallouk, J.M. McKelvey, and W.D. Mohr of *Dupont* in the United States [45–48]. These papers essentially rediscovered the screw pump fluid mechanics of the 1920s [17, 18] and applied it to the screw extrusion of thermoplastics. Also of importance in this period was the 1952 doctoral dissertation of C. Maillefer at the University of Lausanne in Switzerland [49, 50]. Here we find the first treatment of the conveying of solid pellets and its relationship to pressure development in a screw extruder.

The period follwing the end of the war saw the rapid rise of a counter-rotating twin screw extruder industry. In the mid 1940s, *Maschinenfabrik Paul Leistritz GmbH* had produced very large (300 mm screw) intermeshing, counter-rotating twin screw extruders for extruding polyvinyl chloride profiles, but had discontinued this business following the war. In the late 1940s and early 1950s, many companies entered this industry, such as. *Wilhelm Anger* (Austria), *Gerhardt Kestermann Zahnrader und Maschinenfabrik* (Germany), *Krauss-Maffei AG* (Germany), *Mapre SA* (Luxemburg), *Schloemann AG* (Germany), and *Trudex* (France) [51–53]. These machines were used primarily for extruding polyvinyl chloride profiles. In the United States, *Welding Engineers* developed a tangential, counter-rotating twin screw extruder of modular construction that was to be used for continuous mixing and devolatilization [54, 55]. This began the development of a new industry for counter-rotating twin screw extruders.

Following the war, R. Colombo and *LMP's* co-rotating twin screw extruder was patented internationally [56] and licensed to *CAFL* (Clextral) in France, *R.H. Windsor* in Britain, and *Ikegai Corporation* in Japan. It was used around the world for profile extrusion. In the late 1940s, R. Erdmenger, W. Meskat, and others with (now) *Bayer AG* devised various intermeshing, co-rotating twin screw machines, including kneading disk blocks [57] and modular machines [58, 59], for continuous mastication and mixing. Modular machines with long *L/D* ratios were used by *Bayer AG* for reactive extrusion compounding and devoltilization. In the late 1950s, this machine was licensed to *Werner and Pfleiderer GmbH*, who began to manufacture and market it to the industry [10, 25, 26, 60]. The early machines were designated ZSK: System Erdmenger. *Baker-Perkins* also began to manufacture this type of machine in the United States.

1.2.4   1960 to the Present

The decade of the 1960s saw the first major efforts to understand and model the nonisothermal and melting characteristics of a single screw extruder. Simulations of flow in a screw channel of non-Newtonian fluids with viscous dissipation heating appeared [61, 62]. In 1959, Maddock [63] of *Union Carbide* in the USA published the first experimental studies of melting in a single screw extruder. In 1966, Z. Tadmor [64] of the *Western Electric Company* published the first model of the melting process in a single screw extruder. Researchers subsequently with *Western Electric* sought to develop a complete model for a single screw extruder from hopper to die. They combined melt metering, melting, and solid convey models. This model and its comparison with experiment is described in papers by Klein and Marshall [65] and Tadmor et al [66]. Subsequently, Klein and Tadmor resigned from *Western Electric* and formed a new firm, *Scientific Process Research*, which developed computer software for the design of single screw extruders.

The 1960s were a period of new designs of single screw extruders. The most important was the barrier screw of C. Mailieffer [67] of *Mailliefer SA,* which sought to control the solid bed during the melting process. Maillefer's patent led to new screw designs in both the thermoplastics and rubber industry as well as to many legal battles and to patents with alternative designs.

The use of grooves in the barrel of the feed zone of a thermoplastics screw extruder seems to have been introduced by Kautex in Germany in the 1960s [68]. In the late 1960s, during a joint research program an extrusion blow molding of high-density polyethylene involving Kautex and BASF the true values of grooves in the feed zone was realized.

The same period saw the introduction of various new twin screw machines, which competed with the *Buss* Kokneter, the *Werner and Pfleiderer* modular co-rotating twin screw extruder, and the *Welding Engineers* tangential, counter-rotating twin screw extruder. In the 1960s, *Farrel Corporation* introduced a continuous mixer consisting of two separated, machined, counter-rotating rotors containing screw and mixing rotor sections [10, 13, 69, 70]. This machine was originally intended for pelletized or powdered rubber but was found useful for thermoplastics compounding. Similar machines came to be made by *Japan Steel Works*, and under Farrel license by *Kobe Steel,* and by *Pomini SpA. Krauss-Maffei* AG introduced a self-wiping, co-rotating twin-rotor mixer with machined screw and mixing rotor sections [71] which was licensed to *Japan Steel Works. Maschinenfabrik Paul Leistritz GmbH* introduced modular, intermeshing, counter-rotating twin screw extruders for compounding thermoplastics [72].

From the late 1960s, there were major breakthroughs in the design of screw extruders for the rubber industry. New cold-feed rubber extruder screws were devised by G. Menges, J.P. Lehnen and E.G. Harms of the RWTH Aachen and *Uniroyal Englebert Deutschland AG.* The rubber industry had long been dominated by hot-feed (from two roll mills), short length/diameter (*L/D*) screw extruders. As noted earlier, *Paul Troester Maschinenfabrik* had developed cold-feed, long *L/D* screws in the 1930s but they did not produce acceptable extrudates. In the late 1960s, Menges and Lehnen [73, 74] developed various "homogenizing" sections for cold-feed rubber screws. However, significant success came only in the 1970s with the development of the pin barrel extruder for cold-feed rubber extrusion by Menges and Harms [75, 76]. *Paul Troester Maschinenfabrik,* which had worked with Menges and Harms, were the first licensee, and many other companies around the world, including *Berstorff, Farrel, Nakatazoki,* and *Pomini,*became licensees.

Modular, co-rotating twin screw extruders for compounding came to be manufactured by many machinery companies as the original patents expired. In Germany, *Berstorff* and *Leistritz* entered the market. In Japan, *Japan Steel Works, Ikegai Corp., Toshiba Machine,* and *Kobe Steel* became manufacturers. Today more than sixty companies around the world manufacture these machines.

Screw extrusion machines have continued to undergo development. This is most notably seen in recent years through computer-based control systems and increased screw speeds and torques, notably in twin screw extruders.

1.3   Earlier Books on Screw Extrusion

The first monograph on extrusion was published in 1941 by Dr. Ing. Hanns Decker of *Paul Troester Maschinenfabrik* of Hannover–Wulfel, Germany. It was entitled *Die Spritzmaschine (Strangpresse Schneckenpresse): Theoretisch Grundlagen und Ausführung* [77]. It was published by *Troester* and 53 pages long. The monograph discusses flow in screw channels, heating systems, dies, and single screw extrusion technology. Its publication is clearly associated with the rapid growth of the thermoplastics industry and the introduction of electrical heating for thermoplastics screw extruders by Troester.

In 1958, E.G. Fisher published *Extrusion of Plastics* [78], on behalf of the Plastics Institute (of Great Britain). The book is 113 pages long, and discusses in reasonable detail the technology of the time. The emphasis is on single screw extruders, but attention is given to both intermeshing, counter-rotating and co-rotating twin screw machines. The fundamental fluid mechanisms of Newtonian flow in single screw pumps is presented.

The third book on screw extrusion was published in 1959 by Dr.-Ing. Gerhard Schenkel, also of *Paul Troester Maschinenfabrik*. The volume was originally intended as an updating of Decker's *Die Spritzmaschine*. Schenkel's volume became much more than this. It was entitled *Schneckenpressen für Kunststoffe* [79] and was published by *Carl Hanser Verlag* of Munich. A second edition appeared in 1963 with the title *Kunststoff-Extrudertechnik*. This volume is 540 pages long [80]. The Schenkel books emphasize extrusion technology, although the simple Newtonian fluid theory of screw extrusion is discussed. Both single screw and twin screw technology and science is developed.

In 1959, Ernest C. Bernhardt of E.I. DuPont deNemours in Wilmington, Deleware, USA edited a volume entitled *Processing of Thermoplastic Materials* [81], which was sponsored by the Society of Plastics Engineers. The book contains 10 chapters, one of which (Chapter 4) deals with extrusion. This chapter, which is 153 pages long (three times the length of Decker's monograph, longer than Fisher's monograph, but one-third the size of Schenkel's book), is written by five engineers at *DuPont:* J.B. Paton, P.H. Squires, W.H. Darnell, F.M. Cash, and J.F. Carley. The chapter in some part summarized the research efforts of DuPont engineers in the 1950s to understand flow in single screw extruders.

In 1962, James M. McKelvey, who had resigned from *DuPont* and had taken a professorship at Washingon University in St. Louis, Missouri, USA, published a pedagogically oriented advanced textbook entitled *Polymer Processing* [82]. The book is 409 pages long and the chapters on screw extrusion and dies are 124 pages long. This book emphasizes flow fundamentals and neglects technology.

With the publication of the books of Schenkel, Bernhardt, and McKelvey, certain glaring differences between American and German books became clear. American books presented sophisticated mathematical analyses of the screw extrusion process and presented greatly oversimplified technologies. German books present a much more realistic view of existing technology but much more simplified fluid mechanics. As an example, Schenkel's volumes present a solid view of the twin screw extrusion technology of its time. The Bernhardt and McKelvey books do not mention twin screw extrusion.

In 1970, Zehev Tadmor and Imrich Klein of *Scientific Process and Research Inc.* published *Engineering Principles of Plasticating Extrusion* [83]. The book was 500 pages long and devoted to the modeling theories of metering, melting, and solid conveying. It discusses in detail the composite single screw extruder developed by the authors and their coworkers during the previous decade. There is minimal discussion of extrusion technology.

In 1970, Roger T. Fenner of Imperial College of the University of London published *Extruder Screw Design* [84]. This volume, which is based on his PhD dissertation at Imperial College, is 281 pages long. It is devoted to simulation of the metering region in a single screw extruder. Indeed, analyses of solid conveying and melting are missing. There is no discussion of extrusion technology.

The next important book appeared in 1972 by Heinz Herrmann of *Werner and Pfleiderer GmbH* and was entitled *Schneckenmaschinen in der Verfahrenstechnik* [10]. This 179-page monograph describes the historical development of extrusion technology and emphasizes modular twin screw extrusion and its applications. Co-rotating and counter-rotating twin screw machines and the Buss Kokneter are discussed. There is no discussion of fluid mechanical modeling.

In 1974, Willi Dalhoff published *Systematische Extruderkonstruktion* [85]. This monograph was written while the author was Ober-ingenieur at the Institut für Kunststoffverarbeitung at the RWTH Aachen. This 193-page book is a very practical treatment of the detailed mechanical design of a single screw extruder. There is no discussion of modeling.

In 1978, Leon P.B.M. Janssen published *Twin Screw Extrusion* [86]. This book largely represents his doctoral research activities at the Technological University of Delft (Netherlands) on intermeshing, counter-rotating extruders intended for profile extrusion. There is minimal discussion of technology and the emphasis is on the fluid mechanics of pumping and leakage in intermeshing, counter-rotating geometries.

In 1979, Zehev Tadmor (now at the Israel Institute of Technology) and Costas Gogos (Stevens Institute of Technology) published a 736-page textbook entitled *Principles of Polymer Processing* [87]. About 300 pages are devoted to screw and die extrusion. The approach is pedagogical and the book aimed at American graduate student audiences.

Also in 1979, Walter Michaeli of *Freudenberg* published a monograph on die extrusion entitled *Extrusion-Werkzeuge* [88]. This book discusses both technology and flow modeling. A second edition appeared in 1991 [89] by which time Michaeli had become professor at the RWTH Aachen. Both editions were translated in to English [90]. There is no consideration of flow in screws.

In 1981, Helmut Potente of the University of Paderborn (Germany) published a 154-page monograph entitled *Auslegen von Schneckenmaschinen-Baureihen* [91]. The book did not broadly treat screw extrusion technology and science, but was limited to consideration of scale-up.

In 1986, Chris Rauwendaal, a graduate of the University of Twente in the Netherlands, then with *Raychem Corp.* in Menlo Park, California, USA, wrote an extensive book called *Polymer Extrusion* [92]. It is 568 pages long and covers all aspects of screw extrusion technology and flow analysis. It devotes 46 pages to twin screw extrusion and 15 pages to die design. Subsequent editions of this book are essentially reprints of the 1986 edition.

Three years later, in 1989, a two-volume treatise on screw extrusion technology and science was edited by Friedhelm Hensen of *Barmag Barmer Maschinenfabrik*, Reimsheid, Germany, Werner Knappe of the Montan Universitat Leoben, Leoben, Austria, and Helmut Potente of the University of Paderborn [93]. The first volume, *Grundlagen*, has 582 pages. It treats both fundamental flow inside extruder screws and the technology of screw extruders. The second volume, which was translated into English as *Plastics Extrusion Technology* [94], largely treats die and post-die processing. The two volumes, especially the latter, have a heavy emphasis on technology.

In 1990 James L. White of the University of Akron, Akron, Ohio, USA published a 295-page monograph, *Twin Screw Extrusion: Technology and Principles* [26]. This book describes the technology and fluid mechanics for intermeshing, co-rotating, tangential, counter-rotating, and intermeshing, counter-rotating twin screw extruders.

In 1993, Chris Rauwendaal, now heading *Rauwendaal Extrusion Engineering* of Los Altos Hills, California, USA published an edited monograph, *Mixing in Polymer Process-ing* [95]. The monograph is 475 pages long, of which 279 pages relate to extrusion technol-ogy as applied to mixing.

A 1994 volume in the Polymer Processing Society's Progress in Polymer Processing series, edited by Ica Manas-Zloczower (Case Western Reserve University) and Zehev Tadmor (now President of the Israel Institute of Technology), was entitled *Mixing and Compounding in Polymer Processing* [96] of its 867 pages, roughly 330 pages were devoted to continuous extrusion machines.

In 1995, a volume by J.L. White (University of Akron) entitled *Rubber Processing: Technology, Materials and Principles* [97] was published. Of its 586 pages, 190 pages involved screw extrusion or continuous mixing of rubber and dies.

References

1. Usher, A.B., *A History of Mechanical Inventions*, Harvard University Press, 1st ed. (1929), revised ed. (1954), Dover edition (1988)
2. Richter, J.P., *The Notebooks of Leonardo DaVinci* Sampson et al., London (1880), reprinted by Dover (1970)
3. Wright, S. B., Green, H.T., English Patent, 1626 (1855)
4. Murray, C.H., Jennings, M., English Patent, 1057 (1866)
5. Brooman, R.A., English Patent, 10,582 (1845)
6. Bewley, H., English Patent, 10,825 (1845)
7. Lawford, R.A., Nicholson, L.R., *The Telecon Story* (1950) Telegraph Construction and Maintenance Co., Ltd., Fanfare Press, London
8. Sturgis, J.D., U.S. Patent (filed 25 April 1871), 114,063 (1871)
9. Higbie, W.H., U.S. Patent (filed 30 October 1876) 192,069 (1877)
10. Herrmann, H., *Schneckenmaschinen in der Verfahrenstechnik* (1972) Springer, Berlin
11. Gray, M., British Patent (filed 10 December 1879) 5056 (1880)
12. Schenkel, G., in *Kunststoffe, ein Werkstoff macht Karriere*. Glenz W. (Ed.) (1985) Carl Hanser Verlag, Munich
13. White, J.L., *Int. Polym. Process* (1992) 7, p. 110
14. Coignet, F., U.S. Patent 93, 035 (1869)
15. Wiegand, S.L., U.S. Patent (filed 28 April 1874) 155, 602 (1879)
16. Wunsche, A., German Patent (filed 12 September, 1901) 131, 392 (1902)
17. Pearsall, R.H., *The Automobile Engineer* (1924) p. 145
18. Anonymous, *Engineering* (1926) 114, p. 606; Rowell, H.S., Finlayson, D., *Engineering* (1928) 118, p. 249
19. Blau, E., Chem.-*Ztg.* (1930) 54, p. 801
20. Montelius, C., *Teknisk Tidskrift* November (1933) 6, p. 61
21. Anderson, F.B., U. S. Patent (filed 14 November 1930) 1,848,236 (1932)
22. Schenkel, G., *Int. Polym. Process* (1988) 3, p. 3
23. Heidrich, H., German Patent (filed 4 November 1936) 757,255 (1953)

24. Decker, H., German Patent (filed 26 July 1939) 735, 000 (1943)
25. Herrmann, H., In *Kunststoffe, ein Werkstoff macht Karriere.* Glenz W. (Ed.) (1985) Carl Hanser Verlag, Munich
26. White, J.L., *Twin Screw Extrusion: Technology and Principles* (1990) Carl Hanser Verlag, Munich
27. Kiesskalt, S., Tampke, H., Winnacker, K., Weingaertner, E., German Patent (filed 26 July 1935) 652,990 (1931)
28. Leistritz, P., Burghauser F., German Patent (filed 1 December 1935) 682,787 (1939)
29. Kiesskalt. S., *VDI Z.* (1942) 86, p. 752; *Kunststoffe* (1951) 41, p. 414
30. White, J.L., *Int. Polym. Process* (1993) 8, p. 286
31. Kiesskalt, S., Tampke, H., Winnacker, K., Weingaertner, E., German Patent (filed 25 May 1935) 676,045 (1939)
32. Anonymous, German Patent (filed 31 January 1941) 750,509 (1944)
33. Colombo. R., Italian Patent (filed 6 February1939) 370,578 (1939)
34. Quillery, H.P.M., French Patent (filed 8 February 1939) 858,310 (1952)
35. Beck, H., German Patent (filed 16 December 1943) 858,310 (1952)
36. List, H., Swiss Patent (filed 20 August 1945) 247,704 (1947); U.S. Patent (filed 19 August 1946) 2,505,125 (1950); *Kunstoffe* (1950) 40,185
37. White, J.L., *Int. Polym. Process* (1996) 11, p. 2
38. White, J.L., *Int. Polym. Process* (1992) 7, p. 194
39. White, J.L., *Int. Polym. Process* (1995) 10, p. 194
40. Rogowsky, Z., *Proc. Inst. Mech. Eng.* (1947) 156, p. 56
41. Pigott, W. T., *Trans. ASME* (1951) 73, p. 947
42. Meskat, W., *Kunststoffe* (1951) 41, p. 417
43. Meskat, W., *Kunststoffe* (1955) 45,87
44. Riess, K., Meskat, W., *Chem. Ing. Tech.* (1951) 23, p. 205
45. Carley, J.F., Mallouk, R.S., McKelvey, J.M., *Ind. Eng. Chem.* (1953) 45, p. 974
46. McKelvey, J.M., *Ind. Eng. Chem.* (1953) 45, p. 982
47. Carley, J.F., McKelvey, J.M., *Ind. Eng. Chem.* (1953) 45, p. 989
48. Mohr, W.D., Mallouk, R.S., *Ind. Eng. Chem.* (1959) 51, 765
49. Maillefer, C., *British Plastics* (1954) p. 394
50. Maillefer, C., *British Plastics* (1954) p. 437
51. Schaerer, A.J., *Kunststoffe* (1954) 44, p. 105
52. Baigent, K., *Trans. Plastics Inst.* (1956) 134
53. Shultz, F.C., *SPE J.* (1962) p. 1162
54. Street, L.F., *India Rubber World* (1950) 123, p. 58
55. Fuller, L.J., U.S. Patent (field 15 May 1995) 2,615, 199 (1952)
56. Colombo, R., U.S. Patent (filed 7August 1947) 2,563,396 (1951)
57. Erdmenger, R., German Patent (filed 29 September 1949) 813,154 (1951): German Patent (filed 28 July 1953) 940,109 (1956)
58. Meskat, W., Pawlowski, J., German Patent (filed 10 December 1959) 949,162 (1956)
59. Erdmenger, R., U.S. Patent (filed 17 August 1959) 3,122,356 (1964); *Chem. Ing. Tech.* (1964) 36, p. 175
60. Fritsch, R., Fahr, G., *Kunststoffe* (1959) 49, p. 543
61. Griffith, R.M., *IEC Fund* (1962) 1, p. 180
62. Zamodits, H., Pearson, J.R.A., *Trans. Soc. Rheol.* (1969) 13, p. 357
63. Maddock, B.H., *SPE J.* (1959) p. 383
64. Tadmor, Z., *Polym. Eng. Sci.* (1966) 6, p. 185
65. Klein, I., Marshall, D.I., *Polym. Eng. Sci.* (1966) 6 p. 191
66. Tadmor, Z., Duvdevani, I., Klein, I., *Polym. Eng. Sci.*(1967) 7, p. 198
67. Maillefer, C., Swiss Patent (filed 31 December 1959) 363,149 (1962)
68. Boes, A., Kramer, A., Lohrbacher, V., Schnelders, A., *Kunststoffe* (1990) 80, p. 659
69. Ahlefeld, E.H., Baldwin, A.J., Hold, P., Rapetski, W.A., Scharer, H.R., U.S. Patent (filed 15 May 1962) 3,154,808 (1964)
70. Hold, P., *Adv. Polym. Technol.* (1984) 4, p. 281
71. Proksch, W., *Kunststoffe und Gummi* (1964) 3, p. 476
72. Tenner, H., *Kunststoffberater* (1976) 6

73. Menges, G., Lehnen, J.P., *Plastverarbeiter* (1969) 20, p. 31; Lehnen, J.P., *Kunststofftechnik* (1970) 9, pp. 3,90,114,198

74. Lehnen, J.P., Menges, G., Harms, E. G., U.S. Patent (filed 5 March 1970) 3,652,064 (1972); U.S. Patent (filed 16 July 1970) 3,680,844 (1972)

75. Menges, G., Harms, E.G., *Kautschuk Gummi Kunststoffe* (1972) 25, p. 469; (1979) 27, p. 187

76. Harms, E.G., Menges, G., Hegele, R., German Offenlegungsschrift (filed 21 July 1972) 2,235,784 (1984); U.S. Patent (filed 30 January 1978) 4,178,104 (1979); U.S. Patent (filed 21 December 1978) 4,199,263 (1980)

77. Decker, H., *Die Spritzmaschine (Strangpresse, Schneckenpresse): Theorie und Ausführung* (1941) Troester, Hannover-Wulfel

78. Fisher, E.G., *Extrusion of Plastics* (1958) The Plastics Institute, Iliffe, London

79. Schenkel, G., *Schneckenpressen für Kunststoffe* (1959) Carl Hanser Verlag, Munich

80. Schenkel, G., *Kunststoffe-Extrudertechnik* 2d ed. (1963) Carl Hanser Verlag, Munich

81. Bernhardt, E.C., Ed., *Processing of Thermoplastic Materials* (1959) van Nostrand-Reinhold, (1972) New York

82. McKelvey, J.M., *Polymer Processing*, Wiley, New York (1962)

83. Tadmor, Z., Klein, I., *Engineering Principles of Plasticating Extrusion* (1970) Reinhold, New York

84. Fenner, R.T., *Extruder Screw Design* (1970) Illfe, London

85. Dalhoff, W., Systematische Extruder-Konstruktion (1974) Krausskopf, Mainz

86. Janssen, L.B.P.M., *Twin Screw Extrusion* (1978) Elsevier, Amsterdam (1978)

87. Tadmor, Z., Gogos, C.G., *Principles of Polymer Processing* (1979) Wiley, New York

88. Michaeli, W., *Extrusion-Werkzeuge fur Kunststoffe und Kautschuk*, (1979) Carl Hanser Verlag, Munich

89. Michaeli, W., *Extrusion-Werkzeuge für Kunststoffe und Kautschuk* 2d ed. (1991) Carl Hanser Verlag, Munich

90. Michaeli, W., *Extrusion Dies for Plastics and Rubber* (1992) Carl Hanser Verlag, Munich

91. Potente, H., *Auslegen von Schneckenmaschinen Baureihen* (1981) Carl Hanser Verlag, Munich

92. Rauwendaal, C., *Polymer Extrusion* (1986) Carl Hanser Verlag, Munich

93. Hensen, F., Knappe, W., Potente, H., (Eds.) *Kunststoffe Extrusionstechnik* Vol. I, Grundlagen; Vol. II (1988–9) Carl Hanser Verlag, Munich

94. Hensen, F., (Ed.) Plastics Extrusion Technology (1997) Carl Hanser Verlag, Munich

95. Rauwendaal, C., (Ed.) *Mixing in Polymer Processing* (1991) Dekker, New York

96. Manas Zloczower, I., Tadmor, Z., (Eds.) *Mixing and Compounding in Polymer Processing* Vol. 4. In *Progress in Polymer Processing* Series (1994) Carl Hanser Verlag, Munich

97. White, J. L., Rubber Processing: Technology, Materials and Principles (1995) Carl Hanser Verlag, Munich

2 Fundamentals

James L. White

2.1 Introduction

Our purpose in this chapter is to present an overview of extrusion machines and pumping mechanisms. We first describe the basic mechanisms of pumping that have been applied to viscous fluids, including both screw and non screw machines. We then turn to the basic characteristics of screws. Subsequently, we give an overview of the unique characteristics of single screw and the various types of twin screw and multiple screw machines.

The general broad perspective of this chapter has been given in some part in earlier treatises of extrusion, notably the volumes of Schenkel [1], Rauwendaal [2] and Knappe and Potente [3].

2.2 Pumping Mechanisms

2.2.1 General

Many mechanical devices have been invented to pump liquids. These include a wide range of mechanical designs. Some of these devices, such as centrifugal pumps, are suitable only for gases and low-viscosity liquids. Highly viscous liquids are usually pumped by two different mechanisms: (1) positive displacement pumps, in which fluid enters enclosed chambers and is moved forward by the mechanical movement of the solid parts of the machine; and (2) drag flow pumps, in which the fluid enters a region between two surfaces, one of which is in motion. The relative movement of the two surfaces drags the fluid along a channel, gradually pressurizing it and forced it through a die. Other types of pumping machines also exist for viscous fluids. In this section, we describe these various types of machines.

2.2.2   Positive Displacement Pumps

2.2.2.1   Ram Extruders

Positive displacement pumps include a wide range of mechanical machines. The simpliest of positive displacement pumps is the piston pump, which consists of a cylinder containing a liquid to be pumped. Into the top of this cyclinder is placed a piston, which is moved along the cylinder axis and pressurizes the liquid. The pressurized liquid is pushed by the piston through an orifice or die placed at its base. This type of machine was the first extruder used in the polymer industry, being applied by R.A. Brooman (4) in 1845 to extrude gutta percha filament (Fig. 2.1). Machines of this type are generally called ram extruders in the polymer industry.

The piston pump or ram extruder has the disadvantage that it is a batch operation. After the liquid in the cylinder has been extruded through the die, it is necessary to remove the piston and add more liquid (or material to be liquefied). There have been efforts to develop continuous flow ram extruders for thermoplastics. Most notable are the inventions by Westover [5] (Fig. 2.2) and Yi and Fenner [6] (Fig. 2.3).

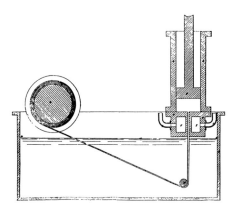

Figure 2.1   Positive displacement piston pump Brooman's 1845 ram extruder

In the Westover machine, which is described in a 1963 paper [5], four plungers interact with a shuttle valve system. Two feed hoppers on opposite ends of the machine connect to two plasticating ram cylinder systems. These plasticating ram cylinder systems melt the polymer and push it past valves into the die head system. Here two additional

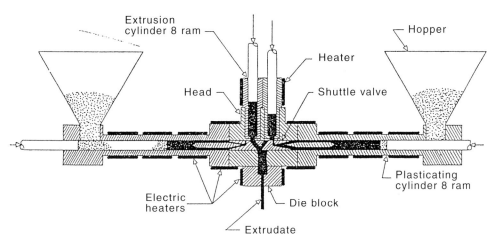

Figure 2.2   Schematic of main parts of Westover's continuous-flow ram extruder

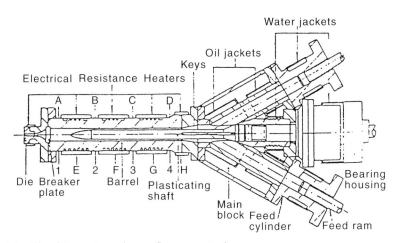

Figure 2.3   Yi and Fenner's continuous flow ram extruder

rams push the polymer melt out through the dies. In the Yi-Fenner machine [6], two recip-rocating rams feed a barrel containing a rotating shaft that plasticates and melts the ther-moplastic.

2.2.2.2   Rotary Positive Displacement Pumps

More practical positive displacement pumps for viscous liquids involve machines with pairs of intermeshing, counter-rotating rotors, such as those shown in Fig. 2.4 [7–9]. Liquid fed into these machines moves into chambers between the two rotors and the pump walls. These pumps have throughputs given by

$$Q = 2NV_c \tag{2.1}$$

where Q is the volumetric throughput, N is the rotor rotation rate (rpm), and V_c is the volume of the filled chamber that moves from the entrance to the exit. These equations represent the behavior of various types of lobe pumps as well as gear pumps. Schmidt's [8] cam pump (Fig. 2.4a) has two open volumes: one in contact with the inlet and the other in contact with the outlet. His second design (Fig. 2.4b) has two similar volumes plus a third enclosed volume. *Gebr. Pintsch's* pump [7] has four lobes and three to four enclosed volumes.

Perhaps the most important of the two intermeshing, counter-rotating rotor positive displacement pumps is the gear pump (Fig. 2.5). Gear pumps are also discussed in the patent literature [10]. The output of a gear pump is given by

$$Q = 2NV_c = 2NnV_t \tag{2.2}$$

where V_c is the volume of empty chambers around the circumference of a gear, n is the number such chambers, and V_t is the individual volume between gear teeth.

Another class of intermeshing counter-rotating, rotor positive displacement pumps is fully intermeshing counter-rotating twin screw pumps [11–16]. Several different designs are shown in Fig. 2.6. The output of such a machine is again

$$Q = 2NiV_c \tag{2.3}$$

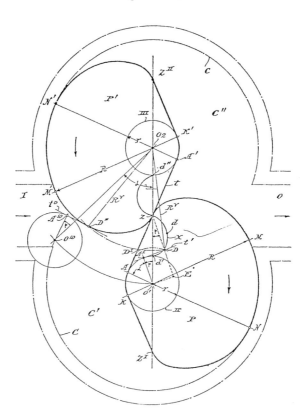

Figure 2.4a   Intermeshing counter-rotating twin rotor pumps: Schmidt cam pump design

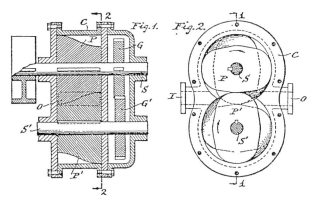

Figure 2.4b   Intermeshing counter-rotating twin rotor pumps: Schmidt design

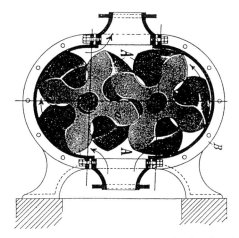

Figure 2.4c   Intermeshing counter-rotating twin rotor pumps: Gebrüder Pintsch design

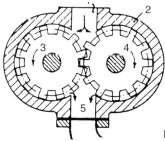

Figure 2.5   Gear pump

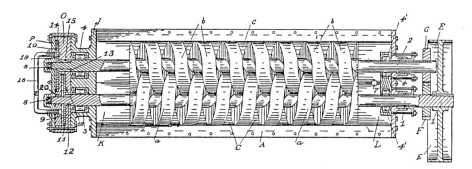

Figure 2.6a   Intermeshing counter-rotating twin screw pumps: Holdaway design

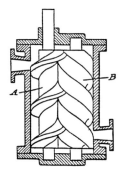

Figure 2.6b   Intermeshing counter-rotating twin screw pumps: Montelius design

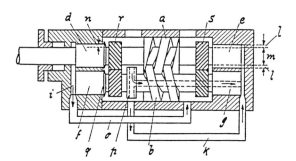

Figure 2.6c   Intermeshing counter-rotating twin screw pumps: Leistritz-Burghauser design

Here i is the number of thread starts and V_c is the volume of a C-chamber. We return to fully intermeshing, counter-rotating twin screw machines in Section 2.5.2.

Fully intermeshed, counter-rotating *multiscrew* pumps are also positive displacement pumps. Examples of these are shown in Fig. 2.7. The output of such a machine is

$$Q = N(\sum_{j} i_j V_{cj})$$

(2.4)

where i_j is the number thread starts and V_{cj} the volume of the C-chamber of the screw j.

(a)

(b)

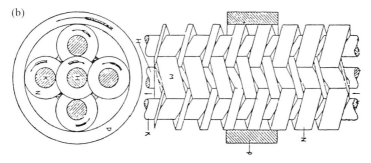

Figure 2.7   Intermeshing counter-rotating multi screw pumps

None of the above machines is a perfect positive displacement pump. There must always be clearances between moving parts and the pressurized viscous fluids being pumped will leak back through them.

2.2.3   Drag Flow Pumps

There are a wide range of drag flow machines that can pump viscous fluids. The simplest machine of this type is the drum flow pump or drum extruder shown in Fig. 2.8 [17, 18]. This very fundamental drag flow machine seems to have been invented by Gabrielli [17] in 1952. A similar extruder is described in a later patent by Beck, Robbins, and Birdsall [18]. Here the material to be pumped is introduced into an annular space between the rotating drum and a surrounding barrel. The rotation of the drum carries the liquid to a position where there is a wiper bar and the entrance to a die from which the liquid exits the machines. The wiper bar diverts the liquid into the die. The die pressurizes the liquid and a pressure gradient develops along the length of the channel between the drum and the barrel.

It is readily possible to present a simple flow model of this machine. If the drum has a linear velocity U, and there is a uniform clearance H between the drum and the barrel, which has length W, the ideal output of the drag flow pump is

$$Q = (1/2)HWU \qquad (2.5)$$

(a)

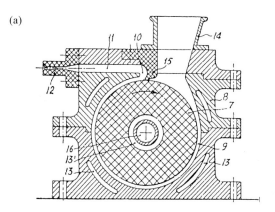

(b)

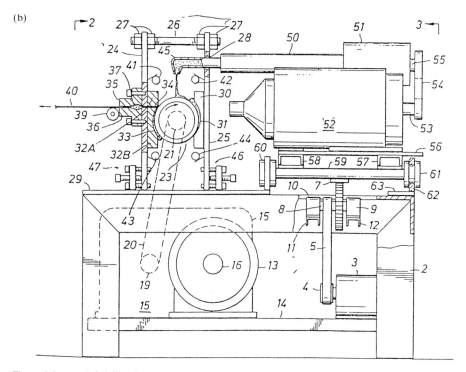

Figure 2.8   (a) Gabrielli's drum extruder; (b) Beck, Robbins and Birdsall's drum extruder

The occurrence of a die at the end of the drag flow pump induces backflows along the annular space, which for a Newtonian fluid are proportional to the pressure and inversely proportional to the shear viscosity:

$$Q = (1/2)HWU - Q_{back} \tag{2.6a}$$

$$Q = (1/2)HWU - \frac{K\Delta p}{\eta} \tag{2.6b}$$

The single screw extruder is a similar drag flow pump in which the fluid being pumped is dragged along the screw helix. Eq. 2.6 again represents the flow, but it is directed along the helix and not around a drum. The fluid is dragged against the die entrance and pressurization occurs. This pressurization leads to backflow along the screw channel, as in the drum flow pump. The single screw pump is of such importance that we devote Section 2.3 to the geometry of screws and Section 2.4 to screw pump principles and design features.

Drag flow extruders have been developed in which a screw spiral is cut into the surface of a disk [19, 20] (Fig. 2.9). The earliest such machine comes from an anonymous German patent application in March 1944 [19], which did not issue until 1954. A second patent was applied for by Keune in 1949 [20] (when the German patent office was reopened after World War II) and issued in 1951. Since both patents emphasize wire coating, Keune might also be the author of the 1944 application.

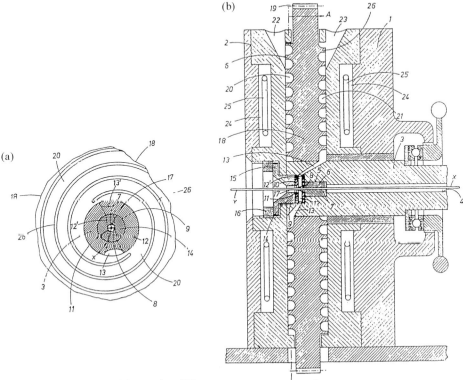

Figure 2.9   Screw spiral disk extruder of Keune

Other types of drag flow pumps based on rotating disks have been proposed. One type of machine is described in a 1962 paper by Westover [21]. This is the slider pad or stepped disk extruder (Fig. 2.10). This machine contains a stepped disk positioned a small distance from a flat disk. The flat disk is rotated and as it turns melt is moved into a section of smaller clearance where it is pressurized. When the pressurized melt traverses into the next depressed region it flows through slots out of the extruder.

Still another type of drag flow pump and processing machine, the diskpack, was developed by Tadmor et al. [22, 23] in the late 1970s (Fig. 2.11). A series of stationary, interconnected, hollow disks are placed on smooth rotating shaft. If we introduce a coordinate system into the surface of the rotating shaft, it will be seen that each hollow disk has three dragging surfaces, not one as in a drum extruder, a single screw extruder, or the disk extruders of the previous paragraphs. The shapes of the disks can be modified to allow for solid conveying, melting, melt conveying, mixing, or devolatilization. The diskpack has been commercially developed by *Farrel Corporation*.

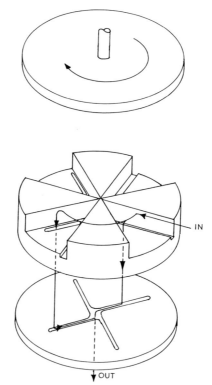

Figure 2.10   Westover's rotary slider pad pump

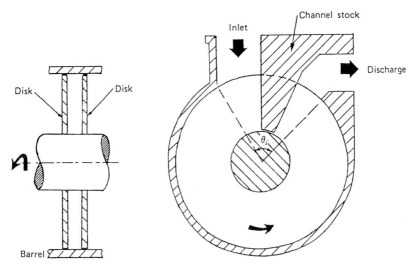

Figure 2.11   Tadmor Diskpack extruder

2.2.4   Normal Stress Pumps

Maxwell and Scalora [24] have proposed that the normal stress effect discovered by Weissenberg [25] for viscoelastic fluids can be used as the basis of a pump. We show in Fig. 2.12a the normal stress effect as described by Weissenberg [25] and in Fig. 2.12b Maxwell and Scalora's normal stress based pump. In this machine a viscoelastic fluid (or thermoplastic pellets to be melted) are added at the outer edge of a rotating disk. The fluid when sheared develops both shearing stresses and normal stresses along and perpendicular to the streamlines. The normal stresses act as tensions along the streamlines. The normal stresses along the streamlines may be shown to have through a force balance a radial inward component. This drives the viscoelastic fluid as shown in Figure 2.12a up the stirring rod and in Figure 2.12b radially inward and into the die. The normal stress pump has been commercialized by *Custom Scientific* as the Mini-Max molder, a laboratory device for molding small quantities of polymers.

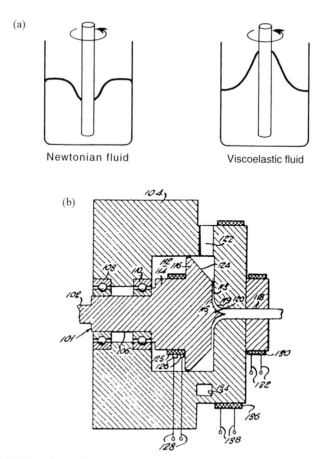

(a)

Newtonian fluid   Viscoelastic fluid

(b)

Figure 2.12   (a) Weissenberg effect for viscoelastic fluid; (b) Maxwell and Scalora's normal stress pump

2.3   Specifications of Screws

The major method of pumping highly viscous fluids involves using screws rotating in barrels. The geometry of a screw located within a barrel is shown in Fig. 2.13. The internal barrel diameter is D_B. The diameter of the root of the screw is D_S and the distance between the root of the screw and the internal surface of the barrel is taken as H. The screw has a helical flight running along its length. The radial clearance between the crest of the screw flight and the barrel is δ_F. The value of H may vary between the flights, due to a curvature of the screw root. It follows that

$$D_B = D_S(z) + 2H(z) \tag{2.7a}$$

At the positions of the screw flights,

$$D_B = D_S + 2\delta_F \tag{2.7b}$$

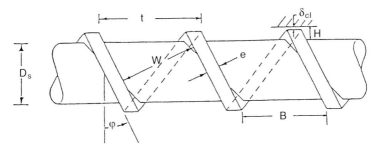

Figure 2.13   Geometry of a screw with a single thread start

The axial distance associated with one full turn of the screw flight is the screw lead or pitch, which is defined as S. The axial distance between the screw flights is B and the perpendicular distance between the flights is W. The flight thickness is designated e.

The angle of the screw helix is ϕ. The screw pitch is related to the helix angle through

$$S = 2\pi r \tan \phi \tag{2.8}$$

Since the screw pitch is independent of radius, the ϕ angle must be radius dependent, $\phi(r)$. Thus, at the screw root

$$S = \pi D_S \tan \phi_S \tag{2.9a}$$

and at the barrel

$$S = \pi D_B \tan \phi_B \tag{2.9b}$$

Angle ϕ decreases as one proceeds from the screw root to the screw barrel.

Many geometric relationships exist between the quantities defined above. The pitch S is related to B and e through

$$S = B + \frac{e}{\cos \phi} \tag{2.10}$$

The channel width W is related to B through

$$W = B \cos \phi \tag{2.11}$$

and to the pitch S through substituting Eq. 2.11 into Eq. 2.10:

$$S = \frac{W}{\cos \phi} + \frac{e}{\cos \phi} \tag{2.12a}$$

and

$$W = \left(S - \frac{e}{\cos \phi} \right) \cos \phi \tag{2.12b}$$

The relationships given above are for screws with a single lead (Fig. 2.13). It is possible for screws to possess multiple leads or multiple thread starts. If a screw has multiple leads, fluid added to it will flow forward along multiple parallel channels. A screw with a double lead (double flighted) is shown in Fig. 2.14a. A screw with three leads (triple flighted) is shown in Fig. 2.14b. The relationships developed in the first part of this section need to be modified for screws with multiple leads. It should be clear from Fig. 2.14 that for a screw with m leads

$$S = m\left(B + \frac{e}{\cos\phi}\right)$$
(2.13a)

$$W = \frac{S}{m}\cos\phi - e$$
(2.13b)

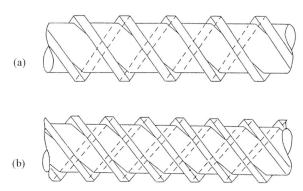

(a)

(b)

Figure 2.14   (a) Geometry of a screw with two thread starts; (b) geometry of a screw with three thread starts

Screws have mirror images that are different from each other as one's left hand differs from his right hand (Fig. 2.15). If a rotating right-hand screw drags fluid toward a screw pump exit, a similarly rotating left-handed screw will drag fluid toward the pump entrance (see Section 2.4.1)

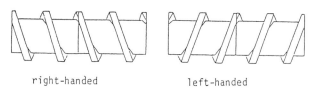

right-handed   left-handed

Figure 2.15   Right-handed and left-handed screws

2.4   Single Screw Pumps

2.4.1   Principles

The single screw pump is, as we have noted, a drag flow pump and basically operates on the same principles as described in Section 2.2.2.2. The liquid between the screw flights adheres to both the barrel and the screw. The rotation of the screw relative to the barrel drags the liquid along the channel formed by the screw flights. By tradition, a right-handed screw drags liquid from the entrance of the pump to its exit and a left-handed screw drags liquid toward the entrance.

The flow rate of liquid along the screw channel may be experessed analytically in terms of the screw geometry described in Section 2.3. First, we think in terms of a flattened-out screw and barrel and erect a coordinate system 123 in the root of the channel of the rotating screw in which 1 is along the channel defined by the flights, 2 is along the radius of the screw shaft, and 3 is transverse to the screw flights (Fig. 2.16). In this coordinate system the screw is stationary and the barrel moves with velocity

$$\mathbf{U} = U_1 \mathbf{e}_1 + U_3 \mathbf{e}_3 \tag{2.14}$$

where the velocity U_1 along the screw channel is positive and the transverse velocity U_3 is negative. U_1 and U_3 may be written in terms of the screw rotation rate and screw diameter as

$$U_1 = \pi DN \cos \phi \qquad U_3 = -\pi DN \sin \phi \tag{2.15}$$

where ϕ is the helix angle of the screw.

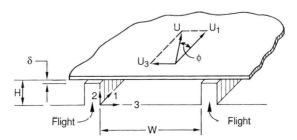

Figure 2.16   Coordinate system in screws channel

To a first approximation the flow through a screw channel is drag flow with a linear velocity profile. If we neglect the drag of the screw root and flights, it follows that for a screw with a single lead

$$Q = (1/2)HWU_1 = (1/2)\pi HWDN \cos \phi \tag{2.16a}$$

and if there are m thread starts

$$Q = m \left[(1/2)\pi HWDN \cos \phi \right] \tag{2.16b}$$

The discussion above neglects the occurrence of backflows induced by pressure gradients in the screw channel. It is possible, however, to describe these backflows and to devise full velocity profiles in the screw radius or 2 direction through setting up force balances involving flow in the 1 and 3 directions. A shear plane across the screw channel at position x_2 is introduced. The shear stress on the barrel is σ_b with components σ_{b1} in the 1 direction and σ_{b3} in the 3 direction. The shear stress on the screw flight is σ_{f1} in the 1 direction. These shear stresses are balanced by pressure gradients. The 1 and 3 force balances have the form

$$W\left(H - x_2\right)dp = W\sigma_{b1}dx_1 + W\sigma_{12}\left(x_2\right)dx_1 + 2\int_{X_2}^{H}\sigma_{f1}dx_1dx'_2 \tag{2.17a}$$

$$L\left(H - x_2\right)dp = L\sigma_{b3}dx_3 + L\sigma_{32}\left(x_2\right)dx_3 \tag{2.17b}$$

where differential slices of thicknesses dx_1 and dx_3 are used and σ_{12} and σ_{32} are shear stresses at positions x_2 in the screw channel in the 1 and 3 directions. L is a length along the screw flights. If friction on the screw flight walls is neglected, Eqs. 2.17a and 2.17b are equivalent to

$$\sigma_{12}\left(x_2\right) = \left(H - x_2\right)\frac{\partial p}{\partial x_1} - \sigma_{bl} \tag{2.18a}$$

$$\sigma_{32}\left(x_2\right) = \left(H - x_2\right)\frac{\partial p}{\partial x_3} - \sigma_{b3} \tag{2.18b}$$

If we presume that the fluid being pumped is Newtonian, the shear stress is proportional to the shear rate and we may write

$$\sigma_{12} = \eta\frac{\partial v_1}{\partial x_2} \tag{2.19a}$$

$$\sigma_{32} = \eta\frac{\partial v_3}{\partial x_2} \tag{2.19b}$$

Substitution of Eqs. 2.19a and 2.19b into Eqs. 2.18a and 2.18b leads to ordinary differential equations for the velocity components v_1 and v_3 as a function of x_2. These equations may be solved using Eq. (2.15) as boundary conditions, which gives

$$v_1(x_2) = U_1\frac{x_2}{h} - \frac{H^2}{2\eta}\frac{\partial p}{\partial x_1}\left[\frac{x_2}{H} - \left(\frac{x_2}{H}\right)^2\right] \tag{2.20a}$$

$$v_3(x_2) = U_3\frac{x_2}{h} - \frac{H^2}{2\eta}\frac{\partial p}{\partial x_3}\left[\frac{x_2}{H} - \left(\frac{x_2}{H}\right)^2\right] \tag{2.20b}$$

In arriving at Eq. 2.20, we have treated σ_{b1} and σ_{b3} as unknown constants and eliminated them to satisfy boundary conditions. This a shown in Fig. 2.17a

The flow rate along the screw channels is

$$Q = W \int_0^H v_1 dx_2 = \frac{U_1 HW}{2} - \frac{H^3 W}{12\eta} \frac{\partial p}{\partial x_1} = \frac{\pi DHWN \cos \varphi}{2} - \frac{H^3 W}{12\eta} \frac{\partial p}{\partial x_1} \tag{2.21}$$

The drag of the barrel due to the rotation of the screw moves the polymer melt forward along the screw channel between the flights.

The velocity component perpendicular to the screw flights, $v_3(x_2)$, is also of interest. Note that

$$\int_0^H v_3 dx_2 = \frac{U_3 H}{2} - \frac{H^3}{12\eta} \frac{\partial p}{\partial x_3} = 0 \tag{2.22a}$$

where

$$U_3 = -\pi DN \sin \phi \tag{2.22b}$$

There is a pressure gradient across the screw channel given by

$$\frac{\partial p}{\partial x_3} = -\frac{12}{H^2} \pi DN \sin \varphi \tag{2.23}$$

The pressure is a maximum at the leading screw flight and then decreases in the normal direction away from the flight in a linear fashion. This results in a velocity field of form

$$v_3(x_2) = -\pi DN \sin \phi \left[3\left(\frac{x_2}{H}\right)^2 - 2\frac{x_2}{H} \right] \tag{2.24}$$

At large x_2/H, the fluid is dragged in a negative direction along the barrel. At small x_2/H, there is a positive pressure. The result is a circulating flow, shown in Fig. 2.17b.

The velocity field in the down-channel and cross-channel directions are shown in Fig. 2.17. The down-channel flow is dominated by forward drag. The transverse motion is circulatory. The melt flows in a helical manner in the screw channel. This is, of course, superposed on the flow screw helix.

The primary results of this section, Eqs. 2.20 to 2.24, are well known and were developed in the literature by various authors in the period 1925 to 1959 [2, 3].

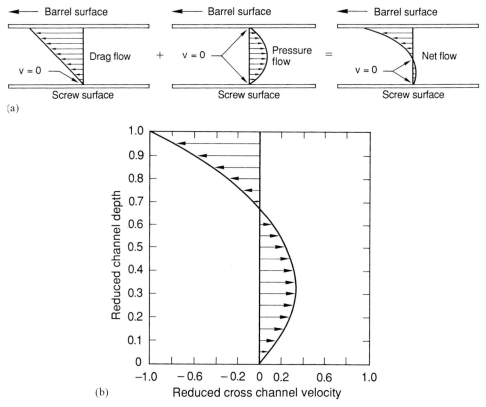

Figure 2.17   Velocity field in a single screw pump: (a) down-channel velocity field in a single screw pump; (b) cross-channel velocity field in a single screw pump

2.4.2   Simple Screw Design Features

It is possible to interpret the influence of screw design variables such as channel depth, width, and helix angle on fluid/melt characteristics such as pressurization using Eq. 2.21, which we may write as

$$Q = \frac{1}{2}\pi^2 D^2 NH \cos\phi \, \sin\phi - \frac{\pi H^3 D}{12\eta}\frac{\partial p}{\partial z}\sin^2\phi \tag{2.25}$$

where we have written W as $\pi D \sin\phi$ and x_1 as $z / \sin\phi$, where z is the screw axis direction.

Let us analyze Eq. 2.25. First, a large viscosity η reduces backflow and makes the pump more efficient. We next consider the influence of screw channel depth H. Clearly, if H is decreased the backflow leakage must fall off more rapidly than the forward drag flow. Thus, small H makes a more effective pump.

In the absence of pressure flow, the maximum output is obtained at $\phi = 45°$. This follows from $\cos\phi \, \sin\phi$ being $(1/2) \sin 2\phi$. Including pressure flow reduces the value of the optimum ϕ.

2.5 Counter-Rotating Twin Screw Machine

2.5.1 Tangential

Tangential counter-rotating (Fig. 2.18) twin screw extruders as envisaged in the work of Fuller [26] are essentially drag flow pumps with the drag being supplied by the fraction of the screw circumference that is in contact with the barrel. These screw machines can have matched screw flight or staggered screw flight configurations (Fig. 2.18).

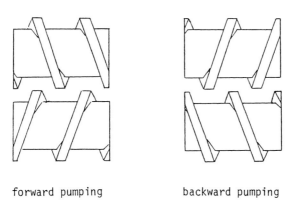

forward pumping   backward pumping

Figure 2.18   Tangential counter-rotating twin screw pump: (a) matched flights; (b) staggered flights

The concept of this flow mechanism was developed by Kaplan and Tadmor [27], whom we follow here. Essentially between the screw and the barrel, we may write, as in a single screw extruder,

$$Q = \frac{U_1 H W}{2} - \frac{H^3 W}{12\eta}\frac{\Delta p_I}{L_I}$$

(2.26)

where Δp_I and L_I refer to pressure drop and channel length in contrast with the barrel. Between the screws (region II) we have only pressure flow for matched screw flights

$$Q = -\frac{H^3 W}{3\eta}\frac{\Delta p_{II}}{L_{II}}$$

(2.27)

where we have considered that in this region the channel dimensions $2H$ by W. The total pressure drop per revolution is

$$\Delta p = \Delta p_I + \Delta p_{II}$$

(2.28)

These machines can have matched screw flight or staggered screw flight configurations (Fig. 2.18). Eq. 2.28 leads to

$$Q = \frac{1}{2}WHU_1\left(\frac{4f}{3f+1}\right) - \frac{WH^3}{12\eta}\frac{\Delta p}{L_{\mathrm{I}}+L_{\mathrm{II}}}\left(\frac{4}{3f+1}\right) \qquad (2.29)$$

where

$$f = \frac{L_{\mathrm{I}}}{L_{\mathrm{I}}+L_{\mathrm{II}}} \qquad (2.30)$$

Since the quantity f is less than unity, drag flow is reduced and backward pressure flow for each screw is increased.

The total flow Q_T for the tangential twin screw extruder encompasses expressions such as Eq. 2.29 for each screw and a backward interscrew leakage, specifically

$$Q_T = 2Q - Q_{leak} \qquad (2.31)$$

or, using Eq. 2.14,

$$Q_T = WH\pi DN\cos\phi\left(\frac{4f}{3f+1}\right) - \frac{WH^3}{6\eta}\frac{\Delta p}{L_{\mathrm{I}}+L_{\mathrm{II}}}\left(\frac{4}{3f+1}\right) - Q_{leak} \qquad (2.32)$$

Here Q_{leak} is the backward flow through the triangular region between the screws. Kaplan and Tadmor estimate Q_{leak} as the backward pressure flow through a uniform triangular section:

$$Q_{leak} = \frac{W_T H_T S}{12\eta}\left(\frac{\Delta p}{\Delta L}\right)_{fl} \qquad (2.33)$$

where W_T is the base of the triangle, H_T, is the height of the triangle, $(\Delta p/\Delta L)_{fl}$ is the pressure gradient flight to flight along the screw axis, and S is a shade factor.

The derivation cited above is valid only for matched screw flights. If the screw flights on the two screws are staggered relative to each other, the behavior will change. Most strikingly, the triangular cross- section through which backward leakage increases in size and the pumping characteristics become worse. Analytical models for flow in staggered flight geometry are given by Rauwendaal [2] and included in the study of White and Adewale [28]. There have also been a number of numerical simulations. This subject is developed in White [29] and in Section 6.4.

2.5.2   Intermeshed

As described in Section 2.3.2, the fully intermeshing, counter-rotating twin screw extruder is a positive displacement pump. The machine consists of alternating thick flights and screw channels on the two parallel opposite screws. As the screws rotate, the open channels, which have C shapes, may be observed to move forward along the screw axis (Fig. 2.19). The volume of material in these C-chambers is pumped forward by the rotation of the screws to the die.

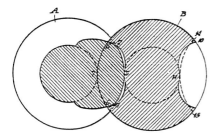

Figure 2.19   Intermeshing counter-rotating twin screw pump and C-chamber; after Montelius [16]

The output of the machine was discussed by Kiesskalt [30] and Montelius [16] in the 1920s and in various books [1–3, 29, 31] since that time. It is given by Eq. 2.3, which we repeat here:

$$Q = 2NiV_c \qquad (2.3)$$

where i represents the number of thread starts and V_c the volume of the C-chamber. The volume of the C-chamber is to a first approximation:

$$V_c = \frac{\pi DHW_m}{\cos \phi_m} \qquad (2.34)$$

where W_m is the mean channel width and ϕ_m is the mean flight angles.

The fluid mechanics of intermeshing, counter-rotating twin screw extruders has generally been simulated based on a model where the coordinate system is fixed on the surface of a screw translating with a C-chamber [1–3, 29–33]. The C-chamber translates along the screw axis at a velocity

$$U_n = SN = \pi \, DN \tan \phi \qquad (2.35)$$

The equations of motion for a Newtonian fluid may readily solved for a Newtonian fluid in a C-chamber under conditions where transverse shearing by the walls of the screw channel is neglected. The 1 and 3 velocity components parallel and transverse to the screw channel directions are of the same form as Eq. 2.20a,b, except for consideration of the translating coordinate system. Specifically, these are

$$v_1(x_2) = -\frac{\pi DN}{\cos \phi} + \pi DN \cos \phi \left(\frac{x_2}{H} \right) - \frac{H^2}{2\eta} \frac{\partial p}{\partial x_1} \left[\frac{x_2}{H} - \left(\frac{x_2}{H} \right)^2 \right] \qquad (2.36a)$$

$$v_3(x_2) = -\pi DN \sin \phi \left(\frac{x_2}{H} \right) - \frac{H^2}{2\eta} \frac{\partial p}{\partial x_3} \left[\frac{x_2}{H} - \left(\frac{x_2}{H} \right)^2 \right] \qquad (2.36b)$$

The total flux in the C-chamber-based coordinate system is zero, so that

$$\int_0^H v_1 dx_2 = 0 \tag{2.37a}$$

$$0 = \frac{\pi DN}{\cos \varphi} HW + \frac{1}{2} \pi DN \cos \phi HW - \frac{WH^3}{12\eta} \frac{dp}{dx_1} \tag{2.37b}$$

The pressure gradient in the C-chamber is

$$\frac{dp}{dx_1} = -\frac{6\eta\pi DN}{H^2} \left(\frac{2}{\cos \phi} - \cos \phi \right) < 0 \tag{2.38}$$

The pressure is the highest at the back of the C-chamber and the lowest in the front.

The pressurized liquid in the C-chamber tends to leak back between the machine clearances. Doboczky [32] and later Janssen et al. [31, 33] and most recently White and Adewale [28] sought to characterize the various leakage flows. Most important is backflow through the calendering gap.

We return to the intermeshing, counter-rotating twin screw extruder in Section 6.3.

2.6   Co-Rotating Twin Screw Machine

The co-rotating twin screw extruder used in modern technology is, as explained in Section 1.2, a modular, intermeshing, self-wiping machine. The basic self-wiping, co-rotating screw machine is shown in Fig. 2.20. This figure is taken from a patent by Wunsche [34] who first discussed such a machine. In the self-wiping, co-rotating twin screw extruder, the channels are open. Forward pumping operates on the principle of drag flow. As described by Erdmenger [35], the pioneer of the modern machine, the fluid moves forward along the screw channel in a figure-eight pattern.

As the screw channel traces the barrel along its path, we may write, if we neglect transverse shearing, the forward velocity profile $v_1 (x_2, x_3)$ in the form of Eq. 2.20a:

$$v_1(x_2, x_3) = U_1 \frac{x_2}{H} - \frac{H^2}{2\eta} \frac{\partial p}{\partial x_1} \left[\frac{x_2}{H} - \left(\frac{x_2}{H} \right)^2 \right] \tag{2.39}$$

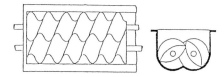

Figure 2.20   Self-wiping, co-rotating twin screw pump

The flux q_1 is

$$q_1 = \int_0^H v_1 dx_2 = \frac{1}{2}U_1 H - \frac{H^3}{12\eta}\frac{\partial p}{\partial x_1} \tag{2.40}$$

The total flow Q is

$$Q = \int_0^{W_{max}} q_1 dx_3 = \frac{1}{2}U_1 A - \frac{1}{12\eta}\left(\int H^3 dx_3\right)\frac{\partial p}{\partial x_1} \tag{2.41}$$

The basic mechanism of flow in a co-rotating twin screw extruder was perhaps first discussed by Erdmenger [35] and more especially Herrmann and Burkhardt [36], who presented a simple fluid mechanical analysis of the flow. The co-rotating twin screw machine is reviewed by Rauwendaal [2] and more especially by White [29].

It should be noted that the above analysis is for a uniform fully filled machine and in practice co-rotating twin screw extruders are operated under starved conditions and are modular in construction. This will be addressed in Chapter 6.

We return to this machine in Section 6.2.

2.7   Multiple Screw Extrusion

While twin screw extrusion machines have become important, there has also been considerable activity through the years on multiple screw extrusion machines. Most attention to multiple screw machines has been for intermeshing, counter-rotating twin screw machines (see Fig. 2.21), which are all positive displacement pumps. An explicit discussion of multiple fully intermeshing, counter-rotating screw pumps is contained in a 1925 patent application of Montelius [14] and in a 1933 paper [37] by the same author. These pumps were clearly being manufactured commercially by the 1930s. The output of a multiple fully intermeshing counter-rotating twin screw machine is clearly given by Eq. 1. Fig. 2.21 shows three screw and five screw intermeshing, counter-rotating screw pumps [38, 39].

Self-wiping co-rotating, multiple screw pumps have been described by Colombo [40] and Meskat and Erdmenger [41] among others (see Fig. 2.22). These are drag flow pumps. Multiple screw machines involving both co-rotating and counter-rotating screws have also been proposed. A four screw machine involving two pairs of self-wiping, co-rotating screws that are tangentially counter-rotating with each other was proposed by Erdmenger and Oetke [42] (Fig. 2.23). This machine was used for devolatilization and operates by drag flow.

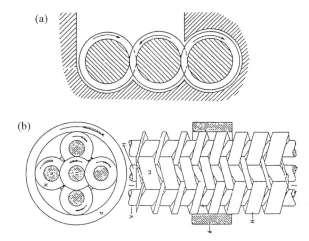

Figure 2.21   Multiple screw, intermeshing, counter-rotating screw pumps: (a) Three screws; (b) Five screws

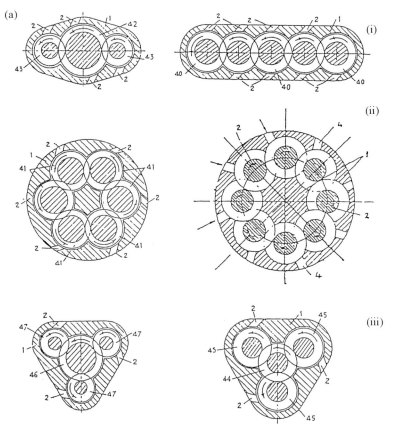

Figure 2.22a   Self-wiping, co-rotating multiple screw pumps: multiple co-rotating screw extruders after Colombo [40] indicating (i) 3- and 5-screw linear arrangements, (ii) 6- and 8-screw circumferential arrangements, and (iii) 4-screw arrangement with contact 65 spars

(b)

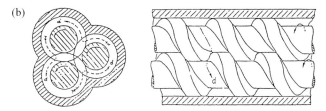

Figure 2.22b   V-shaped 3 co-rotating screw arrangement after Meskat and Erdmenger [41]

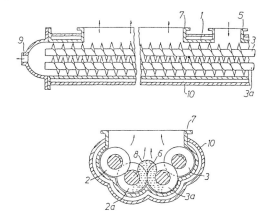

Figure 2.23   Four-screw machine of Erdmenger and Oetke [42] showing two pairs of self wiping, co-rotating twin screws that rotate in a counter-rotating manner tangential to each other

References

1. Schenkel. G., *Schneckenpresse für Kunststoffe* and *Kunststoff-Extrudertechnik* (1959, 1963) Carl Hanser Verlag, Munich
2. Rauwendaal, C., *Polymer Extrusion* (1986) Carl Hanser Verlag, Munich
3. Knappe, W., Potente, H., (Eds.) *Kunststoff-Extrusionstechnik* (1989) *I. Grundlagen,* Carl Hanser Verlag, Munich
4. Brooman, R.A., English Patent 10,582 (1845)
5. Westover, R.F., *Mod Plastics* (1963) March, p. 130
6. Yi, B., Fenner, R.T., *Plast. Polym.* (1975) p. 224
7. Gasapparate- und Maschinenfabrik Gebrüder Pintsch, German Patent 75,506 (1892)
8. Schmidt, R., U.S. Patent (filed 6 June 1930) 1,846,692 (1932)
9. Ungar, G.A., U.S. Patent (filed 11 June 1930) 1,846,700 (1932)
10. Kullmann, A., German Patent (filed 29 December 1931) 603,303 (1934)
11. Wiegand, S.L., U.S. Patent (filed 28 April 1874) 155,602 (1874)
12. Holdaway, W.S., U.S. Patent (filed 2 June 1915) 1,218,602 (1917)
13. Societe Anonyme des Establishements, Olier, A., British Patent (filed 21 May 1921) 180,638 (1922)
14. Montelius, C.O.J., U.S. Patent (filed 20 March 1925) 1,698,802 (1929)
15. Leistritz, P., Burghauser, F., German Patent (filed 24 April 1926) 453,727 (1927)
16. Montelius, C.O.J., U.S. Patent (filed 1 March 1929) 1,965,557 (1934)
17. Gabrielli, E., British Patent (filed 29 December 1952) 759,354 (1956); German Ausgeschrift (filed 29 December 1952) 1,129,681 (1962)

18. Beck, E., Robbins, A.L., Birdsall, J.C., U.S. Patent (filed 16November 1972) 3,880,564 (1975)
19. Anonymous, German Patent (filed 5 March 1944) 909,821 (1954)
20. Keune, W., German Patent (filed 7 June 1949) 822,261 (1951)
21. Westover, R.F., *SPE J.* (1962) p. 1473
22. Tadmor, Z., Hold, P., Valsamis, L., *SPE ANTEC Tech Papers* (1979) 25, pp. 193, 205 (1979)
23. Tadmor, Z., Hold, P., Valsamis, L., *Plast. Eng.* November, p. 26; December, p. 30 (1979)
24. Maxwell, B., Scalora, A. J., *Mod. Plast.* (1959) 37, p. 107
25. Weissenberg, K., *Nature* (1947) 159, p. 310
26. Fuller, L.J., U.S. Patent (filed 15 May 1945) 2,675,199 (1952)
27. Kaplan, A., Tadmor, Z., *Polym. Eng. Sci.* (1974) 14, p. 58
28. White, J.L., Adewale, A.O., *Int. Polym. Process* (1993) 8, p. 210
29. White, J.L., *Twin Screw Extrusion: Technology and Principles* (1990) Carl Hanser Verlag, Munich
30. Kiesskalt, S., *Z. Ver. deut. Ing.* (1927) 71, p. 453
31. Janssen, L.B.P.M., *Twin Screw Extrusion* (1978) Elsevier, Amsterdam
32. Doboczky, Z., *Plastverarbeiter* (1965) 16, p. 57
33. Janssen, L.B.P.M., Mulders, L.P.H.R.M., Smith, J.M. *Plast. Polym.* (1975) 43, p. 93
34. Wunsche, A., German Patent 131,396 (1901)
35. Erdmenger, R., *Chem. Ing. Tech.* (1964) 36, p. 175
36. Herrmann, H., Burkhardt, U., 5th Leobener Kunststoff Kolloquium Doppelschnecken-Extruder (1978) Lorenz Verlag, Vienna
37. Montelius, C., *Teknisk Tidskraft* (1933) June, p. 61
38. Burghauser, F., Erb, K., German Patent (filed 5 February 1938) 690,990 (1939)
39. Farbenindustrie, I.G, Italian Patent (filed 9 April 1938) 373,183 (1939)
40. Colombo, R., Italian Patent (filed 6 February 1939) 370,578 (1939)
41. Meskat, W., Erdmenger, R., German Patent (filed 7 July 1944) 868,668 (1953)
42. Erdmenger, R., Oetke, W., German Auslegeschrift (filed 16 March 1960) 1,111,154 (1961)

3   **Screw Extrusion Technology**

3.1　Rubber Extrusion

James L. White

3.1.1　Introduction

The screw extrusion of rubber compounds is one of the older areas of extrusion technology, dating back well into the 19th century. Rubber extruders have distinctive differences from thermoplastics extruders. First, they operate at much lower temperatures (130 °C maximum). Second, they are usually fed with rubber compound strips (and only in exceptional cases with pellets), which undergo no phase changes or major densification in the screw extruder system. This is unlike thermoplastics, which are processed from 180 to 300 °C (and more) inside screw extruders, and where low-density assemblies of solid pellets are usually fed, which melt and densify as they move along the screw.

Rubber extruders are usually primarily classified according to hot-feed and cold-feed machines. In hot-feed extruders, a rubber compound heated by the mechanical action of a two-roll mill is stripped off and fed into the extruder as a continuous heated strip. In a coldfeed screw extruder, the extruder is fed by a room-temperature rubber strip. Rubber extruders are also often classified by whether or not they are intended for degassing.

There have been few other reviews of screw extruders for the rubber industry. Of recent vintage, we can cite only the review articles of Gohlisch et al. [1] and several chapters in a book by the present author [2].

3.1.2　Extruder Technology

3.1.2.1　Hot-Feed and Cold-Feed Extruders

The oldest extruders of the rubber industry were hot-feed extruders, where hot rubber compounds at 80 to 100 °C are stripped off two roll mills and fed into the extruder. In some cases preheated slabs are discontinuously fed into the extruders.

The requirements of the hot-feed rubber extruder are small, i.e., simply reshaping a rubber compound under pressure accompanied by little temperature change. Not surprisingly, hot-feed rubber extruders are relatively simple devices. They generally have screw length–diameter ratios of order 4 to 6. The screw channels are of constant depth. Because of their short lengths, the output of hot-feed rubber extruders are almost immediately affected by variations in feed rate, which translate into changes in the geometry of the extrudate profiles.

Cold-feed rubber extruders are a more recent development. They were first investigated by *Paul Troester Maschinenfabrik* in the 1930s but with unsatisfactory results. It was only in the 1960s with innovations in screw design that they were commercially introduced. In these machines, cold rubber strip or even rubber pellets are introduced in the feed section. Screw length–diameter ratios are in the ranges of 10 to 18.

The functions of a cold-feed screw extruder are more complex than those of hot-feed extruders. They may be considered to operate in four stages: In the first stage the machine must accept the feed and convey and compact the rubber compound. In the second stage it must heat and plasticate the rubber to make it a more fluid mass. In the third section the extruder must mix and thermally homogenize the rubber compound. In the fourth stage it must build up pressure to force the rubber compound through dies. For hot-feed rubber extruders, the second and third stages are provided by the preheat mill.

In recent years, cold-feed screw extruders for rubber compounds have become widely adopted in most areas of rubber extrusion. They are significantly more economic, because preheating mills need not be purchased.

3.1.2.2   Special Screw Designs for Cold-Feed Extruders

3.1.2.2.1   *Plastication and Homogenization*

Many cold-feed screw extruder designs have been proposed in the patent literature. It is not always clear to what extent these designs act to plasticate or homogenize the rubber compounds being extruded. We will thus treat them together.

The earliest plasticating extruders were developed by W.A. Gordon [3–5] and became part of the Gordon plasticator, which was marketed by the *Birmingham Iron Foundry* (now *Farrel Corp.*) to prepare by continuous mastication natural rubber that had been in storage for compounding internal mixers. Similar machines were developed by V. Royle [6, 7] These screw extrusion machines have conical kneading chambers with fluted walls through which the rubber must pass [4, 7]Some machines also have serrations or threads cut into their barrels.

In 1952, C.M. Parshall and P. Geyer [8] of the *U.S. Rubber Company*, Tire Division filed a patent application for a "mill". This describes a single screw machine with helical grooves in the barrel (see Fig. 3.1). As the screw rotates, rubber will pass from the screw to the barrel and back again. This machine was used as a continuous kneading unit to enhance carbon black dispersion following incorporation of black in an internal mixer. This machine was later called a "Shear-Mix" within *U.S. Rubber* [9] and in later years was commercially marketed as a "Transfer-Mix" (e.g., [9, 10].

A new concept in screw design was proposed in a December 1959 Swiss patent application by C. Maillefer [11] of *Maillefer SA* and in an April 1961 American patent application by P. Geyer [12] of the *U.S. Rubber Company*. The screw designs involved the development of a new screw flight out of the leading flight. This flight gradually moves across the screw channel and eventually joins the neutral flight. The patent drawings are shown in Fig. 3.2. This design is often called the barrier screw. The motivations of Maillefer and

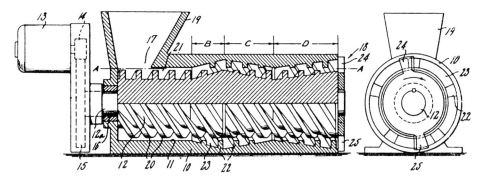

Fig. 3.1   Transfer mix of Parshall and Geyer [8–10] of U.S. Rubber

Geyer were different, as is made clear by reading their patents. Maillefer was concerned with the screw extrusion of thermoplastics and apparently knew of the experimental studies Maddock [13], which had shown that the bed of solid pellets during melting was usually pressed against the neutral flight. Maillefer sought to control this solid bed. Geyer expresses concern for scorched rubber gel particles in the compound being extruded and sought to use the additional screw flight to prevent their movement to the die.

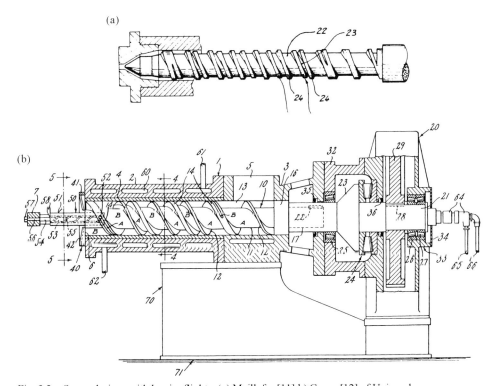

Fig. 3.2   Screw designs with barrier flights: (a) Maillefer [11] b) Geyer [12] of Uniroyal

The barrier screw design was licensed by *U.S. Rubber (Uniroyal)* within the United States as the Plastiscrew in 1963. It was incorporated into the design of cold-feed rubber extruder screws, where it was found to improve extrudate quality. This is remarked upon in a paper by Christy [14] of *NRM*, who describes the improved cold-feed rubber extruders of the 1970s that used Plastiscrew designs. Christy [15] also notes the superior economics of the cold-feed extruder, because it is not necessary to purchase preheat mills.

In succeeding years, there were many efforts to improve the plastication and homogenization characteristics of cold-feed rubber extruders. In a June 1968 German patent application, K. Koch [16] of *Paul Troester Maschinenfabrik* proposed a screw section with low helix angle and four thread starts (Fig. 3.3). It is perhaps better to say that this section consists of four axially aligned shear gaps intended to provide significant shearing of the rubber compound. This is often placed at the end of the screw. The device is called a Troester shear section.

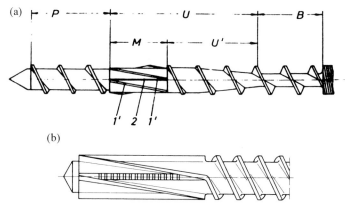

Fig. 3.3   Troester shear section

In this same period, a joint activity developed between *Uniroyal Engelbert Deutschland AG* based in Aachen and Prof. G. Menges of the Institut für Kunststoffverarbeitung at the Rheinisch-Westfälische Technische Hochschule of Aachen (RWTH Aachen). Involved for *Uniroyal Engelbert* were J.P. Lehnen and E.G. Harms, who would use the results of this program in their Dr.-Ing dissertations with the RWTH Aachen. This program has two stages: The first involves patent applications [17–21] and publications [22–26] in 1969–70 describing special screw designs; the second involves the development of the pin barrel extruder. Menges et al. in one patent [20] describe a special homogenizing section consisting of a backward-pumping, left-handed screw element with grooves in its flights. This produces backward drag, which is overcome by forward pressure flow (Fig. 3.4a). In a second design by Lehnen et al. [17, 19] the screw contains two plasticating/homogenizing zones. The first zone is again a left-handed, backward-pumping section with grooves in its flights. The second zone contains a barrier section similar to the Maillefer [11]–Geyer [12] designs of Fig. 3.2. This design is shown in Fig. 3.4b. In a third design, Menges et al. [18, 21] describe complex screws of the type shown in Fig. 3.4c in which there is continuous division and rejoining of material streams. This screw design is also intended to promote homogenization. The screw design can be varied by making removable sections that can be interchanged (Fig.3.4d). The work described here became the Dr.-Ing. dissertation of J.P. Lehnen at the RWTH Aachen.

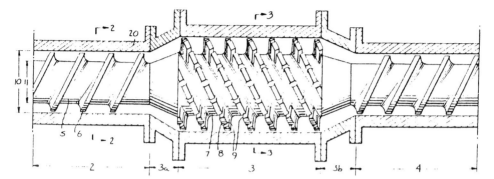

Fig. 3.4a   Special backward pumping homogenizing section of Menges et al. [20]

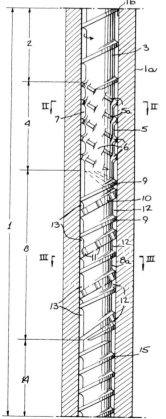

Fig. 3.4b   Lehnen et al., screw design with backward pumping section and barrier section [17, 19]

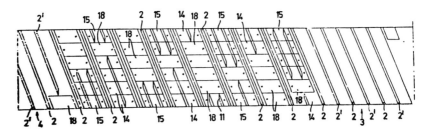

Fig. 3.4c   Menges et al., screw design with continuous division and rejoining of streams [18, 21]

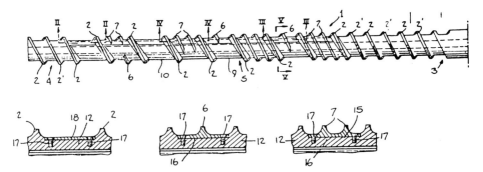

Fig. 3.4d   Menges et al. interchangeable screw sections [20]

Menges and Harms continued to work together, which led to the development of the pin barrel extruder for cold-feed rubber extrusion. The idea of pin barrel extruders is actually old [2]; the application to cold-feed rubber extrusion was new. This development is described in a series of patents [27–29] and papers [30–34] that were filed and appeared from 1972 to 1979. The pin barrel extruder design is shown in Fig. 3.5. Combining the pins with the slices in the screw flights leads to continual subdivision and recombining of the flows of the rubber compounds, which results in greatly enhanced thermal homogenization. One was able to achieve higher output rates of good quality extrudates than with earlier cold-feed extruder designs. This could be accomplished either by using higher screw speeds or by cutting the screw channels deeper. The pin barrel extruder for cold-feed rubber extruders became the Dr.-Ing dissertation of Harms at the RWTH Aachen. *Paul Troester Maschinenfabrik GmbH* had worked with Menges, Harms, and Hegele during the early development of the pin barrel extruder. They received the first license and became the first manufacturer. Many other machinery manufacturers, throughout not only Germany and Europe but the world, have since taken out licenses to manufacture pin barrel, cold-feed rubber extruders. With the acquisition of *Uniroyal Englebert*, by *Continental Gummiwerke*, the ownership of the patents passed to the Hannover-based tire firm. There have been continuing efforts by extruder manufacturers to improve on the design of pin barrel extruders. Thus H. Brusehoff of *Paul Troester Maschinenfabrik* in a 1985 German patent application [35] describes a machine with control devices in the pins.

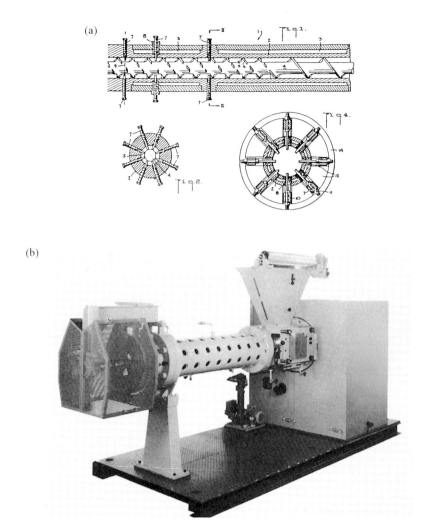

Fig. 3.5   (a) Pin barrel extruder design of Menges, Harms, and Hegele [27–29] (b) Paul Troester Maschinen-fabrik QSM pin barrel extruder

Other designs aimed at achieving better plastication and homogenization have been proposed, including the Troester mixing section [1] (Fig. 3.6). Another example of a mixing section is the cavity mixer developed by G.M. Gale [36], of the UK-based Rubber and Plastics Research Association (RAPRA) (Fig. 3.1). In this section, which is generally placed at the end of a screw, an intensive exchange of volume elements of the rubber compound take place. A screw extruder with a variant cavity transfer mixer has been developed by Fukumizu, Inoue, and Kuriyama [37] of *Kobe Steel* (Fig. 3.8). There have also been efforts to combine, Transfer Mix, pin barrel and other mixing sections in single screws.

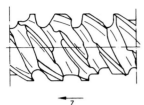

Fig. 3.6   Troester mixing section

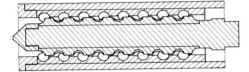

Fig. 3.7   Gale's RAPRA cavity transfer mixer [36]

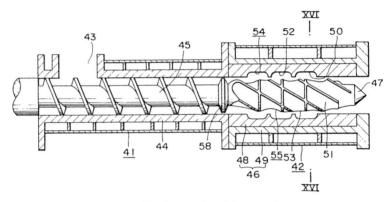

Fig. 3.8   Fukumizu et al. [37] Kobe Steel cavity transfer mixing extruder

3.1.2.2.2   *Vacuum Extruders*

It is necessary to remove all volatile components from rubber compounds during the extrusion process if it is intended to carry out continuous pressureless vulcanization of profiles and tubes. This is achieved by essentially dividing the extruder screw into two parts. The first section functions as a cold-feed screw extruder that passes the rubber compound through a vacuum zone to a second extruder that carries out the degassing and pumping the melt against the die resistance.

There is considerable patent literature associated with the degassing of rubber compounds and thermoplastics. Much of it is old. A March 1913 patent application by R.B. Price [38] of the *Rubber Regenerative Company* describes a screw extruder essentially divided into two sections with an intermediary barrier for this purpose (Fig. 3.9). Degassing of rubber compounds in the extrusion of rubber hose is described in a 1915 patent application by Price with W.J. Steinle [38]. An improved degassing extruder was presented by Steinle [39] of the *Rubber Regenerating Company* in a second 1915 patent application. This involves converting the rubber compound into sheet form in the extruder before it enters the vacuum zone (see Fig. 3.10). Another method that has been used for thermoplastics is depressurization and starvation by rapidly decreasing the root diameter of extruder screws and then repressurization by increasing the root diameter and compress the material. This is discussed in an August 1959 patent application by A.J. Palfrey [40] of *Dow Chemical*. A combination of using a barrier and decreasing screw root diameter is contained in a devolatizing screw developed by Tedder [41] of *Shell Oil* for removing water and similar impurities from polyolefins.

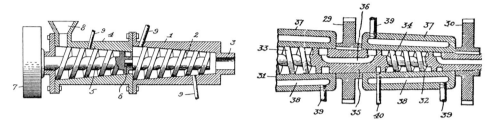

Fig. 3.9   Price's degassing extruder [38]

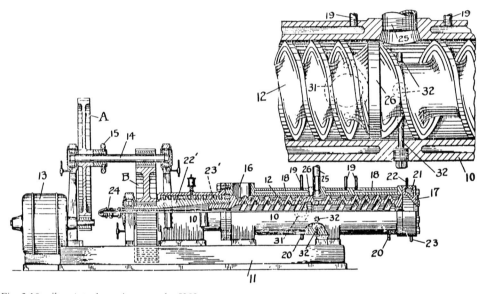

Fig. 3.10   Steinle's degassing extruder [39]

A typical vacuum screw for rubber compounds is shown in Fig. 3.11 [1]. It contains a barrier zone through which the rubber compound passes. When a rubber compound passes through this region, it leads both a large pressure drop and the division of the rubber compound into a large number of thin strands, each with a high surface area.

Fig. 3.11   Typical rubber vacuum screw

3.1.2.3   Overall Screw Extruder Design

The overall mechanical design of rubber processing extruders has been described by Gohlisch et al. [1] of *Paul Troester Maschinenfabrik.* We will try to summarize the key issues here. Generally, leading screw extruder manufacturers have developed modular systems for the basic components of their machines. The advantage to the equipment manufacturer is efficient manufacture. The advantages for the user include easy disassembly and thus reduced downtime because of the interchangeable parts.

Different drive systems have been used on rubber screw extruders. Thyristor-controlled dc drives have been successfully used on modern extruders. Three phase shunt wound or variable frequency ac drives are also used, as well as hydraulic motors.

These extruder drive motors can be coupled to the extruder through different drive transfer elements. The most common is power transmission to an extruder gearbox by a high-performance V belt or toothed belt. On very large cold-feed extruders, with screw diameters of 200 mm and more, the motor is usually directly connected.

Rubber extruders require a mechanical gearbox unless they are direct-driven by a slow-running hydraulic motor. Generally, the machine frame is used as the gearbox housing. Reducing gears, which are usually ground and polished, helical gears, and the oil storage tank are accomodated within the machine frame. Some rubber extruders use gearboxes of proprietary design and construction. These may be modular systems for spur, bevel, or worm gearboxes. In this case the machine frame for the basic extruder must be specially designed to accomodate the gearbox. Vertically mounted extruders are often fit with such gearboxes.

Screw extruders also contain a thrust bearing. In most designs, the thrust bearing that supports the drive shaft axially is built into the gearbox. This has the advantage of very compact construction, but it significantly reduces the flexibility needed for different extrusion products. Axial roller bearings are usually employed for the thrust bearing, with self-aligning taper rollers being preferred because of their self-centering behavior.

The most important feature of a screw extruder is the extruder screw. The geometry that extruder screws should have is discussed elsewhere in this chapter and throughout the book. It is of more concern here to state that the screws must be made out of wear-resistant steels and possess high torsional strength, especially at small screw diameters. To achieve high output rates, rubber processing extruders have very deep screw channels, thus making this concern very important. High-strength steels must be used to meet this requirement. Generally, extruder screws require surface treatments, such as nitriding, to produce wear-resistant flights.

We now turn to extruder barrels. The effective barrel length is 4 to 6 diameters with normal hot-feed extruders, 10 to 18 diameters for cold extruders, and 14 to 24 diameters for degassing extruders. There are three different kinds of extruder barrels: (1) barrels with wet liners, (2) jacketed barrels, and (3) peripherally drilled barrels. The various types of extruder barrels require different designs. Depending on the length of the extruder barrel, it is made in one piece or in several sections that are bolted together. The overall length is divided into thermal control zones, thus permitting individual control of the different sections.

Removable barrel liners provide a rapid and inexpensive means for repair when worn out ofter long periods of use. Wet liners provide a good basis for optimum temperature control of the extruder barrel. Usually, the liners have smooth bores and are made from steel with nitrided surfaces. For the processing of very abrasive compounds, extremely wear-resistant, special alloy, bimetallic sleeves with wave-resistant alloy thickness 2 to 3 mm are used.

As mentioned above, the barrels of screw extruders are divided into separate temperature control zones. Electrically heated high-pressure water devices are usually used for thermal control. Oil-based units are used much less frequently.

Feed housings and spiral undercuts in the feed housing are provided in rubber extruders with the intention of achieving trouble-free feeding. The feed roll that forms a feed-roll pair with the conveying screw is the most usual material feeding aid in cold-feed extruders. Generally, the feed roll is designed so that its cylindrical surface projects into the feed opening in the extruder. The feed roll is in most cases driven by gears attached to the main shaft of the extruder. Usually, only the feed housing is cooled, not the remainder of the barrel.

Rubber extruders are fitted with extrusion heads that contain dies that control the output of the extruder. Generally, extrusion dies contain the highest surface finish quality of the flow channel surfaces and are designed to avoid dead spots and to have flow proceed by the shortest possible channel length. The extrusion head must be designed to have good temperature control and to allow for easy dissembly and cleaning of dies.

3.1.3   Basic Experimental Studies

3.1.3.1   General

The literature on screw extrusion of rubber compounds is quite sparse compared to that on thermoplastics. The earliest papers, by Vila [42] and Pigott [43], appeared in 1944 and 1951. The first paper is part of the American Rubber Reserve Program to develop butadiene–styrene synthetic rubber (GR-S) for tires during World War II. The second of these papers by Pigott contains a fundamental analysis of the fluid mechanics of a screw extruder, plus many fundamental observations of flow in a simple cold-feed rubber extruder. It is with Pigott [43] that the engineering science of screw extrusion of rubber compounds begins.

More recently, general studies of the cold-feed screw extrusion of rubber compounds have been published by Menges and Lehnen [22] Brzoskowski et al. [44, 45], and Kubota et al. [46]. Menges and Lehnen in a 1969 paper [22] made many basic observations. Their study involved feeding red and white strips of similar rubber compounds with curing agent to a cold-feed rubber extruder (45-mm *Schwabenthan Maschinenfabrik*). The rubber strips were extruded for less than 30 min. The screw rotation was then stopped and the extruder heated to 180 °C, where the rubber compound was cured for 30 min. Subsequently, the extruder was cooled down and the screw removed. The spiral of rubber compound around the screw channel was sectioned. Menges and Lehnen observed that rubber compounds exhibited circulating flows in screw channels (Fig. 3.12). Such circulating flows had been predicted by earlier investigators [47] to arise from the circumferential drag of the direction and barrel and the angular configuration of the screw channel (Eq 2.24). This leads to crosschannel as well as down-channel drag flow components. These circulating flows in screw channel cross sections were confirmed by Brzoskowski et al. [44, 45] using similar experimental techniques but with different rubber compounds.

In later investigations of flow in cold-feed rubber extruders, Brzoskowski et al. [44, 45] and Kubota et al. [46] described the existence of a state of starved flow in which the screw in the region in front of the die was fully filled, but through most of the hopper end of the extruder there is only a thin strip of rubber compound along the leading flight. The extent of fill was found to increase with the die pressure, the feed strip width, and the use of power feed. Similar behavior was found by these investigators on three different cold-feed extruders. The increase in fill level with feed strip width and power feed can be interpreted as associated with an increase of a metered input flow rate Q.

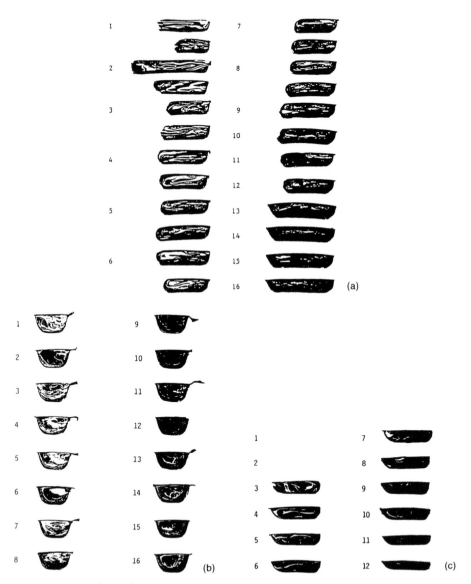

Fig. 3.12 Cross sections of vulcanized rubber strip indicating circulating flows in screw channel: (a) 37.5-mm 20 L/D NRM screw; (b) 48.82-mm 10 L/D Monsanto screw; (c) 32-mm 12 L/D boy injection molding screw from the work of Kubota et al. [46]

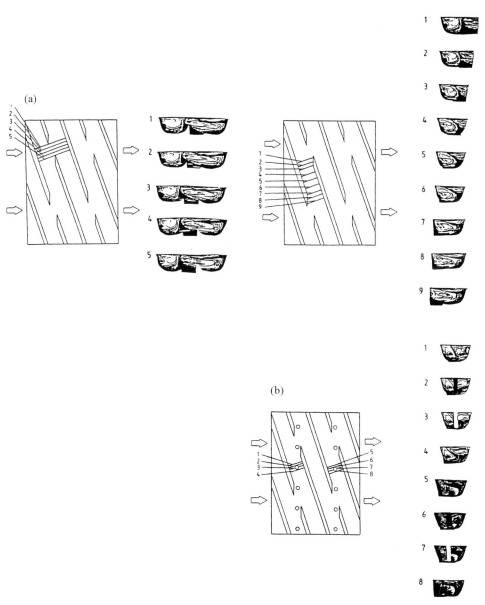

Fig. 3.13   (a) Cross sections of vulcanized rubber compound strip from screws indicating circulating flows and backflows through screw slices; (b) Cross sections of vulcanized rubber compound strip from screws; indicating circulating flows and disruptions of flow by pins and screw; slices

Screw pumping characteristic curves relating throughput Q to pressure rise Δp at fixed screw speeds were determined by Brzoskowski et al. [44] and Kubota et al. [46] under various types of operating conditions. They found that for fully filled screws at any screw speed, the throughput Q was a decreasing function of increasing die pressure. For all conditions studied, the throughput at fixed die pressure increased with screw speed. For starved screws, the output Q was independent of Δp at fixed screw speed, but increased with screw speed.

3.1.3.2   Plasticating and Homogenization Screws

The homogenization of rubber compounds in cold-feed rubber extruders was described in a series of papers by Menges and Lehnen [22, 26] and Lehnen [23–25] published in 1969–70. They prepared cross sections of the spiral of rubber compound wrapped the screw and compared the observed mixing quality in them with a scale of "mixing goodness" they developed. Special attention was given to comparing the effect of screw design and studying the variation of mixing goodness along the screw axis.

Menges and Lehnen made many specific observations of the influence of screw design on mixing quality. The barrier screw or Plastiscrew design of Maillefer [11] and Geyer [12] (Fig. 3.2) did a much better job of inducing mixing than any screw made up of all right-handed screw elements with varying pitch. However, introducing special screw elements of the type described in the Menges–Lehnen patents [17–21] along the screw axis leads to further improved mixing.

3.1.3.3   Pin Barrel Extruders

As described previously, pin barrel extruders were introduced into rubber extrusion technology by the efforts of Menges, Harms, and Hegele [27–34] in the early 1970s. They since have achieved a position of major importance in cold-feed rubber extruder technology. The first basic experimental studies of this machine are found in the published papers of these authors [30–34]. Menges et al. describe flow marker studies using techniques essentially the same as those of Menges and Lehnen [22]. Similar flow marker studies involving pin barrel extruders have been reported by Yabushita et al. [47] and by Shin and White [48]. These experiments show the role of the pins in enhancing mixing. The results of the latter authors [47, 48] are of additional interest because they begin with a simple screw and then successively modify it by first introducing slices in the screw flights and then subsequently introducing rows of pins. Fig. 3.13 from the work of Yabushita et al. shows the results of marker experiments, indicating (1) the backward movement of rubber compound through screw flight slices and (2) the disruption of flow by the presence of pins in the screw channel.

Shin and White [48] compare the homogenizations of three different rubber compounds in a cold-feed rubber extruder using the above three screw designs. One compound is based on butadiene–styrene copolymer and polybutadiene and is a typical traditional passenger tire tread; the second, based on natural rubber, is a traditional truck tire tread; and the third, based on nitrile rubber, is a mechanical goods compound. The best mixing is found in the traditional passenger tire compound and the worst in the mechanical goods compound, but there is a steady improvement for all three compounds in mixing and homogenization as one introduces slices in screw flights and then subsequently introduces pins into the barrel.

3.1.3.4   Vacuum Extruders

There have been experimental studies of the flow of rubber compounds in vacuum extruders. The only investigation we are aware of is that of Brzoskowski et al. [49]. As described in Section 3.1.2.2.3, vacuum screws for rubber compounds contain a barrier in a central section of the screw. Brzoskowski et al. observe that screws removed from such extruders are fully filled in front of the die and in front of the barrier. There is a starved region between the barrier and the fully filled region in front of the die (Fig. 3.14). It is clear

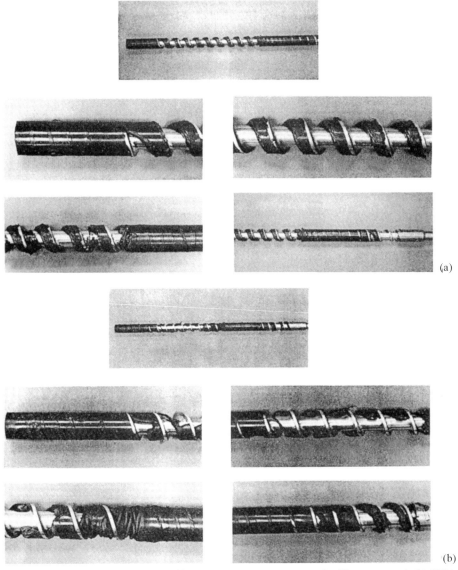

(a)

(b)

Fig. 3.14   Distribution of rubber compound in a two-stage screw: (a) natural rubber compound; (b) SBR/ BR compound

that this starved region is unpressurized during the operation of the extruder. It is thus a suitable position to attempt to remove volatiles by means of a vacuum. The rubber compound would not surge out the vent port as might be expected in a pressurized region of the extruder.

The characteristics of the rubber compound in the starved region appear to differ depending their rheological properties. The butadiene–styrene copolymer/polybutadiene (SBR/BR)-based compound breaks up into pieces (Fig. 3.14b), while the natural rubber compound maintains itself as a continuous strip (Fig. 3.14a). This appears to be due to the occurrence of strain induced crystallization in the natural rubber compounds.

Another interesting study of devolatilization of rubber compounds in a screw extruder is by Fellenberg [50].

3.1.4   Flow Simulations

3.1.4.1   Rheological Properties of Rubber Compounds

To successfully model the flow of rubber compounds in a screw extrusion machine, it is necessary to have a basic knowledge of their rheological properties. Generally, we are not dealing pure elastomers or elastomers with low levels of particle fillers, but rather with compounds having 50 to 150 parts by weight of small particles, such as carbon blacks, silicas, and clays, as well as oils and curatives, per hundred parts of rubber. Not surprisingly, the rheological properties of these compounds are quite complex. There has been a long history of investigations of rheological properties of rubber–carbon black compounds [51–63]. In steady shear flow, rubber compounds exhibit a shear viscosity that rapidly decreases with increasing shear rate. At very low shear rates, the shear viscosity becomes unboundedly high. It has been shown in recent years using creep measurements that there are stresses below which there is no flow in many rubber compounds [61–63]. Typical shear viscosity shear stress data for rubber compounds is shown in Fig. 3.15.

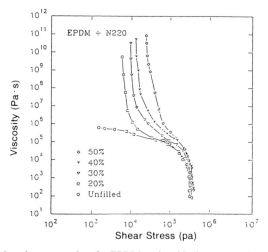

Fig. 3.15   Shear viscosity–shear stress data for EPDM-carbon black compounds

Some investigations [52, 58, 61–63] have additionally shown that rubber compounds exhibit complex time-dependent properties. Rubber compounds with high carbon black loadings that remain at rest for extended times exhibit a build up in their transient startup flow behavior.

The shear stress σ_{12} of rubber compounds has been represented by equations of the form

$$\sigma_{12} = Y + \frac{A\dot{\gamma}}{1 + B\dot{\gamma}^{1-n}} \tag{3.1}$$

where σ_{12} is the shear stress, Y is the yield value, n is the power law exponent, and A and B are material constants. Eq. 3.1 predicts a yield value as the shear rate goes to zero and highly non–Newtonian behavior at higher shear rates. At high shear rates Eq. 3.1 becomes

$$\sigma_{12} = \left(\frac{A}{B}\right)\dot{\gamma}^{n} = K\dot{\gamma}^{n} \tag{3.2}$$

which represents power-law behavior between the shear stress and shear rate.

3.1.4.2   Simple Screws

The first mathematical modeling flow in a screw extruder that could be considered a rubber extruder is that of Pigott [43], which we have already cited. However, the first full modeling that specifically considers non–Newtonian shear viscosity, viscous dissipation heating with reference to rubber compounds is by Zamodits and Pearson [64]. The mathematical model of this paper does not differ from the earlier paper of Griffith [65], which was aimed at much different materials. In the work of these authors, the material being processed is considered to be a non-Newtonian fluid along the length of the screw. There are no first-order or glass-transition thermal transitions, such as occur in thermoplastics, that transform from beds of pellets to homogeneous melts. The rheological behavior is represented by Eq. 3.2, the power-law model. Zamodits and Pearson [64] account for both flow along the screw channel and transverse flows that give rise to the circulating flow described in Section 3.1.3.1.1 and Fig. 3.12. The model uses equations describing the balances of shear stresses and pressure gradients as well as an energy balance that includes viscous heat dissipation and heat conduction.

One important result of the above calculations is that the effect on non-Newtonian viscosity is to deteriorate screw pumping characteristics. If the shear viscosity decreases with increasing shear rate, then the effect of pressure-induced backflows from the die will be larger than for a Newtonian fluid. Second, viscous heating also deteriorates screw pumping characteristics. This is because viscous heating also reduces the shear viscosity and causes enhanced backward pressure flow.

3.1.4.3   Screw Sections with Slices

Screw sections with slices in screw flights are common in screw designs intended for homogenization. The flow of rubber compounds in right-handed screws with slices in the flights has been modeled by Brzoskowski et al. [66, 67] and Shin and White [68]. The simulation of Brzoskowski et al. [66] is for Newtonian fluids and that of Shin and White [68] for power law non–Newtonian fluids. The results of these authors show that introducing slices in screw flights leads to backward leakage flows. These leakage flows deteriorate screw pumping ability as compared to simple screw sections.

3.1.4.4   Pin Barrel Extruders

The flow in pin barrel extruders was also modeled by Brzoskowski et al. [66, 67, 69, 70] and by Shin and White [68, 71]. The results show that the pressure-induced backflows and pumping characteristics are dominated by the screw slices. The pins appear to play only a secondary role. Again it is found that non-Newtonian viscosity reduces screw pumping characteristics [68] (Fig. 3.16), as does viscous heat generation [71].

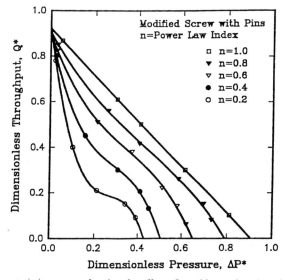

Fig. 3.16   Screw characteristics curves showing the effect of non-Newtonian viscosity and viscous heating on screw pumping characteristics for a pin barrel extruder

3.1.5   Multilayer Extrusion Lines

Generally, all rubber extruders have dies that determine the extrudate profiles. The combination of different extruders together to a single head that produces a combined multilayer profile is something special. Multilayer extrusion is common in rubber extrusion. Frequently today extruders are mounted in a piggyback arrangement (Fig. 3.17) to accom-

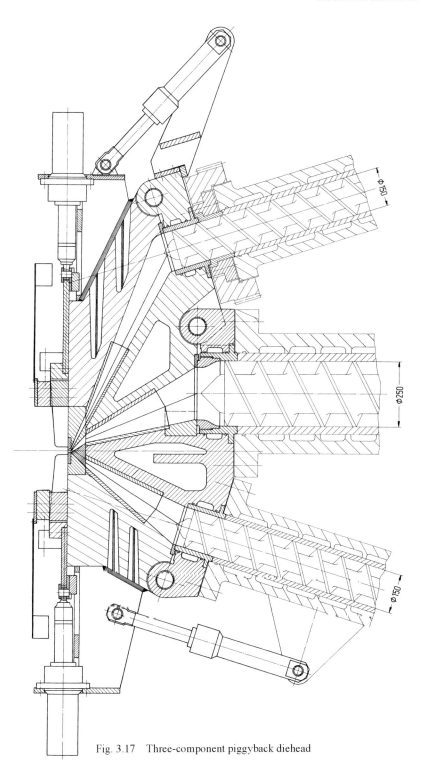

Fig. 3.17   Three-component piggyback diehead

plish this. The newer piggyback arrangements of extruders have replaced the earlier Y configurations. Such extrusion systems are used for the manufacture of flat profiles from two, three, or four rubber compounds.

The tire industry is the largest user of such multilayer extrusion systems. There is a long history of patents dating to the 1930s [72–77] for extruding combined treads sidewalls and liners. Multilayer extrusion lines in this industry have special extrusion heads known as tread heads. These are specially subdivided dies fitted with hydraulic clamping and opening systems.

3.1.6   Continuous Vulcanization Lines

Continuous vulcanization of extrudates of rubber extruders has always been preferred to discontinuous processes since the beginning of vulcanization. Such technologies date to Parkes 1846 invention [78] of rapid curing of thin sheets of natural rubber and gutta percha with solutions of sulfur dichloride and the subsequent applications of this by Brockedon and Hancock [79, 80]. The reasons for the desirability of continuous production of vul-canized extrusion profiles involve economy as well as quality. The economic reasons are (1) high-speed lines, (2) no intermediate conveying of the extrudate, (3) saving on man-power, (4) smaller space requirement, (5) smaller overall investment, and (6) energy saving as extrudate heat is used so that only a small amount of energy is needed to raise the com-pound temperature to the vulcanization temperature. Reasons involving quality include a greater tendency to constant profile geometry, extrudate length, and better quality control.

Various technologies for continuous vulcanization have been developed through the years. Among these are (1) hot air, (2) fluidized beds, (3) salt baths, (4) infrared heating, (5) helicure method, and (6) shear head lines. Typical continuous vulcanization lines contain vacuum extruders, heating and cooling sections, and haul off (or wind-up or cutting) apparatus.

Hot air continuous vulcanization lines are the oldest found in industrial plants. Today they are mostly used in combination with UHF or shear head lines. The pure hot air lines, which are today often used with highly expanded foam rubber tubes or sheeting, can be as much as 100 to 150 m long.

Fluidized bed vulcanization uses a pseudo-fluid of air-suspended glass microsheres of 0.2 to 0.4 mm in diameter as a heat transfer medium. These lines consist of a tray with steel wire mesh through which alternatively hot air or superheated steam can be blown.

Salt bath vulcanization is by far the most common and universally used continuous vulcanization method for extrusion profiles. This method was introduced by DuPont in 1960. The salt bath consists of a eutectic mixture of salts consisting of 53% potassium nitrate (KNO_3), 40% sodium nitrite ($NaNO_2$), and 7% sodium nitrate ($NaNO_3$). This mixture has a melting point of about 140 °C and a density of 1.9 g/cm^3 at 250 °C. It can be used at the vulcanization temperatures of all well-known compounds. In more recent years, salt baths based on nitrite free mixtures of lithium nitrate ($LiNO_3$) and potassium nitrate (KNO_3) have been used.

References

1. (a) Gohlisch, H.J., May. W., Ramm, F., Ruger, W., Extrusion of elastomers. In *Plastics Extrusion Technology.* Hensen, F. (Ed.) (1988) Carl Hanser Verlag, Munich; translated from Extrusion von Elastomeren. In *Handbuch der Kunststoff Extrusions Technik* Vol. 2. Knappe, W., Potente, H., Hensen, F. (Ed.) (1988) Carl Hanser Verlag, Munich. (b) Gohlisch, H.J., May, W., Ramm, F., Ruger, W., Extrusion of Elastomers. In *Plastics Extrusion Technology,* 2d ed. Hensen, F., (Ed.) (1996) Carl Hanser Verlag, Munich
2. White, J.L., *Rubber Processing Technology, Materials and Principles* (1995) Carl Hanser Verlag, Munich
3. Gordon, W.A., U.S. Patent (filed 1 October 1919) 1,364,549 (1921)
4. Gordon, W.A., U.S. Patent (filed 18 August 1925) 1,608,980 (1926)
5. Gordon, W.A., U.S. Patent (filed 29 November 1930) 1,935,050 (1933)
6. Royle, V., U.S. Patent (filed 4 June 1929) 1,904,884 (1933)
7. Royle, V., U.S. Patent (filed 29 January 1937) 2,200,997 (1940)
8. Parshall, C.M., Geyer, P., U.S. Patent (filed 24. September 1952) 2,744,287 (1956)
9. Parshall, C.M., Saulino, A.J., *Rubber World* (1967) May, p. 78
10. Stansfield, M.D., *SPE ANTEC Tech. Papers* (1971) 17, p. 282
11. Maillefer, C., Swiss Patent (filed 31 December 1959) 363,149, (1962); U.S. Patent (filed 20 December 1960) 3,358,327 (1967)
12. Geyer, P., U.S. Patent (filed 13 April 1961) 3,375,549 (1968)
13. Maddock, B.H., *SPE J.* (1959) May, p. 383
14. Christy, R.L., *Rubber World* (1979) July
15. Christy, R.L., *Rubber World* (1979) September
16. Koch, K., German Offenlegungschrift (filed 1 June 1968) 1,778,770 (1970)
17. Lehnen, J.P., Menges, H.G.L., Harms, E.G., U.S. Patent (filed 19 March 1969 / 5 March 1970) 3,652,064 (1972)
18. Menges, H.G.L., Harms, E.G., Lehnen, J.P., German Offenlegungschrift (filed 30 June 1970) 2,032,197 (1972)
19. Uniroyal Engelbert Deutschland, British Patent (filed 15 July 1970) 1,318,913 (1973)
20. Menges, H.G.L., Lehnen, J.P., Harms, E.G., U.S. Patent (filed 16 July 1970) 3,680,884 (1972)
21. Menges, H.G.L., Harms, E.G., Lehnen, J.P., U.S. Patent (filed 18 June 1971) 3,946,998 (1976)
22. Menges, H.G.L., Harms, E.G., Lehnen, J.P., *Plastverarbeiter* (1969) 20, p. 31
23. Lehnen, J.P., *Kunststofftechnik* (1970) 9, p. 3
24. Lehnen, J.P., *Kunststofftechnik* (1970) 9, p. 114
25. Lehnen, J.P., *Kunststofftechnik* (1970) 9, p. 198
26. Menges, G., Lehnen, J.P., *Kunststofftechnik* (1970) p. 352
27. Harms, E.G., Menges, H.G.L., Hegele, R., German Offenlegungschrift (filed 21 July 1972) 2,235,784 (1974)
28. Menges, H.G.L., Harms, E.G., Hegele, R., U.S. Patent (filed 30 January1978) 4,178,104 (1979)
29. Menges, H.G.L., Harms, EG., Hegele, R., U.S. Patent (filed 16 July 1973 and again 30 January 1978) 4,199,263 (1979)
30. Menges, G., Harms, E.G., *Kautsch Gummi Kunstst* (1972) 25, p. 469
31. Menges, G., Harms, E.G., *Kautsch Gummi Kunstst* (1974) 27, p. 187
32. Harms, E.G., *Rubber Age* (1977) 109 (6), p. 33
33. Harms, E.G., *Kautsch Gummi Kunstst* (1977) 30, p. 735
34. Harms, E.G., *Eur. Rubber J.* (1977) 160 (5), p. 26
35. Brusehoff, H., German Patentschrift (filed 25 January 1985) DE3,502,437 C2 (1988)
36. Gale, G.M., U.S. Patent (filed 31 March 1981) 4,419,014 (1983)
37. Fukumizu, S., Inoue. K., Kuriyama, A., U.S. Patent (filed 14 October 1986) 4,695,165 (1987)
38. Price, R.B., U.S. Patent (filed 26 March 1913) 1,156,096 (1915); Price, R.B., Steinle, W.J., U.S. Patent (filed 24 April 1915) 1,211,370 (1917)
39. Steinle, W.J., U.S. Patent (filed 24 April 1915) 1,283,947 (1918)
40. Palfrey, A.J., U.S. Patent (filed 3 August 1959) 3,023,456 (1962)
41. Tedder, W., U.S. Patent (filed 30 April 1962) 3,115,675 (1963)
42. Vila, G.R., *Ind. Eng. Chem.* (1944) 36, p. 1113

43. Pigott, W.T., *Trans ASME* (1951) 73, p. 947

44. Brzoskowski, R., Kubota, K., Chung, K., White, J.L., Weissert, F.C., Nakajima. N., Min, K., *Int. Polym. Process* (1987) 1, p. 130

45. Brzoskowski, R., White, J.L., *Int. Polym. Process* (1987) 2, p. 102

46. Kubota, K., Brzoskowski, R., White, J.L., Weissert, F.C., Nakajima, N., Min, K., *Rubber Chem. Technol.* (1987) 60, p. 924

47. Yabushita, Y., Brzoskowski, R., White, J.L., Nakajima, N., *Int. Polym. Process* (1989) 4, p. 219

48. Shin, K.C., White, J.L., *Rubber Chem. Technol.* (1993) 65, p. 121

49. Brzoskowski, R., White, J.L., Weissert, F.C., Nakajima, N., Min, K., *Rubber Chem. Technol.* (1986) 59, p. 634

50. Fellenberg, K., *Kautsch Gummi Kunstst* (1965) 18, p. 665

51. Dillon, J.H., Johnston, N., *Physics* (1933) 4, p. 225

52. Mullins, L., Whorlow, R.W., *Trans. IRI* (1951) 27, p. 55; also Mullins, L., *J. Phys. Chem.* (1950) 54, 539

53. Vinogradov. G.V., Malkin, A.Ya, Plotnikova, E.P., Sabsai, O.T., Nikolayeva, N.E., *Int. J. Polym. Mat.* (1972) 2, p. 1

54. Lobe, V.M., White, J.L., *Polym. Eng. Sci.* (1979) 19, p. 617

55. Toki, S., White, J.L., *J. Appl. Polym. Sci.* (1982) 27, p. 3171

56. Montes, S., White, J.L., *Rubber Chem. Technol.* (1982) 55, p. 1934

57. White, J.L., Wang, Y., Isayev, A.I., Nakajima, N., Weissert, F.C., Min, K., *Rubber Chem. Technol.* (1988) 8, p. 431

58. Montes, S., White, J.L., Nakajima, N., *J. Non-Newtonian Fluid Mech.* (1988) 28, p. 183

59. Song, H.J., White, J.L., Min, K., Nakajima, N., Weissert, F.C., *Adv. Polym. Technol.* (1988) 8, p. 431

60. Shin, K.C., White, J.L., *Kautsch Gummi Kunstst* (1990) 43, p. 181

61. Osanaiye, G., Leonov, A.I., White, J.L., *J. Non-Newtonian Fluid Mech.* (1993) 49, p. 87

62. Osanaiye, G., Leonov, A.I., White, J.L., *Rubber Chem. Technol.* (1995) 68, p. 50

63. Li, L.L., White, J.L., *Rubber Chem. Technol.* (1996) 69, p. 628

64. Zamodits, H., Pearson, J.R.A., *Trans. Soc. Rheol.* (1969) 13, p. 357

65. Griffith, R.M., *Ind. Eng. Chem. Fund.* (1962) 1, p. 180

66. Brzoskowski, R., White, J.L., Szydlowski, W., Nakajima, N., Min, K., *Int. Polym. Process* (1988) 3, 134

67. Brzoskowski, R., White, J.L., *Int. Polym. Process* (1990) 5, p. 238

68. Shin, K.C., White, J.L. *Rubber Chem. Technol.* (1997) 70, p. 264

69. Brzoskowski, R., Kumazawa, T., White, J.L., *Int. Polym. Process* (1990) 5, p. 191

70. Kumazawa, T., Brzoskowski, R., White, J.L., *Kautsch Gummi Kunstst* (1990) 43, p. 688

71. Shin K.C., White, J.L., *Kautsch Gummi Kunstst.* (2000) 53, p. 434

72. Snyder, R.W., Haase, J.J., U.S. Patent (filed 17 January 1931) 1,952,469 (1934)

73. Lehman, P.W., U.S. Patent (filed 21 March 1936) 2,096,362 (1937)

74. Fay, P., U.S. Patent (filed 21 September 1948) 2,569,378 (1951)

75. Weston, R.D., Barns, F.K., U.S. Patent (filed 24 February 1958) 2,897,543 (1959)

76. Eilerson, J.E., U.S. Patent (filed 6 June 1961) 3,099,859 (1963)

77. Greenwood, A., Hale, J.T., U.S. Patent (filed 18 January 1967) 3,486,195 (1969)

78. Parkes, A., English Patent (filed 25 March 1846) 11,147 (1846)

79. Brockedon, W., Hancock, T., English Patent (filed 19 November 1846) 11,455 (1846)

80. Hancock, T., *Origin and Progress of the Caoutchouc or India-Rubber Manufacture in England* (1857) Longman, London

3.2 Extrusion of Thermoplastics

Ulrich Berghaus

3.2.1 Trends in the Development of Single-Screw Extruders

The extruder is the key component of all extrusion lines. Homogeneous plastification and uniform conveying by the extruder screw have a decisive effect on the nature of the endproduct. For the production of extruded semifinished and finished articles single-screw extruders are used almost exclusively.

Due to its outstanding cost/effectiveness ratio and versatility the single-screw extruder is one of the most frequently used continuous plastics processing machine. The single-screw extruder mainly has the function of conveying, plasticizing, and homogenizing the material. With respect to dimensioning and technical design, the following requirements must be met [1]:

- High torques to realize high outputs
- High and surge-free melt throughput
- High output rates at low screw speeds, i. e., minimum material changes due to degradation or cross-linking
- Optimum melt temperature
- High flexibility as a result of a large processing range
- Reduced wear

The development of single-screw extruders is determined by the following objectives [2]:

- Enhancement of melt and product quality
- Increase in efficiency

A higher efficiency of a plastizising extruder of a given screw diameter can be realized only by an increase in melt throughput. Fig. 3.18 shows the development of melt throughput from 1955 to 1995 for single-screw extruders of various screw diameters used for LDPE processing into blown film [2].

With conventional plasticizing extruders, i. e., extruders provided with a smooth feed bush, a high melt throughput can be reached only by large-diameter screws. The essential benefits of a smooth-bore extruder are minimum wear and easy processing of regrind. These features contrast with a low specific output, higher melt temperatures, and a conveying action depending on the back pressure. The requirement of high output with small extruder size could be met by the development of grooved feed extruders in the early 1960s [3].

The advantages of a grooved feed extruder are a high specific melt throughput, a conveying action that is almost independent of the back pressure, and low melt temperatures. An extruder concept developed in the late 1980s, based on a highspeed extruder fitted with a melt pump, combines the advantages of a smooth-bore extruder and the grooved feed design [4–6].

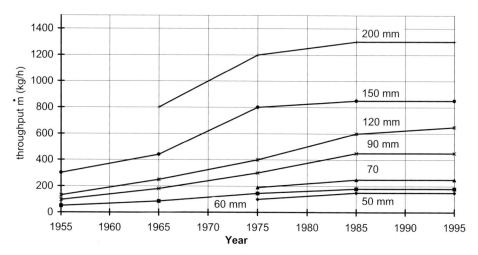

Figure 3.18   Development of melt throughput in LDPE blown film extrusion as a function of the screw diameter

3.2.1.1   High-Speed Extruders with Melt Pumps

The extruders used are of 27D length and operate adiabatically. A melt pump allows the extruder to be run at low pressures. Thus, the geometry of the screw can be optimized to meet the requirements placed on its feeding, melting, and homogenizing efficiency. Even high screw speeds can be reached at low melt temperatures and with low temperature fluctuations.

A comparison of the three above-mentioned extruder systems used for the production of PP film tapes [6] (Fig. 3.19) shows that a high-speed adiabatic extruder (90 mm/27D) with melt pump (180 cm³/rev) can realize outputs similar to the rates reached by a grooved feed extruder (120 mm / 33D). The high-speed extruder, however, operates at the most favorable melt temperature and develops the smallest pressure variations. A comparison to the smooth bore extruder (120 mm/33D) shows that this extruder is the most inefficient solution for this type of application.

The capital investment in a high-speed extruder with melt pump is of the same order as for a smooth-bore extruder of comparable output [6]. So far, in industrial practice the use of the high-speed extruder has been limited to the extrusion of PP film tape.

3.2.1.2   Grooved Feed Extruders with 30D or 33D Screws

The 25D to 30D extruders, which were regarded as sufficient for a long time, developed into extruders of 30D to 33D length. This development entailed a proportional increase in the motor outputs and screw torques (Fig. 3.20). As a result of this development, over the last twenty years the output of a typical extruder of 90 mm screw diameter increased up to the quadruple (Fig. 3.20).

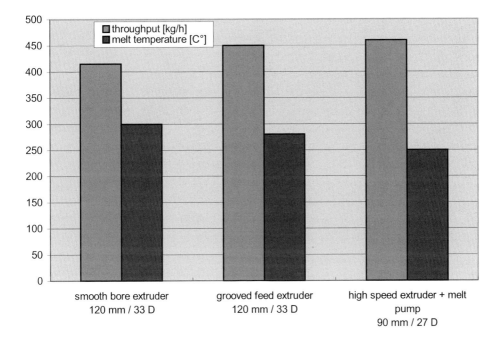

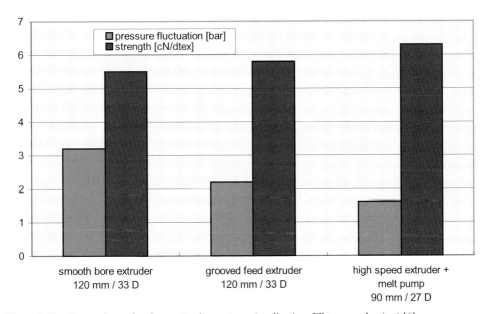

Figure 3.19   Comparison of various extrusion systems (application: PP tape production) [6]

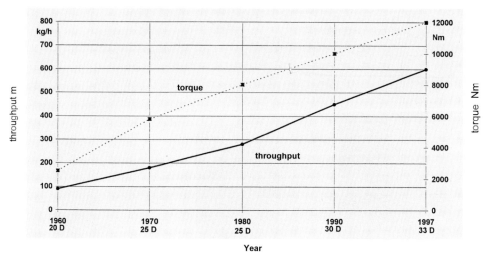

Figure 3.20   Development of the throughput and torque of a 90-mm single-screw extruder (application: polyolefin pipe extrusion)

3.2.1.3   Feed Section Design

To obtain a conveying effect that does not depend on the back pressure, the grooves and the geometry in the feeding zone must be precisely geared to each other [7, 8]. The number and cross section of the grooves, flight depth, pitch, and geometry of feed pockets are determined as a function of the machine size and raw material to be processed in accordance with the required output. With a precise geometry the material is conveyed at a constant specific output rate through the heated or unheated grooved feed section (Table 3.1).

Table 3.1   Performance of a Grooved Bush Extruder (70 mm/30D) at Various Grooved Bush Temperatures (raw material: LLDPE)

Grooved bush temperature (°C)	cold	40	60	80	100
Speed (min⁻¹)	60	60	60	60	60
Output (kg/h)	117	114	114	116	116
Specific output (kg · min/h)	1.92	1.90	1.90	1.93	1.93

For about ten years the grooved bushing has been lined with hard metals [9]. The solid hard-metal linings are made with high precision and shrunk into the prepared barrel sections. As an alternative, bushings made by hot isostatic pressing of metal powders can be welded into the extruder barrels. The increase in sales since the introduction about ten years ago shows that this design has prevailed on the market.

3.2.1.4   Screw Designs

Whereas in Europe extruders with grooved feed bushes and screws of conventional design have been developed and used for the most part so far, the extruder development in North America has concentrated on the consistent advance of barrier screws with smooth feed bushes [2]. Nevertheless, in Europe high-capacity extruder designs combining the advantages of barrier screws and grooved feed extruders have been developed for a few years.

3.2.1.5   Increased Output Without Increase in Wear

As the rate of wear grows with the pressure within the screw channel, the main objective of a grooved feed extruder optimization is to minimize the grooved bush pressure. The request for higher output in combination with low wear can be met only by using a two-stage screw. The two-stage screw has different pitches in the feed section than in the other screw sections and is optimally geared to the highly efficient grooved feed system [8]. Due to the different pitches (feed section < 1D, downstream sections > 1D), the pressure is markedly reduced along the barrel, especially in the feed section (Fig. 3.21). The screw runs at a constant specific output throughout the whole speed range (Fig. 3.22) and is unaffected by pressure and speed. Melt temperature control is possible only if the specific output is constant over the complete speed range.

If the specific throughput decreases, for instance, because the feed section is not optimized, the melt temperature inevitably increases (Fig. 3.23). At production speed, the

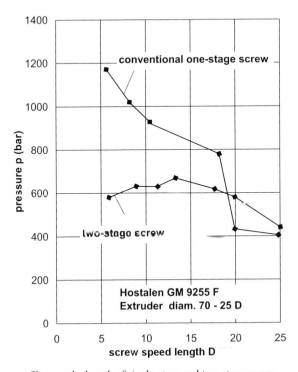

Figure 3.21   Pressure profile over the length of single-stage and two-stage screws

specific throughput of the two-stage screw is considerably higher than the specific through-put of a single-stage screw. This means a lower melt temperature with the same throughput when using the two-stage screw. Low pressures in the grooved bushing and high outputs at low screw speed cause appreciably less wear on the screws and barrels.

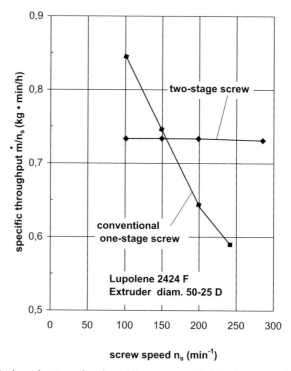

Figure 3.22   Specific throughput as a function of the screw speed of single-stage and two-stage screws

3.2.1.6   Homogenizing, In-Line Dispersion

All high-capacity extruders that are capable of plasticizing large amounts of material require shearing and mixing sections for homogenizing and dispersing [2, 10]. In conventional extruders, mixing action results from opposite pressure and drag flows, in contrast to grooved bush extruders. The pressure and drag flows of the latter type of lower conveying efficiency run in the same direction. The use of shearing and mixing sections more than compensates the lower homogeneity of the melt produced in the grooved bush extruder. Depending on the requirements that are placed on the raw material and product, shearing sections are used for dispersing the material and mixing sections for mixing it both axially and transversely.

Highly effective dynamic mixers with stator and rotor (Figs. 3.24 and 3.25) are used in combination with shearing and mixing sections for all applications that require high homog-enizing efficiency. Because the cavities in the screw and barrel dynamic mixers combine separation and distribution, they are particularly suitable for homogenizing the color concentrates or liquid additives contained in the melt. They comply even with critical standards such as British standard (Figs. 3.26 and 3.27).

Technically interesting and proven applications of dynamic mixers can be found in the industrial production of

- Stretch film (polyisobutylene (PIB) injection into LLDPE melt)
- Pipes and sheets (dispersion of color concentrates in polyolefins)
- Monofilaments (dispersion of hydrolysis stabilizers in polyolefins)
- Blown film (dispersion of color concentrates in polyolefins)
- Nonwoven and meltblown (dispersion of color concentrates and additives in PE, PP, PA, PET, PBT, PPS, PS).

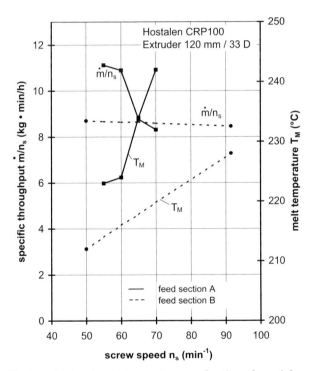

Figure 3.23   Specific throughput and melt temperature as a function of speed for various feed section geometries

As shown in Fig. 3.27 for the example of polyolefin sheet extrusion, this system has proven in service in extruders of up to 200 mm screw diameter.

The cost effectiveness of this dynamic mixer consisting of stator and rotor is plain from the fact that the raw material cost can be reduced by using a virgin material and color concentrate. A comparison of two plants that both produce colored HDPE pressure pipes in a four-shift operation reveals that the processing of HDPE virgin material and color concentrate reduces the production cost by approximately 15% as against the processing of a

Figure 3.24   STAROMIX stator and rotor (photo: Reifenhäuser)

Figure 3.25   STAROMIX stator (photo: Reifenhäuser)

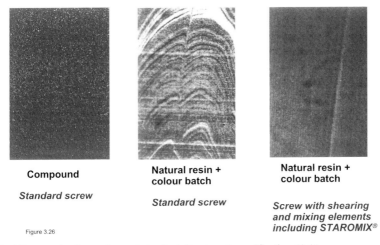

Compound

Standard screw

**Natural resin +
colour batch**

Standard screw

**Natural resin +
colour batch**

*Screw with shearing
and mixing elements
including STAROMIX®*

Figure 3.26

Figure 3.26   Microscopic views of extruded polyolefin pipes (magnification 40:1)

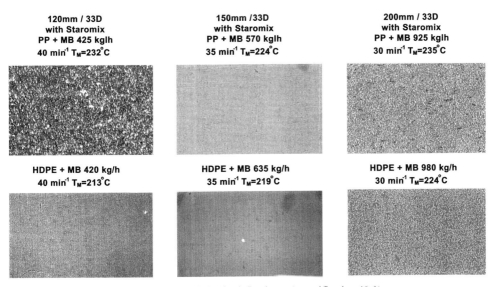

120mm / 33D
with Staromix
PP + MB 425 kglh
40 min⁻¹ T_M=232°C

150mm /33D
with Staromix
PP + MB 570 kglh
35 min⁻¹ T_M=224°C

200mm / 33D
with Staromix
PP + MB 925 kglh
30 min⁻¹ T_M=235°C

HDPE + MB 420 kg/h
40 min⁻¹ T_M=213°C

HDPE + MB 635 kg/h
35 min⁻¹ T_M=219°C

HDPE + MB 980 kg/h
30 min⁻¹ T_M=224°C

Figure 3.27   Microscopic views of extruded polyolefin sheets (magnification 40:1)

HDPE ready-to-use compound, which is DM 0.25/kg cheaper (see exemplary calculation in Table 3.2). A first approach already shows that capital investment in modern technology leads to a lasting economic success within a very short time.

Table 3.2   Economics of HDPE Pipe Extrusion

		HDPE ready-to-use compound Single-screw extruder	HDPE virgin material + color concentrate Single-screw extruder with dynamic mixer
Output	(kg/h)	800.00	800.00
Effective production	(h)	6800.00	6800.00
Raw material cost	(DM/kg)	1.70	1.45
Production cost	(DM/kg)	1.89	1.63
Production cost	(%)	100.00	86.00
Saving due to in-line dispersion as compared to the use of a ready-to-use compound: DM 1,418,800 per year			

3.2.1.7   Barrier Screw

The special feature of a barrier screw is a second screw flight in the barrier zone. Due to the slightly higher pitch of the second flight (type: *Maillefer*) and the decreasing flight depth of the solid-phase channel (type: *Barr*) the produced melt is separated from the solid material and conveyed along the shearing part of the second flight into the melt channel [2, 11, 12]. The barrier system offers the following advantages:

- High melting efficiency as a result of solid/melt separation and controlled melt temperatures
- Good homogeneity of the melt due to defined shear stress to the raw material
- Neutral pressure profile along the screw and, consequently, reduced susceptibility to wear

Some examples of applications for which barrier screws have proven themselves in practice due to their specific advantages over conventional screws are

- Sheathing lines for PE, PP, PVC, TPU, TPE, and PA (grooved feed extruders)
- Thermoforming sheet lines for PET, PP, and PE (smooth bore extruders, grooved feed extruders)
- Cast film lines for polyolefins (smooth-bore extruders, grooved feed extruders)
- Spinbonding lines for PP, PA, and PET (smooth-bore extruders)
- Blown film extrusion lines for polyolefins (grooved feed extruders).

A summary of the possible applications of a barrier screw and its specific advantages and disadvantages as against a conventional screw is presented in Table 3.3.

Table 3.3   Comparison of a Conventional Screw and a Barrier Screw

	Conventional screw	Barrier screw
Use in smooth bore extruders	Yes	Yes
Use in grooved feed extruders	Yes	Yes
Use in vented extruders	Yes	Yes
Melting performance	–	+
Homogenising efficiency	–	+
Conveying performance	0	0
Uniformity of pressure	–	+
Melt temperature development of highly viscous polymers	+	–
Processing range for various raw materials	–	+
Production flexibility	+	–
+, better; –, worse; 0, no difference		

3.2.1.8   Vented Extruders with 33D or 36D Screws

Economic processing of filled thermoplastics or plastics, such as ABS, PS, SB, PMMA, and PC, without predrying requires the use of vented screws. Such screws are generally provided with smooth or slightly grooved feed sections (longitudinal grooves). Feed sections with a large number of rectangular or crescent grooves are not employed, because, on the one hand, the screw section downstream from the vented zone cannot convey the large melt flow coming from a system of positive conveying action at a sufficiently high speed, and, on the other hand, the processing of regrinds leads to the reduction in the throughput of high capacity grooved feed extruders, especially in sheet and thermoforming sheet lines. This means that the installation of a smooth or slightly grooved feed section allows better matching of the first screw section, which plasticizes the raw material, and the second section, which vents the melt and builds up the pressure. Present-day conventional vented ex-truders are provided with processing units of 33D and 36D length.

The development of the vented extruders for the production of sheet and thermoforming sheet is particularly oriented to the improvement of extruders of >120-mm diameter. At present, there is a trend toward larger extruders that ensure a substantially higher productivity and efficiency.

The processing of PET without predrying into films or fibers on vented single-screw extruders or co-rotating twin-screw extruders has been propagated since the beginning of the 1990s [13, 14]. In PET processing, polymer chains degrade by thermal and shear stress as well as by hydrolysis. PET can be easily processed on conventional extruders if it is intensively dried first. Without predrying an equilibrium reaction occurs during extrusion, resulting in lower molecular weights (degradation reaction). If the PET is dehumidified by a vacuum venting system during extrusion, a degradation reaction can be almost completely avoided (Table 3.4).

Table 3.4   Intrinsic Viscosity (IV) and Degradation in G-PET Processing Without Predrying

PET type	Moisture (%)	Initial IV value	Final IV value	Degradation (%)	Note
G-PET	0.10	0.75	0.74	1.3	With vacuum venting
G-PET	0.10	0.75	0.71	5.6	Without vacuum venting
G-PET	0.28	0.75	0.72	4.2	With vacuum venting
G-PET	0.28	0.75	0.70	7.1	Without vacuum venting
A-PET	0.22	0.80	0.75	6.7	With vacuum venting
A-PET	0.22	0.80	0.69	15.9	Without vacuum venting
Extruder of 100-mm screw dia., length 33D, $\dot{m}$ = 500 kg/h					

G-PET can be processed in an undried condition without any qualification. In A-PET processing it is advisable to predry the material at 80 °C for 1 or 2 h. As against conventional crystallization at 100 °C for 1 h and A-PET drying by air at 175 to 180 °C for about 5 h, PET processing on vented extruders offers economic advantages since the capital investment and energy costs are lower. At present, only some limited test results are available for C-PET. Considering the pronounced degradation behavior of C-PET, its processing on vented extruders seems critical. A loss of film quality, e.g., in terms of transparency, may occur, particularly in the processing of PET regrind (for the most part bottle regrind). But especially with regard to the recycling of material, for which not only ecological but also economic aspects must be taken into account, PET processing without predrying on a vented extruder is an interesting process.

3.2.1.9   Reduced Wear as a Result of Appropriate Material Selection

Extruder wear is a subject that requires differentiation. Altogether, five wear inducing factors can be listed:

- Polymer to be processed
- Additives contained in the compound
- Processing parameters
- Design of screw and barrel
- Material used for the plasticizing unit

Increasing demands on high-capacity extruders led to the improved wear resistance of screw and barrel materials [9, 15].

Table 3.5 gives a survey of suitable screw and barrel materials for the processing of various raw materials into finished and semifinished products. To be able to observe the required quality standards, the materials used must be processed by special production methods, subject to thermal treatments, and meet detailed specifications.

Because the specific wear of plasticizing units may become very complex, customer-specific solutions must be found. It has proven to be a great benefit if everything comes from one single source, from the selection of the alloys, to the powder production and coating process, up to the completion of the plasticizing unit.

Table 3.5   Screw and Barrel Materials for Single Screw Extruders

Raw material	Finished / semifinished product	Screw	Barrel
HDPE, LLDPE, blends and fillers/ additives	Blown film, cast film	Through-hardened, with or without hard metal coating, or armored flights, with or without hard metal coating	With hard-metal linings or in bimetal design
HDPE or PP compound	Pipes, sheet, profiles, other	High-grade hardened and tempered steel	High-grade hardened and tempered steel
HDPE, PP and fillers / additives	Pipes, sheet, profiles, other	Through-hardened, with or without hard metal coating, or armored flights, with or without hard metal coating	Bimetal
PMMA, PS, SB, PC	Film, sheet, profiles, other	High-grade hardened and tempered steel, long-time gas-nitrided	High-grade hardened and tempered steel, long-time gas-nitrided
ABS	Film, sheet, profiles, other	Armored flights	Bimetal
Fluoropolymers	Miscellaneous	Hastelloy C4	Bimetal
Corrosive polymers or fillers/additives	Miscellaneous	Hastelloy C4 or Inconel 625, armored flights or X 35 CrMo17, armored flights	Bimetal

3.2.2   Single Screw Extrusion Technologies

3.2.2.1   Blown Film – A Classic Product with Good Properties

Blown film is a commodity product in the best sense of word, for which there is great demand worldwide in a very large number of fields. Despite the enormously high production figures involved, the trend is moving toward ever-improving product quality standards.

- *Industry packaging.* This range includes shrink and stretch film, bag film, and container liners.
- *Consumer packaging.* Here we find film for frozen products, shrink film for transport packaging, food wrap, and every kind of form, fill and seal packaging.
- *Laminating film.* This range of high-grade film is applied in a follow-on production process to another material, such as aluminum or paper, which is then used to package, for example, milk or coffee or products in the photographic industry.
- *Agriculture film.* In addition to manufacturing greenhouse film, crop promotion, and silage film, this group also includes lines for the manufacture of silage stretch film.
- *Barrier film.* This film incorporates special raw materials, such as polyamides and EVOH, which act as an aroma or oxygen barrier. It is used for packaging food, e.g., cold meats and cheese, and for medical products. It is manufactured on blown-film lines with typical structures ranging from 5 to 8 layers.

For all these blown film products, blown film lines are available with a blown film width from 50 to 20,000 mm and a throughput range between 20 and 2000 kg/h. Employing modular design principles, the individual subassemblies

- Extrusion
- Forming and cooling
- Sizing and haul-off
- Winder
- Line automation

enable blown filmblown film lines to be adapted to any requirement and application, thereby offering a maximum flexibility (Fig. 3.28).

The blown film process [16] is the method that implies the highest potential with regard to film dimensions. It is used for flexible, semirigid, and rigid mono- or multilayer film. The key feature is the adjustable biaxial orientation of the melt from the die to the film tube by the so-called blowup ratio (BUR). A blowup ratio from 1:1 to approximately 1:5 allows to influence specifically the strength, elongation at break, and shrinkage of the film both in longitudinal and transverse direction. Film composites of 2 to 8 layers can be produced by coextrusion, but the number of layers has to be determined in accordance with the principle "as many as necessary" for the intended application and required functionality of the film composite. This is all the more important because one melt distributor channel for each layer has to be integrated in the coextrusion die so that it becomes increasingly complex with an increasing number of layers.

In blown film coextrusion, primarily two systems are used:

- Disk-type system with horizontal melt distributor (Fig. 3.29)
- Ring-type system with vertical melt distributor (Fig. 3.30)

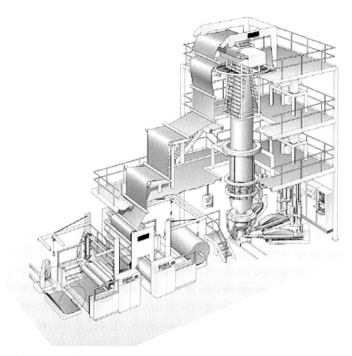

Figure 3.28   3-layer blown film line

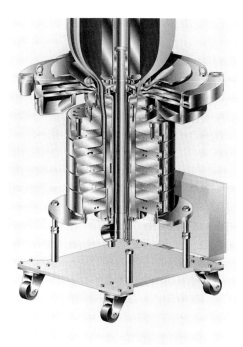

Figure 3.29   Coextrusion blown film die with horizontal melt distributor

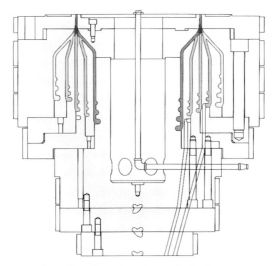

Figure 3.30   Coextrusion blown film die with vertical melt distributor

State of the art for blown film dies are spiral mandrel dies with rheological and thermal optimized flow channels. A minimum length of the flow channel guarantees short purging periods and increased service life of the die. Optimum rheological design offers a high flexibility in producing different film structures with all common resins in blown film extrusion.

In comparison with traditionally spiral mandrel dies, the horizontal stack die system has the advantage that the polymer melts are thermally isolated until they join in the die. In addition, it is possible to change the number and combination of layers by simply exchanging the individual modules. This offers the film producer a high degree of flexibility and the possibility of meeting new market requirements at any time.

3.2.2.2   Double-Bubble Process for Biaxially Oriented Blown Film

Biaxially oriented film that is high in strength, is high in yield, and offers excellent optical clarity is the product of a double bubble line. This type of machine combines blown film extrusion with the double-bubble process for in-line production of triple-layer, coextruded shrink films from polyolefin compounds (LDPE, LLDPE, PP, mPE). The double-bubble process is divided into two steps (Fig. 3.31). The first step is the perfect extrusion of a primary tube, which is shock-cooled with a wet sizing unit. This primary tube is preheated by infrared radiators in the second step and finally simultaneously biaxially oriented in the downstream stretching unit, which is heated, too.

The quality of a double-bubble line is reflected both in the stability of production sequence and in the ability to switch more quickly and more flexibly to varying raw materials and film compounds. The in-line production method for simultaneously biaxially oriented shrink film is the alternative to known off-line production methods for biaxial film, in which the cast film is stretched on large-area, two-stage draw stands.

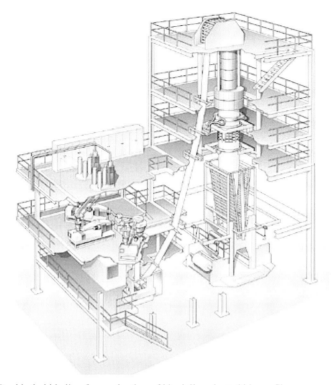

Figure 3.31   Double-bubble line for production of biaxially oriented blown films

Typical applications are

- Packaging film for books and newspapers
- Packaging film for audio and video cassettes
- Packaging film for cosmetic products
- Packaging film for toys and games
- Packaging film for confectionery and foodstuffs, e.g., bread
- Capacitor films
- Tapes
- Vending machine shrink films

3.2.2.3   Cast Films – Quality of the Highest Standard

Precision films of high quality are typically produced by the chill roll process. The key feature of this technology is the cooling of the melt, leaving the die on a chill roll. This ensures the production of films with outstanding optical and physical properties suitable for a large range of applications. The increasing demand for high-quality cast films ensures a constant growth market. The range of applications for these products, made on cast film lines, is constantly expanding. The required room to manenver in this development of new products calls for flexible plant technology and wide-ranging facilities for the processing of various raw materials. The main applications for cast film lines cover the following single-layer or multi-layer products:

- Packaging of textiles
- Film liners
- Office organization material
- Flower packaging
- Food packaging
- Pharmaceuticals packaging
- Metallized foils
- Sanitary product packaging
- Stretch films
- Industrial films

The reliable, trouble-free manufacture of high-quality films can be ensured only by production lines that are made to the highest quality standards, are product related, incorporate advanced technology, and are supported by efficient after-sales service. Today, different series of cast film lines are available with a width range between 900 and 4000 mm and throughputs up to 1500 kg/h. The following modular components help to fulfill the customer's wishes or production requirements and allows a tailor-made line concept (Fig. 3.32):

- Extruder
- Cast films die with multilayer adapter system
- Casting station with cooling system, air knife, vacuum box
- Preliminary haul-off
- Slitting equipment
- Winder
- Winding shaft or reel withdrawal unit
- Edge trim recycling system
- Preembossing heating station
- Embossing unit
- Stretching unit
- Line automation

The cast films extrusion process [16] is limited by the gauge and width range of the cast films, which depends on the width and cooling capacity of the chill roll. The chill roll process is characterized by higher flexibility than the blown film method because the chill roll covering can be provided with surface textures, enabling the production of slightly embossed or matte film surfaces of approximately 180 μm thickness. These films are used for the production of office organization material, such as transparent covers. Additional embossing units are necessary if thicker films of up to approximately 400 μm are to be textured with strong patterns.

Compared to the blown film method coextrusion of various melt layers is much easier since a so-called coextrusion adaptor can be used in combination with a slot die (Fig. 3.33). There are a few systems available on the market:

- Dow feed block
- Cloeren plug adaptor
- Reifenhäuser slide valve feedblock (Fig. 3.33)

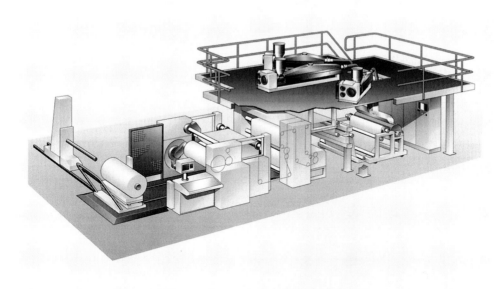

Figure 3.32   Cast films extrusion line

The slide valve feedblock is the only one that allows adjusting the velocity of the single flow as well as the distribution of each layer without changing inlets of the feedblock and all this during production. With divided or profiled slide valves which are similar to the behavior of a restrictor bar, it is possible to compensate high throughput and viscosity differences. Quick interchangeable cassettes enable a different structure of the polymers. Indicators on all sliders guarantee the reproduction of different structures. Typical products are films with up to 7 layers.

The die design is characterized by a rheologically optimized flow channel in form of a coat hanger distributor and a flexible lip. The layout of a die is based on many years of experience, especially in balancing the die head with the manifold. Systems with an additional restrictor bar are not common. The hard chromium plated surface of the flow channels gives a high resistance against corrosion due to the resin and also a minimized tendency for degradation.

Heater cartridges are used to warm up the die body and to compensate for heat losses. The position of the heater cartridge has to be selected very accurately to get a uniform heat distribution of the flow channel surface. This yields in a thermal equilibrium to achieve a minimum of gauge tolerances. Optimized insulation will support this. The flat film dies can be equipped with steplessly variable deckle plates on both sides.

State of the art are flat dies with automatic flexlips. Actuators are available as thermal bolts or as "translators" (Fig. 3.34) that use the piezo effect. The translators have the advantage of a short response time and a very low energy consumption. These automatic flexlips are incorporated in a gauge control system and keep the gauge tolerances within a minimum range.

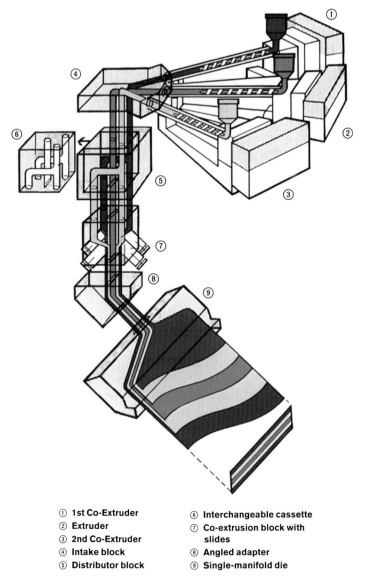

① **1st Co-Extruder**	⑥ **Interchangeable cassette**
② **Extruder**	⑦ **Co-extrusion block with**
③ **2nd Co-Extruder**	**slides**
④ **Intake block**	⑧ **Angled adapter**
⑤ **Distributor block**	⑨ **Single-manifold die**

Figure 3.33   Adaptor coextrusion system (patented by Reifenhäuser)

Another advantage of the cast film method lies in the fact that with comparable film composites and film dimensions, which can also be produced by the blown film process, it is possible to achieve a higher film transparency and higher production capacity due to more intensive cooling of the melt. Here, film composites with PP outer layers are predominant. The extrusion technology using slot dies, however, allows only monoaxial orientation, which may involve some disadvantages in further processing into industrial packaging because of a tendency to split.

Figure 3.34   Flat film die with automatic slot control (ASC) (photo: Reifenhäuser)

3.2.2.4   Extrusion Coating – Improved Quality for Composites

New composites of high quality are created by the durable bonding of material webs like, e.g., paper, aluminum foils, plastic film, textiles, nonwovens, sheet steel, etc. After lamination or coating with thermoplastic materials suiting the specific product properties of the goods to be filled in, these composites find a broad application in the packaging industry. Milk packs and food packaging paper are classical examples.

Composites (substrates) with the following properties are obtained by lamination with thermoplastic materials:

- Improved mechanical properties (tear strength, elongation, thermoformability)
- Improved barrier effects against water, gases, fats, aromatic substances, etc.
- Improved printability and optical properties (gloss and transparency)
- Lamination is also a prerequisite for improved heat-sealing properties

Other essential advantages of the extrusion coating technology are that

- Production of plastic films (base film) and conversion in one operation step.
- This technology does not use any pollutant solvents and adhesives as required by pollution control regulations becoming more and more stringent.

Some typical product examples for extrusion coating are

- *Paper/PE (polyethylene)*. Paper/PE coated composites are used for all kinds of packagings; e.g., for meat products:
    - A thinner coating (7–20 g/m²) features good adhesion, good printability, and a certain barrier effect; e.g., packaging of sugar in portions
    - A thicker coating (25–100 g/m²) shows improved mechanical strength, good barrier effect, hot-sealing property, surface gloss, and smoothness.
- *Cardboard/PE (polyethylene), cardboard/PE/aluminium foil*. This coating has a positive effect on appearance, excellent barrier properties and stiffness; e.g., milk packs.
- *Cardboard/PP (polypropylene)*. PP is resistant to boiling water and is sterilizable; therefore, these composites are used for ready meals and square-bottom packs.

- *PP nonwovens/PP (polypropylene)*. PP nonwovens coated with PP plastics films are used for roofing applications requiring mechanical strength and water impermeability but water vapor permeability. They are also used for sanitary applications.
- *PP tape fabric/PE (polyethylene)*. PP tape fabric coated with PE is used for heavy-duty sacks, tarpaulins, and waterresistant decorative laminates; e.g., wipe-clean wallpapers.

A variable line concept (Fig. 3.35) suitable for a wide range of products provides the following essential features:

- Extruder with movable extruder undercarriages
- Coating die with steplessly variable deckling system and multilayer adapter system (Fig. 3.33)
- Laminator
- Unwinding and winding systems for nonstop operation at maximum speed
- Line automation

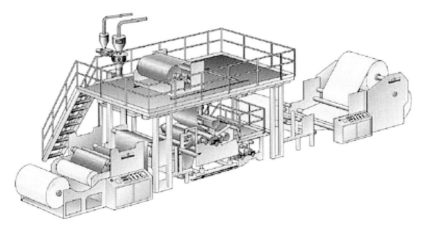

Figure 3.35   Extrusion coating line

Generally, the process of extrusion coating [16] of different laminates (composites) with a film of thermoplastic material is as follows: In the extruder, the thermoplastic polymer is processed into a homogeneous melt and formed in the flat die head into a melt web. The material web (substrate) is fed into the gap between laminator roll and pressure roll. The melt web applied to the substrate sticks as a foil to the surface of the substrate. The contact pressure of the pressure roll and the required temperature of the melt effect an intensive joining of the plastics foil and the substrate. Depending on the structures being combined (coextrusion), certain bonding agents may be required. The coated structure is then cooled down by the constant temperature rolls and transported via a thickness gauge and a post-cooling unit to the winder.

The adhesion between the substrate and the thermoplastic polymer, e.g., PE melt, depends on the degree of oxidation. For coating, a temperature between 230 and 300 °C is required at the coating die head, depending on the resin being processed. If this temperature range cannot be observed, the following procedures may improve the adhesion between melt and substrate:

- Electrostatic Corona pretreatment of the substrate
- Preheating of the substrate (especially paper and cardboard)
- Use of a bonding agent in coextrusion
- Additional oxidation by ozone-enriched air blown against the melt

3.2.2.5   Plastic Sheets and Thermoforming Sheets – In Highest Quality for a Host of Applications

We encounter plastic sheets in all walks of life. In most cases, the starting product is a sheet that is turned into the final product on vacuum-forming lines. Typical sheet and thermoforming sheet products include:

- Rigid packaging films from PS, PP, or PET for food and nonfood applications
- Office organization material (PP)
- Filled semifinished products for the automotive industry or food packaging (PP + mineral fillers)
- Highly transparent protective glazings (PMMA, PC, CAB)
- Internal refrigerator components (SB, ABS)
- Suitcase shells (ABS)
- Side panels for shower cubicles (PS)
- Sanitary products, such as bath tubs and washbasins (ABS/PMMA)
- Apparatus, laboratory, and tank construction (PP, HDPE)
- Transparent or colored flexible sheets (soft PVC)

Economic extrusion is always governed by a variety of factors. The requirements placed on the product must be harmonized with production. Extrusion lines for the production of sheets and thermoforming sheets [16] fully satisfy this demand, thanks to their modular design, extensive range variants, and advanced technology. The main components are

- Extruder
- Slot dies with multilayer slide adapter or multilayer slot die
- Various polishing stack versions (steel belt, vertical or horizontal version)
- Cooling tables
- Haul-off units
- Cutting units (for longitudinal cutting and transverse cutting)
- Stacking systems
- Winder (single or multiweb)
- Embossing station
- Laminating devices for protective films
- Automation systems

Line systems can be tailored to suit the requirements placed on the product. Thin thermoforming sheets polished on both sides can be produced with the steel-belt technology (Fig. 3.36).

The steel-belt system is capable of producing superior optical surface qualities to double-polished PP sheets in the critical thickness range from 0.1 to 0.4 mm. This thickness range of film products cannot be produced in an optimum way, either with a chill roll line or with a thermoforming sheet line (Fig. 3.31). It is also the ideal choice for the production of sheet in highest optical quality from PS, PET, PC, and PMMA. Typical applications are

- Blister packs
- Pharmaceutical blister packs
- Transparent folding boxes
- Stationery sheet > 0.150 mm
- Polished sheet with high filler content for index dividers and file covers
- High transparency thermoforming sheet
- Packaging sheet and sheet for technical applications from PS, PET, PC, and PMMA

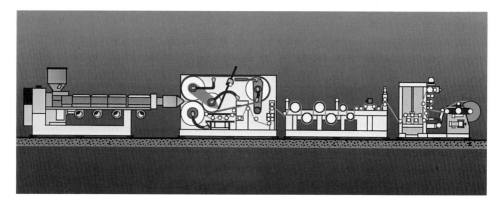

Figure 3.36   Steel-belt extrusion line

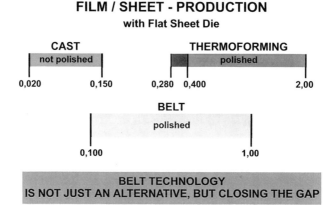

Figure 3.37   Thickness range for film products, produced on chill roll, steel belt, and thermoforming sheet lines

With the steel-belt system the sheet is double polished between a roll and the steel belt, which is in contact with a segment of the roll. Double polishing with this new system produces a much better effect than linear polishing in a three-roll polishing stack, which is the conventional method used for polishing thermoforming sheet. The use of the steel-belt system allows a considerable improvement in sheet quality to be achieved, especially in respect of transparency and gloss.

Thermoforming sheets are mainly extruded by vertical extrusion (Fig. 3.38), which offers the following advantages:

- Controlled melt flow into the polishing nip, also for thin film with low melt stability
- Perfect handling, allowing good insight into the polishing nip
- Feed and web position easily adjustable
- No condensation on the film
- Optimum chill roll operation for thin film production

Figure 3.38   Thermoforming sheet line (vertical extrusion) (photo: Reifenhäuser)

Sheet extrusion is mainly done by horizontal extrusion (Fig. 3.39). Advantages of horizontal extrusion are

- Particularly suitable for film of high melt stability
- Simple line assembly
- Easy access
- Also suitable for combined thermoforming sheet / standard sheet lines

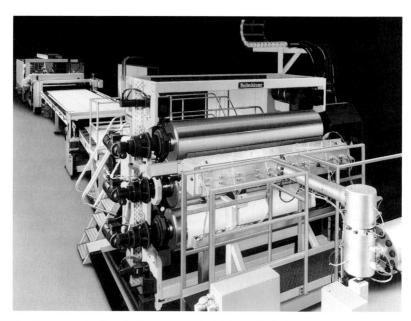

Figure 3.39   Sheet extrusion line (horizontal extrusion) (photo: Reifenhäuser)

3.2.2.6   Plastic Pipes and Profiles – The Ideal Components for Civil Engineering and Technological Applications

The diversity and continual development of the plastic pipe and profile market place exacting demands on extrusion line systems. The following explanations are mainly focused on extrusion systems with single-screw extruders. The very important market of PVC pipes and profiles is not put into consideration, because these products are produced on production lines, equipped with counterrotating twin screw extruders processing PVC dry blend in powder form. The production of pipes [16] can be divided into following products:

- Polyolefin pipes up to 2000 mm diameter; e.g., pressurized pipes for gas and water, drainage pipes, and cable protection pipes
- Coextruded pipes and tubes, comprising up to 5 layers, for special applications
- Textile-reinforced, plasticized PVC tubes; e.g. water and compressed-air hoses
- Biaxially orientated pipes from HDPE for cartridge cases
- Polyamide tubes; e.g., for the car industry (single and multilayer)
- Tubes and hoses for the medical sector (single and multilayer)

The economic extrusion of pipes is always determined by a host of factors. The requirements placed on the product must be matched with production capacity. A cable protection pipe intended for domestic installation is, for example, manufactured with a reduced wall thickness and from lower-grade material but at a high extrusion speed, whereas piping designed for underwater installation is produced with a thicker wall of high-grade material at a slower extrusion speed. Similar criteria apply to hydraulic pipelines, pipes for under-floor heating systems, and pressurized pipes for gas, water, and chemicals.

Extruded profiles [16] of all kinds have increased their market share, and in certain areas it is hard to imagine how to do without them, since they have completely replaced some traditional raw materials. Furthermore, there are almost no boundaries to creativity regarding the shape and design of profiles. Typical non-PVC products are as follows:

- Profiles made from polystyrene (PS), styrenebutadiene (SB), and acrylonitrile butadiene styrene (ABS) are used for furniture profiles, e.g. as main body parts and corner connection pieces. PS profiles are used for fluorescent light-tube covers.
- Profiles from PMMA are common because of their very good light stability and resistance to aging. Light covers, with or without embossing, have particularly proven themselves, as have double- and triple-walled sheets for greenhouse, wind protection on terraces, and pergola covers; or as solar collector elements, where, particularly in the greenhouse area, the outstanding translucency and insulating property of the multishell building elements offer enormous advantages.
- The range of applications for PC profiles is similar to that of PMMA, though PC possesses an inherently higher flexural strength; it is, however, subject to aging after long exposure to sunlight, and it yellows slightly. This can be improved with a coextruded or lacquered PMMA top layer. Thereby, it is possible to combine the advantages of both raw materials.
- Areas of applications for PP profiles are in the construction of solar collectors, as cardboard substitute where moisture is a problem, as a thermal break for aluminium windows, and in the construction of ventilation shafts.
- Although polyethylene (PE) profiles are difficult to calibrate, certain mass products are on the market, e.g., heat shrinkable collars made of cross-linked PE for joining or repair of telephone cables, or multiapplication profiled pipes as cable-protection conduits for broadband communication systems.
- Cellulose acetobutyrate (CAB) has established itself for trim strips, because of its outstanding aging resistance and good mechanical properties.
- Solid rods from thermoplastics like PA, POM, PE, PC, and PMMA are extruded in the same way as profiles – through sizing sleeves. Quality requirements regarding homogeneity and freedom from blowholes are more stringent than with normal extruded profiles and place special demands on the die and the downstream equipment during production. Solid rods from PA and POM are used as machining stock for gearwheels, bearing bushes, etc.

The basic components of a line for production of pipes (Fig. 3.40) and profiles are

- Extruder
- Profile extrusion dies
- Pipe extrusion dies (single and multilayer)
- Vacuum tank sizing units, profile sizing units
- Cooling facilities
- Wall thickness measuring units
- Haul-off units
- Cutting equipment
- Winders
- Automation systems.

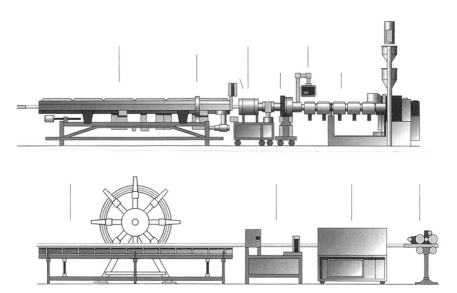

Figure 3.40   Pipe extrusion line

3.2.2.7   Synthetic Filaments – Monofilaments, Slit Film Tapes, and Strapping Tapes

3.2.2.7.1   Monofilaments

The numerous applications of monofilaments [16] with a currently possible diameter range of 0.06 mm to 4 mm are the reason for the great market importance of monofilament lines, which will be increasing in future due to specially developed polymers or blends that allow to open up new possibilities or to extend existing applications. The most important applications and endproducts are

- Bristles for broom and brush applications (PVC, PP, HDPE, PA, PET)
- Technical applications as filter meshes, filter fabrics, and conveyor belts (PET, PA, PPS, PVDC, PP, HDPE)
- Fishing lines and nets (PA, HDPE)
- Woven fabrics (HDPE, PP, PVC)
- Packaging products (HDPE, PP)
- Zip fastener (PET, PA)
- Optical fibers (PMMA, PC)

3.2.2.7.2   Slit Film Tapes

In the past many of slit film tape products were made of natural fibers, such as jute, hemp, and sisal. These fibers, however, have some disadvantages that can be eleminated by the use of plastic fibers: Woven plastic sacks have higher resistance to chemicals. Plastic sacks are fungistatic. Water is practically not absorbed. Compared with a jute sack, a plastic sack that has the same carrying capacity weighs less. Typical slit film tape products made from PP, HDPE, and LLDPE include

- Carpet backing
- Woven sacks for fertilizers, raw materials, and other bulk goods; e.g., 50-kg sacks
- Heavy-duty sacks for packaging
- Binder twine
- Raschel-knitted sacks for nets
- Sewing thread for sacks
- Concrete reinforcing material
- Simulated turf, etc.

For the production of slit film tapes there are three different processes:

1. Film cooled in a water bath
   Advantages
   - Low capital outlay, as, among other things, less winders are required
   - Simple to operate
   - More effective cooling, enabling the production of thicker tapes
   - Higher takeoff speeds, thus higher output rates
   Disadvantage
   - More water is left sticking on the film if PP with UV stabilizers or especially HDPE is processed
2. Film cooled by chill roll casting
   Advantages
   - High takeoff speeds for HDPE and PP film additives
   - Thin tapes up to 0.022 mm can be produced
   - No water is left sticking on the film
   Disadvantages
   - High capital investment
   - Thick tapes cannot be produced because they are cooled on one side only
3. Blown film cooled by air
   Advantage
   - Only in the production of tapes from HDPE that shall be converted on raschel-knitting machines that require extremely soft tapes
   Disadvantages
   - High capital investment
   - Low throughput and line speed due to air cooling
   - Thick tapes cannot be produced

3.2.2.7.3 Strapping Tapes

Prior to the discovery of the plastic strapping tape, steel hoops were used almost exclusively. After the first use of the plastic tape, people quickly recognized the advantages it offers over steel hoops. The tapes made from PP and PET are used for strapping goods such as newspapers, cardboard boxes, or other products that are to be packaged. Monofilament and strapping tape lines (Fig. 3.41) are mainly composed of modules, such as [16]

- Extruder
- Spinning heads

- Cooling tanks
- Draw stands
- Hot air or hot water stretching bath
- Spray cooling systems for tape production lines
- Embossing unit for strapping tapes
- Winding machines
- Automation systems

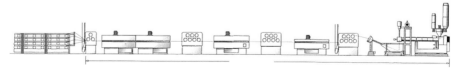

Figure 3.41   Monofilament extrusion line

3.2.2.8   Nonwoven Textiles – The Cost-Effective Alternative

PP and PE spunlaid nonwovens are a basic material for numerous applications. As their name implies, nonwoven textiles are not made by conventional weaving or knitting processes, but are composed of ultrafine continuous fibers made from thermoplastics, which are randomly laid to a web onto a conveyor belt and subsequently bonded together. Nonwovens are capable of straightforward further processing; e.g., they can be welded, sewn, printed, or secured with adhesives. In addition, they are washable, can be dry cleaned, and are abrasion-resistant. Special properties, such as water repellence, permeability, or UV stability, can be obtained by mixing additives to the raw material or applying agents to the fabric. The multitude of potential uses for nonwovens opens up entirely new markets.

The entire range of today's nonwoven materials made of polypropylene and poly-ethylene can be manufactured by the spinbonding method. Typical applications for non-wovens include:

- Personal hygiene
- Garments and shoes
- Medicine applications
- Specific industrial apllications
- Agriculture and horticulture
- Motor vehicles
- Coating substrates
- Packaging

Microfibers nonwoven fabrics are produced by the melt-blown method. One of the advantages of the melt-blown process is that it is insensitive to the raw material being processed. It is not limited to any thermoplastic polymer. Typical melt-blown resins are

- Polypropylene
- Polyesters (PBT and PET)
- LLDPE, LDPE, HDPE
- EVA, EMA, EVOH

- Polyurethane (TPU)
- Polyester elastomers (TPE)
- Nylon 6, 6.6, 11, 12
- Copolymers

Products made from melt-blown nonwovens, or composite structures including melt-blown fabrics, continue to grow in both number and diversity. The best known among them include

- Hot-melt adhesive fabrics
- Industrial wipes
- Oil cleanup products
- Cable insulation
- Filters (air/liquid)
- Patient drapes
- Thermal insulation for outer wear
- Tablecloths
- Artificial suede and leather
- Battery separators
- Diapers
- Sanitary products
- Surgeons´ masks and gowns

Spunbonding lines and melt-blown systems are composed of the following components:

- Suction conveyor and metering station for the polymer and additives
- Spinning extruder and edge trim recycling extruder
- Metering pump
- Spinning beam with integrated die pack, melt blowing die
- Air control system for the entire process
- Integrated web-forming system
- Conveyor belt machine with suction-air duct
- Thermo-embossing calender
- Automatic fabric winder with longitudinal and crosscutting devices
- Automation system

State-of-the-art spunbonding systems are using a heated coat hanger die with integral spinneret. The melt is uniformly distributed across the spinning beam and the one die pack only, which produces a filament curtain composed of thousands of single filaments. Cooled air on both sides against the filament curtain accelerate the single filaments through the stretching zone. The stretched force is applied to the filaments without using compressed air. The same air stream guides the filaments into the web-forming unit, where the very uniform and homogeneous web structure is created. The nonbonded web is transferred to a thermobonding calender, where the filaments are welded together, then to a winder where the nonwoven web is wound.

The melt-blowing process, developed by Exxon, is a one-step process for the production of fibrous webs directly from resin, by attenuating filaments with high-velocity hot air. Melt blowing is able to produce finer fibers than other filament-producing processes.

The intermingling of the filaments forces a random lay down with a self-bonding effect. Melt blown seems to be an easy process, but there has to be much experience in the linkage of both air and polymer process. In the melt-blown process, the melt metering system is comparable to the spinbonding system. Melt spinning is done through a single line die with 25 up to 50 orifices/inch and a diameter of each between 0.25 and 0.4 mm. The hot air stream embraces the filaments and accelerates them rapidly, so that ultrafine fibers are achieved. Fibers and powders can be incorporated into the melt-blown fabric to tailor it to specific end use demands.

The web is formed by interlacing and thermal-self bonding while the filaments are blown onto the conveyor belt. For controlling web lay down an exhaust fan is used. Afterwards, the web is wound up and/or slitted in the same way as used in the spinbonding system.

By the incorporation of one or more melt-blown units in a spinbonding line (Fig. 3.42), the spinning technology has been brought to highest perfection. Spunbond/melt-blown (SM), SMS, or SMMS fabric with superior characteristics for hygienic, medical, and industrial applications are the result. While the spunbonded layer gives the necessary strength and stability, the melt-blown layer offers superb absorption, filtration, and barrier properties. SM, SMS, or SMMS fabrics also have superior optical properties. Because of the melt-blown systems, microfiber webs of only 2 g/m² can be applied onto or into the spunbonded fabric. For some applications, the weight of the microfiber web is up to 12 g/m².

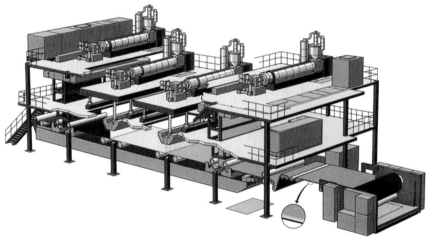

Figure 3.42   SMMS nonwoven extrusion line

3.2.2.9   Outlook

Increasing demands on the performance of single-screw extruders give rise to further efforts to increase the melt throughputs and to enhance the melt and product quality.

3.2.2.9.1   Design Conditions

The basic condition for the manufacture of efficient single-screw extruders is a well-thought-out modular principle of design. The modular principle allows an optimum layout and adjustment of the machine components in accordance with the processing requirements. This results in individually tailored machines with standardized single components, which, nevertheless, do not represent nonstandard lines and ensure proven quality and high performance.

3.2.2.9.2 Processing Conditions

Powerful gearings are the basis for high specific and absolute outputs. High specific outputs that are stable throughout the entire speed range are the condition for processing at low melt temperatures. The consistent technical improvement of extruders with grooved feed bushes, e. g., for pipe and blown-film extrusion, shows that the required increases in output can be reached only by extended plasticizing units. The increasingly heavy requirement to improve the economics induces extruder manufacturers to attempt to further enlarge the processing range (range of raw materials and output) of their machines.

More and more new types of raw materials, such as thermoplastics of higher molecular weight, plastics of bimodal molecular weight distribution, metallocenecatalyzed polyolefins, and or high-temperature thermoplastics, require not only defined melt throughputs of the plasticating extruders but also homogeneous melts up to a processing temperature of 400 °C.

The development of hard-wearing screw and barrel materials will continue to be pursued in the future.

References

1. *Der Extruder im Extrusionsprozeß – Grundlage für Qualität und Wirtschaftlichkeit* (1989) VDI Verlag, Düsseldorf
2. Wortberg, J., Extruderkonzepte unter Berücksichtigung internationaler Entwicklungen und neuer Rohstoffe. In [1]
3. Boes, D., Krämer, A., Lohrbächer, V., Schneiders, A., *Kunststoffe* (1990) 80, p. 659–664
4. Langhorst, H., *Praktische Entwicklung von Schnecken für Hochleistungsextruder* RWTH Aachen (College of Advanced Technology), (1989) Dissertation
5. Menges, G., Bolder, G., Langhorst, H., *Kunststoffe* (1988) 78, 3, p. 261–268
6. Stausberg, G., Improved flexibility in tape production by advanced extrusion technology. *Polypropylen '94 Zürich* (September 1994)
7. Krumm, K., Schroeter, B., Betriebserfahrungen mit axial genuteten Zylindereinzugszonen und die speziellen Anwendungen für die Extrusion unterschiedlicher Rohstoffe und Produkte. SKZ Fachtagung (trade conference) "Feststoff Förderung in Einschneckenextrudern", 20.–21. 10. 1987
8. Schroeter, B., Einfluß der Einzugszone auf das Arbeitsverhalten des Extruders. In [1]
9. Stommel, P., Extrudieren. In: *Verschleiß in der Kunststoffverarbeitung* (1990) Carl Hanser Verlag, München, Wien
10. Feistkorn, W., Gleichmäßigkeit: Statistische und dynamische Mischer bestimmen die Homogenität plastifizierter Schmelzen. *Maschinenmarkt 95* (1989) 45, p. 20–23
11. Limper, A., Barriereschnecken für die Hochleistungsextrusion. In [1]
12. Fischer, P., Wortberg, J., *Plastics No. One* (1995) 8, p. 21–27
13. Wiedmann, E., Wohlfahrt-Laymann, H,. Kunststoffe (1992) 82, p. 934–938
14. Nissel, F. R., *V-PET PET-Sheet Extrusion by Vented Extrusion*, PET '96, Zürich, October 1996
15. Lülsdorf, P., Verschleißschutz für Zylinder und Schnecken.In [1]
16. Hensen, F., *Plastics Extrusion Technology* (1997) Carl Hanser Verlag, München, Wien

3.3   Screw Design

Volker Schöppner

3.3.1   Introduction

The screw is an essential component of extrusion lines. The task of an optimal design of the screw geometry confronts even experienced process engineers with considerable difficulties. This is due to the very complex screw geometry of today, which comprises a great number of degrees of freedom. Since it is difficult to predict the interaction of these geometrical variables and the process behavior, a concrete product-oriented screw optimization always requires for many experiments. Fig. 3.43 shows the usual process, where the precise choice of the single screw modification steps is simplified by means of simulation software. The screw may become the bottleneck of the entire plant if the downstream equipment has large-scale dimensions. The factor that determines the throughput is the homogeneity of the melt. Besides this, an increase in throughput is possible in almost all extrusion lines if the melt temperature can be reduced by a better screw design. Thus, the efficiency of an extrusion plant can be improved in almost all cases by an optimization of the screw.

The screw design is mainly done by the machine manufacturers, who wish to achieve the highest throughput possible with a fixed machine diameter to remain competitive on the market. Here it is important to achieve the throughput that has been agreed upon with the machine operating customer in the first attempt because subsequent changes of the screw

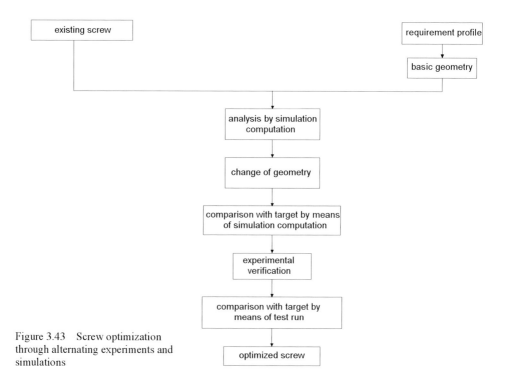

Figure 3.43   Screw optimization through alternating experiments and simulations

will cause downtime in the extrusion plant and will thus entail higher costs. The screw must be able to operate materials from different raw material producers to allow the highest degree of flexibility for the purchase. In contrast to this, many raw material producers offer to participate in the screw design for very special types of raw materials.

This chapter deals with the boundary conditions and the evaluation criteria of screws. Different screw concepts are presented, setting forth the application-specific advantages and disadvantages. A design systematic is presented by means of which the steps from the requirements profile to the screw geometry are illustrated.

3.3.2   Evaluation and Specification

Fig. 3.44 shows a scheme of the boundary conditions of the screw in the barrel. The crucial input variable is the material that enters the screw channel. The quality of the raw material is of great importance for the product quality. The second input variable is the machine control, i.e., the temperature control and the screw drive. This also critically influences the product quality. The third input variable (drawn on the output side according to the material flow) is the downstream equipment. From the screw's position the downstream equipment starts directly beyond the tip of the screw. The flow resistance influences the pressure at the tip of the screw and thus the behavior within the screw. For the screw designer it is essential to exactly specify the input variables to obtain reliable boundary conditions. This is especially important if the machine control and the drives are not comprised in the delivery range.

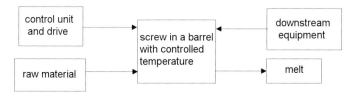

Figure 3.44   Input and output variables

The melt is the output variable of the screw. Regarding the features that determine the quality, the guaranteed values generally have to be indicated at the corresponding junction point of the delivered module. A direct specification of these values at the end of the screw is needed only if the downstream equipment is provided by another supplier.

3.3.2.1   Specification of the Input Variables

3.3.2.1.1   Material

The following information must be acquired in advance to detect risks in determining the screw geometry:

- The polymer types to be processed have to be known.
- The viscosity range of the materials to be processed must be indicated. Not only the viscosity range of new materials but also that of the mixture with low-viscosity regranulated scrap must be determined.

- The additives to be incorporated have to be specified because abrasive filling materials influence the choice of the steel and the surface treatment.
- The content of the regranulated scrap and of the nongranulated recycling material must be specified. Here, the determination of the fines content has to be considered.
- The minimum bulk density, especially in case of a high content of film chips, affects the selection of channel depth of the feeding section.
- The temperature of the input raw material, e.g., the drying temperature must be known.
- For hygroscopic polymers the maximum content of humidity that is admissible has to be determined.
- For the guarantee runs a reference material has to be determined.

Table 3.6 shows an example of a material specification.

Table 3.6   Resin Specification

Material	Polypropylene-homopolymer
Viscosity range	MFI 3 – 6 (230 °C, 2.16)
Melting point	165 to 168 °C
Melt Flow Index variation	± 0.5
Density	0.9 g/cm^3 to 0.905 g/cm^3
Pellet size	2 to 4 mm, regular
Water content	max. 0.005%
Maximum content of regranules	15%
Maximum content of scrap	15%
Minimum content of virgin	70%
Minimum bulk density of mixture	0.4 g/cm^3
Temperature of the resin	5 to 35 °C

3.3.2.1.2   Drive

The necessary torque and motor power that results from the maximum speed of the screw have to be determined by the screw designer. Besides, the consistency of the drive speed directly affects the product quality, because a temporal speed variation will show in the thickness of the extrudate that results from the corresponding throughput variation. The constant speed depends on the quality of the drive motor and on the consistency of the voltage. This is quite a problem in many countries of the world. Table 3.7 shows an example of a specification of the drive situation.

3.3.2.1.3   Temperature Control

The screw operates in a temperature-controlled barrel, which in most cases is heated with electric band heaters and is air-cooled. There are special constructions where the screw temperature is controlled by means of a constant temperature fluid, or the air cooling system is omitted, too which is frequently the case in injection molding. For the screw design it is important to know where the temperature at the barrel wall can be set. This mainly depends on the design of the barrel, the thermocouple position, and the type of temperature control, which affects the value of the heat-transfer coefficient that can be obtained at the barrel liner. Table 3.8 indicates estimated values.

Table 3.7   Drive Specification

Torque	5500 Nm in the whole speed range
Screw speed	210 rpm
Controller accuracy (screw speed)	± 0.1%
AC frequency	50 Hz ± 1%
Main voltage	380 V ± 10%

Table 3.8   Heat-Transfer Coefficients

	Heating (Wcm²)	Cooling (W/cm²)
Water cooling	–	3.50
Oil heating/cooling	0.8	0.80
Air fan with ribs	–	0.40
Air fan without ribs	–	0.15
Electric band heater	0.5	–

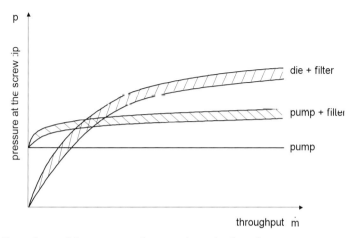

Figure 3.45   Dependency of the pressure at the screw tip on the throughput

Temperature control by means of constant temperature fluid is characterized by a constant temperature that is kept over a long time because the heating and cooling of the corresponding medium is done in a separate cooling device. The regulators of the electronic band heaters are mostly not equipped with a constantly working phase control but with self-optimizing, three-point controllers. Today, a constant temperature of $\pm 1\,°C$ can also be obtained with these regulators and is absolutely satisfactory.

3.3.2.1.4   Downstream Equipment

Here, only the relation between the pressure at the screw tip and the throughput has to be mentioned, which must be known for the screw design. Fig. 3.45 shows possible profiles. Frequently, only a maximum back pressure for the operation of extruders is mentioned. However, this bears the risk of an insufficient homogenization of the melt in case of back pressure that is too low.

3.3.2.2   Quality Criteria of the Output Values

At the end of the screw the melt quality can be evaluated by means of the following factors:

- The melt throughput should be as high as possible.
- The melt temperature distribution over the cross section should be as constant as possible (temperature homogeneity). Special measuring devices are required that are equipped with thermocouples at several positions at the same time. Alternatively, there are thermocouples whose measuring depth is adjustable.
- Additives and colorants should be distributed very uniformly and finely (mechanical homogeneity).
- The pressure pulsation at the end of the screw should be very low. On the one hand, this is a measure for constant throughput; on the other hand, it is also a measure for the homogeneity of the melt. If there are unmolten particles in the melt and poor distributions of filling material, or if the melt temperature varies strongly, these can be measured in the pressure profile.
- If possible, the extruder throughput should be independent of the back pressure so that the extruder speed and the output do not vary strongly if the pressure at the screw tip changes (e.g., filter contamination).
- The melt temperature should be constant over a long period of time.

Whereas some of the mentioned variables can be specified quite exactly it is especially difficult to characterize the mechanical homogeneity by means of data. Frequently, the expression "Melt suitable for further processing to ..." summarizes the melt characteristics. This formulation indicates that the quality requirements vary according to the product. This mainly depends on the thickness of the extrudate and the kind of the further processing. Thick-section, colored, nonoriented products, such as thermoforming sheets and tubes, require the lowest quality. With these products a much higher throughput can be achieved successfully and efficiently with the same machine size than with thinner, transparent, and oriented products, such as spin fibers and stretching films. Table 3.9 shows an example (polypropylene) of the required melt qualities for different products.

Table 3.9   Required Melt Quality

	Pipe	Profile	Blown film	Cast film	Sheet	Fiber
Temperature (°C)	230	230	210	260	230	260
Variation transverse (°C)	± 3	± 3	± 3	± 3	± 3	± 3
Variation longitudinal (°C)	± 1	± 1	± 1	± 1	± 1	± 1
Pressure fluctuation (bar)	± 1.5	± 1.5	± 1	± 0.5	± 1.5	± 1
MFI – Loss	1:2	1:2	1:2	1:1.8	1:2	1:1.8
Thickness tolerance	50% of DIN	–	± 8%	± 3%	± 5%	–

3.3.2.3   Quality Criteria Within an Extruder

Besides the melt quality, other crucial quality criteria regarding can be observed within the screw. First, the energy requirements of the extruder are considered:

- The drive power of the screw should be as low as possible to keep the motor small and to reduce the operating costs.
- The drive torque should be as small as possible to keep the price of the machine low through a small gearbox. Therefore, the so-called high-speed machines are interesting, which, with the same throughput and the same drive power, operate with a low specific throughput and at a high speed.
- Points with an extremely high local energy conversion prevent the adjustability of the temperature at the barrel wall because at these points the air cooling system runs at 100%. The temperature at the barrel wall can no longer be influenced by the regulator's temperature. This status may occur at the end of the screw if the melt is too hot. Constrictions in the conveying section where the material friction heats up the barrel wall are also to be avoided (e.g., torpedoes that are arranged too much ahead or an early sharp compression).
- The process should reduce the machine wear to a minimum. This has to be considered for the pressure profile. High absolute pressure and high pressure gradients that deflect the screw from its middle position contribute to the machine wear.

3.3.2.4   Quantification of the Quality Criteria

To satisfy all these criteria, a quantification of the quality criteria and summary of a total quality are necessary. By doing this, one has to distinguish the necessary from the desirable features by a corresponding weighting.

$$Q = \left(\prod_i q_i^{a_i} \right)^{\left(\sum a_i \right)^{-1}} \tag{3.3.1}$$

Where Q is total quality, q_i is single quality, and a_i are weighting factors.

It is useful to standardize the quality features in a single quality value, which is indicated as 0 if the feature is not realized and 1 if it is fully realized. Fig. 3.46 shows simple possibilities for the evaluation of the melt temperature and the drive power.

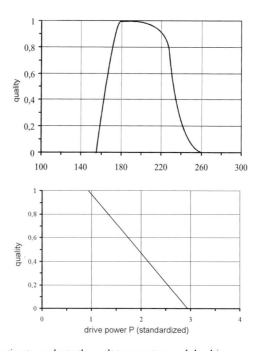

Figure 3.46   Quality function to evaluate the melt temperature and the drive power

When evaluating the screw, the pressure profile and the melting process are very important. Fig. 3.47 shows an "ideal" pressure profile versus the axial position in the three-
section screw with the torpedo and the mixing section. The pressure increases constantly
over the length of the screw; the torpedoes and the mixing sections are pressure-neutral.
The following advantages result:

- The air in the bulk material is vented backward in the direction of the hopper because
  there are no pressure gradients promoting forward flow.
- The pressure gradient is low.
- The are no pressure peaks that contribute to the machine wear.
- By means of the opposing direction of the pressure flow and the drag flow an acceptable
  effect can be yielded.

Fig. 3.48 shows an "ideal" melting process. The ascending line indicates the molten
material fraction at the corresponding axial position of the screw. This line starts at 0 at the
beginning of the screw and reaches 1 after the end of the melting. The descending line
shows the relation between the width of the solids bed and the channel width. This "ideal"
profile is characterized by a constant melting rate that is completed early enough before the
beginning of the torpedo.

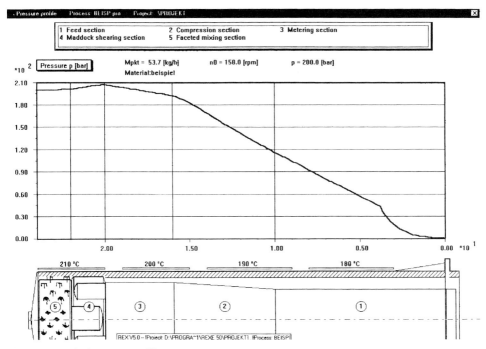

Figure 3.47   Ideal pressure profile

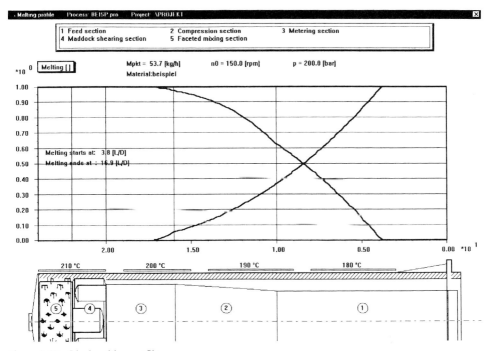

Figure 3.48   Ideal melting profile

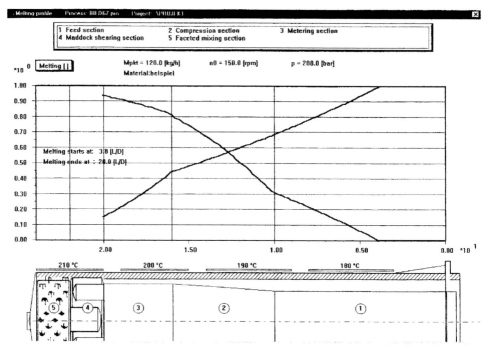

Figure 3.49   Poor melting profile 1

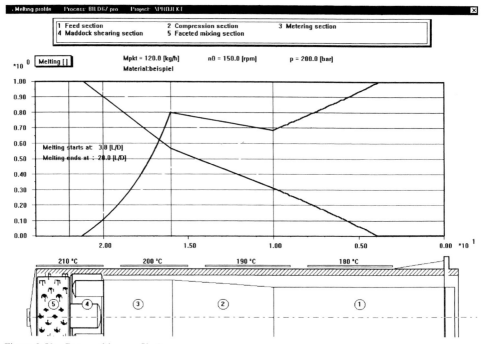

Figure 3.50   Poor melting profile 2

In contrast to this, Figs. 3.49 and 3.50 show an insufficient melting behavior. First, the melting end is not reached due to a too high specific throughput or a too high speed, so the melt still contains unmolten particles. In the other figure, the solids bed ratio increases in the transition zone. Physically, this means that the compression goes faster than the melting of the solids bed at the barrel wall. As a consequence the quality is reduced by a blocked screw channel, which reduces the throughput and may cause an increased wear at the screw flights at this point; this is due to an increasing pressure that results from the blocked channel.

3.3.3   Subdivision into Operating Zones and Types of Construction

In the polymer's flow direction one can distinguish the following operating zones of the screw:

material feeding – melting – homogenization of the melt

If possible, all zones should contribute to the conveying of the melt against the die's resistance so that a special pressure buildup zone, as is frequently mentioned in older literature, is not presented here. In the following, the functions of the single zones and the common types of construction are identified. This can be regarded as a modular construction of the units that compose the complete screw.

3.3.3.1   Feeding Section

In the feeding zone, the material that is poured into the feed hopper is seized by the screw and conveyed forward. The feeding section ends with the beginning of the melting process at the barrel wall. Since the pressure already increases gradually in the feeding section because of the backflow from the end of the screw, the feeding section generally shows a compacted solids bed where the granules are conveyed through the channel without a relative countermovement (plug flow). Therefore, the feeding section has to meet the following conditions:

- The flow resistance against the fed material must be as low as possible.
- The transport has to be quite independent of the die back pressure.
- Effective transport, even against a pressure gradient, must occur to contribute to the pressure buildup.

3.3.3.1.1   Smooth Barrel Extruder

The feeding section is constructed as a deep-flighted screw channel without any change in geometry. The pitch and the flight depth are the degrees of freedom of this zone.

3.3.3.1.2   Grooved-Barrel Extruder

The feeding section is constructed as a shallow-flighted screw channel without any change in geometry. Because of the much higher throughput of a grooved-barrel extruder with the same channel depth, the feeding sections below the grooves must be accordingly shallower to achieve an acceptable specific throughput. At the end of the grooves, a decompression zone is useful to start the melting process in a zone with a low flow rate (Fig. 3.51), [1]. The length of the shallow feeding section has to be equal to the groove length; the degrees of freedom are again the pitch and the flight depth.

Figure 3.51   Feeding zone of a grooved-barrel extruder

3.3.3.2   Melting Section

For designing the melting section of a plasticating extruder, three types of construction meet the following conditions:

- Complete melting of the throughput over a short melting length.
- Venting of the air in the granular bed in backward direction through the hopper.
- Preventing material damage through a careful melting process.
- Transport of the material against a pressure gradient to contribute to the pressure buildup.

The easiest way is to use a classical transition zone. In contrast to this, there are variants, such as the barrier screw, that are quite difficult to construct.

3.3.3.2.1   Transition Zone

The most common type of construction is the transition zone that is presented in Fig. 3.52. The compression in the transition zone takes into account the difference between the bulk density and the solid density or the melt density, respectively. The usual compression ratio lies between 1.8 and 3. In the transition zone, the melting behavior shown in Fig. 3.53 is obtained. The degrees of freedom comprise the pitch, the compression ratio, and the zone length.

Figure 3.52   Transition zone

When manufacturing the transition zone of a screw, there are three possibilities to change the flight depth (Fig. 3.54). The variant that is most similar to the screw channel model is the one below where the lines of the same channel depth are vertical to the screw flight. According to their production procedures and CNC-programming possibilities, different screw manufacturers favor different variants; so far, a technical comparison has not yet been published.

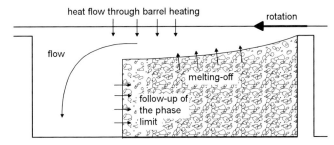

Figure 3.53   Melting model for transition zones

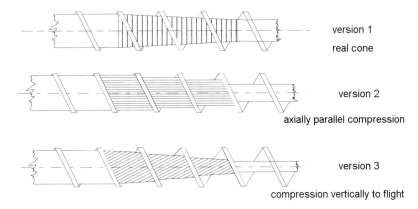

Figure 3.54   Different types of compression

3.3.3.2.2   Barrier Zone

Fig. 3.55 shows the channel cross section of a barrier zone where the solids and the already existing melt are separated by a barrier flight. The barrier flight has the following functions:

- The barrier flight acts as a filter for granules, so that the material in the melt channel is free of unmolten particles.
- When overflowing the barrier channel, the produced melt is separated and homogenized in the melt channel.
- Due to the vortex flow in front of the flights, which is increased in comparison with the normal screw channel, an increased energy input can be obtained.

Figure 3.55   Barrier screw

The melting behavior is illustrated in Fig. 3.56. The barrier zone has a multitude of degrees of freedom because the width as well as the height in the melt channel and in the solids channel can be varied.

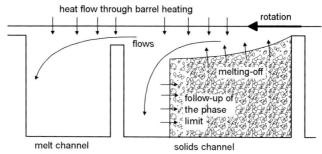

Figure 3.56   Melting behavior in barrier screws

3.3.3.2.3   *Dispersive Melting*

In transition zones as well as in the solids channel of barrier screws the polymer is molten at the interface with the compacted solids bed. If distruptions are made in the melting section, the solids bed breaks up. The single granules are then melted by the thermal conduction of the surrounding melt (Fig. 3.57). This effect can be achieved by means of the screw types that are presented in Fig. 3.58. However, the usefulness of this mechanism is controversial. Screws of this type are successfully used, especially in the processing of polyolefins in high-speed machines. A quite low melt temperature is yielded. So far, a theoretical analysis that compares these empirically developed geometries with respect to compactly melting systems in a systematical manner and with consideration of high speed and material data of polyolefins has not been made.

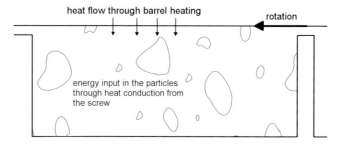

Figure 3.57   Dispers melting

3.3.3.3   **Metering Zone**

In general, the metering zone is designed as a normal screw channel. It has the following functions:

- Melting of the remaining solids if the melting section requires it.
- Homogenization of the melt. This requirement, however, can better be fulfilled by the torpedoes and mixing sections that are described in the following if they exist.
- Pressure buildup and transport. It is especially important to achieve a stable transport

over a longer period of time and to balance possible transport variations in the pressure gradient of the preceding zones so that the throughput pulsation of the screw is minimized.

In three-section screws the flight depth of the feeding section, the compression ratio, and the flight depth of the metering zone cannot be chosen independently. This is, however, possible in the five-section screw illustrated in Fig. 3.59, where the compression ratio and the flight depth of the metering zone are decoupled.

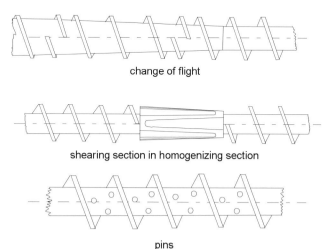

change of flight

shearing section in homogenizing section

pins

Figure 3.58   Installations to achieve a disperse melting

Figure 3.59   Five-section screw

3.3.3.4   Torpedoes

Torpedoes improve the melt homogeneity by the precise input of shear loads. It is a common feature of all construction types that the melt must overflow a narrow shear flight. The locally high shear load at a short dwell time ensures to a good dispersion, e.g., the dispersion of the filling material can be improved. Besides, the narrow slot acts like a filter. It keeps back unmolten material that may still exist. Fig. 3.60 shows some common construction types.

Concerning the positioning of the torpedoes – and of the mixing sections that are dealt with in the following chapter – there are different points of view. Whereas formerly the torpedoes were arranged at about 2/3 of the screw length to "calm" the melt in the subsequent metering zone, today one runs the risk of having too much solids at this point because of higher specific throughput. Therefore, the torpedoes are frequently installed behind the metering zone, in front of either the end of the screw or a terminating mixing section.

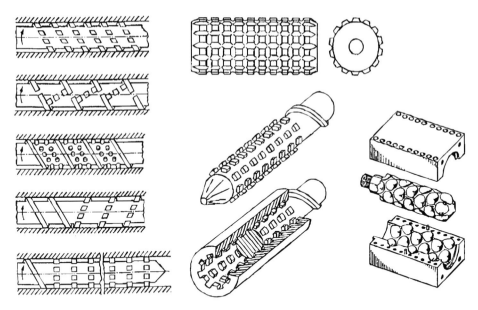

Figure 3.60   Torpedoes

3.3.3.5   Mixing Sections

By means of mixing sections the thermal and mechanical homogeneity of the melt is improved by splitting and remixing of the melt. Therefore, the common characteristics of the different mixing section geometries is the use of disruptions the melt must bypass (Fig. 3.61). Fig. 3.61 shows typical types of construction. The influence, especially on the thermal homogeneity, is considerable (Fig. 3.62) so that it is not recommended to eliminate the mixing sections. With all geometries, a complete cleaning of the internal barrel wall is desired to avoid deposits that decompose slowly.

3.3.3.6   Special Geometries

The term "special geometries" comprises all common screw designs that are possible and have not yet been enumerated and that cannot be allocated to the previously mentioned function zones.

3.3.3.6.1   Combination of a Metering Zone with Shear or Mixing Elements

The metering zone combined with a shear flight or installations to bypass this zone can take the function of a shear or mixing zone [3]. The screw is thus shortened because the combination zone can be kept shorter than a series arrangement of screw flights and a torpedo/mixing section (Fig. 3.63).

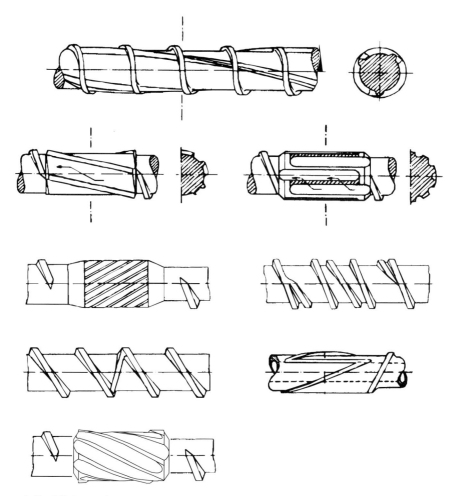

Figure 3.61   Mixing sections

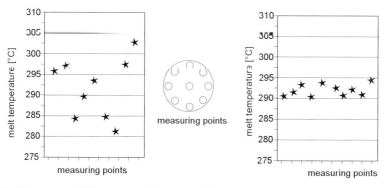

Figure 3.62   Temperature difference over the cross section with and without mixing section (values measured by Barmag AG, D-Remscheid)

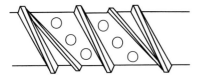

Figure 3.63   Screw channel with
Maillefer flight and torpedoes

3.3.3.6.2   Mixing Screws

The screw types discussed here will achieve a higher heat transfer coefficient at the barrel wall by intensifying vortex flow processes and by a better mixing of the melt (Fig. 3.64) [4, 5].

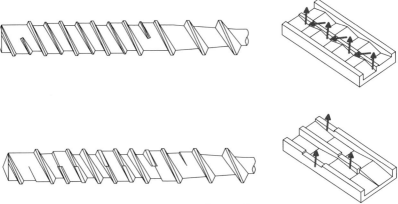

Figure 3.64   Mixing screws (wave screw, energy transfer screw)

3.3.3.6.3   Dynamic Mixers

Dynamic mixers are very efficient but costly method to mix melt. They possess depressions or cavities in the barrel through which the melt is transported several times from the rotating screw to the barrel and back Fig. 3.65). Thus, the principle of flow splitting and remixing is quite efficiently used for homogenization [6]. Applications of this large-scale technique are used to processing colored masterbatch, e.g., in fiber production.

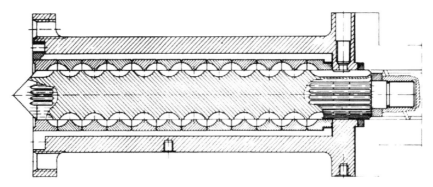

Figure 3.65   Three-dimensional mixer (Barmag Abi)

3.3.3.7   Screw Construction Types and Applications

In the following, solutions for assembling the single-function zones are presented to point out their advantages and disadvantages.

3.3.3.7.1   Plasticating Extruders

3.3.3.7.1.1 Three-Section Screw

The three-section screw in Fig. 3.66 is a very simple screw, as is frequently found in injection molding machines or in extruders, where the throughput requirements are quite restricted (e.g., coextruders in film extrusion lines). There, it is used as a universal screw for a broad range of materials. The addition of a shear torpedo is helpful, the torpedo being in charge of the complete plastification. If these screw has to be used for different materials, the channel depth has to be low for achieving an acceptable homogeneity for each material. Due to these shallow channels, the specific throughput is low.

Figure 3.66   Three-section screw

3.3.3.7.1.2 Barrier Screw with Mixing Section

In extrusion, the barrier screw with a mixing section (Fig. 3.67) is preferred. The high energy input in the barrier zone allows a greater channel depth and thus a higher specific throughput than the three-section screw. Because of the shear flight in the barrier zone, a torpedo is not obligatory; instead, the thermal homogeneity is improved by the mixing section. In the literature one can find different opinions on whether the barrier screw is applicable as a universal screw for processing a broad range of materials or rather as a material-specific optimized solution for processing a few raw materials [7, 8]. Unfortunately, a clear comparison cannot be found in the literature, so a statement that can be verified by experiments is not available. The screw in Fig. 3.67 is preferred for smooth-barrel extruders. For flat film applications, a grooved-barrel extruder is only rarely used for a limited range of materials because of the edge trim that has to be recycled.

Figure 3.67   Barrier screw with mixing section

3.3.3.7.1.3 Screw with Topedo and Mixing Section

In a typical design, the grooved-barrel extruders will have an obviously higher specific throughput than smooth-barrel extruders, so it is useful to provide the former with a torpedo and a mixing section to achieve the desired homogeneity. The screw in Fig. 3.68 has a shallow feeding zone to adjust the desired specific throughput, followed by a deeper zone to reduce the pressure. The metering zone is separated by a decompression so that the compression ratio and the flight depth of the metering zone can be chosen separately. This screw is characterized by a constant specific throughput even up to a high speed. With a corresponding design this screw could be used in all grooved-barrel extruders; i.e., in polyethylene extruders in film-blowing lines and blow-molding lines as well as in polyolefin pipe extrusion lines.

Figure 3.68   Screw with torpedo and mixing section for grooved-barrel extruders

The corresponding variant for smooth-barrel extruders can be seen in Fig. 3.69. With a corresponding design it can be used for all applications, e.g., for the extrusion of technical plastics. By changing the channel depth, it can be adjusted to the corresponding processing behavior of the material, e.g., as a very shallow screw for the processing of polyamide.

Figure 3.69   Five-section screw with torpedo and mixing section

3.3.3.7.1.4 Barrier Screw with Torpedo and Mixing Section for Grooved-Barrel Extruders

The use of a barrier screw is sensible for grooved-barrel extruders, too. Due to the intense energy input in the barrier section and in the torpedo, the screw presented in Fig. 3.10 is especially appropriate for process high specific throughput.

Figure 3.70   Barrier screw with torpedo and mixing section for grooved-barrel extruders

3.3.3.7.2   Melt Extruders

Melt extruders are used to damp pulsations and the thermal homogenization of the melt [9]. This can be achieved with the screw in Fig. 3.71, which is equipped with a mixing section at the beginning as well as at the end.

Figure 3.71   Melt extruder screw

3.3.3.7.3   Vented Extruders

Screws for vented extruders (Fig. 3.72) are characterized by their two-stage structure, consisting of a melt stage and an output stage. To avoid flooding of the vent, the conveying capacity of the second stage must be greater than that of the first stage. Since the first stage must be deep enough to feed the granulate, this stage is restricted by means of a restriction ring.

Figure 3.72   Vented screw

Usually the second stage is designed as a three-section screw, with torpedoes or mixing sections if need be and with a greater channel and flight depth. Usually, the first stage is composed of a short three-section screw. It can also be constructed as a barrier screw, but this is difficult because of the short length available.

The vented screw has a severely restricted operating window because it may not flood. By means of filters that are arranged downstream it is confronted with changing back pressures – according to the degree of filter contamination and the mesh width – which influence the backflow length. Since this length may vary only slightly for a low-pulsation throughput without flooding, vented screws are not flexible enough to be used for a multitude of raw materials and in a wide throughput range. The situation can be simplified by a melt pump that makes the pressure at the screw tip an adjustable value.

3.3.4   Geometrical Design

3.3.4.1   Design Strategy and Means

A graphical systematic scheme of the geometrical design procedure can be seen in Fig. 3.73 [10]. The requirements profile has to be defined. In the first step, Concept definition, the extruder or screw type must be selected. The following decisions have to be made:

- *Shall a grooved-barrel extruder be used?* This question should be affirmed if the machine exclusively processes polyolefines which can be entered as granulate, granules, or granular regrind. The design of grooved-barrel extruders is possible for other applications, too, but demands special know-how.
- *Shall a vented extruder be used?* This question must be answered under economic aspects because a predrying unit can be used alternatively. With the polycondensates PA and PET, a drying system is recommended to avoid a reduction of the molecular weight; with PS the predrying offers the advantage of a greater operating window in extrusion. PMMA is frequently processed in vented extruders because an atmospheric vent may suffice.
- *Shall shearing torpedoes be used?* This question has to be affirmed for products with high quality requirements. Barrier screws can do without a shear torpedo to keep the product temperature at a lower level. It is also useful to do without torpedoes in low-performance screws that have a low specific throughput, with thermally sensitive materials and if an extremely low output temperature is desired.
- *Shall mixing sections be used?* This question has to be affirmed if filling material, masterbatch, or regranulate is processed. To improve the thermal homogeneity, mixing sections can be recommended for the processing of pure material, too.

- *Shall a barrier screw be used?* It is difficult to answer this question. Frequently, the customer makes the decision. With a corresponding design, the barrier screws can almost always increase specific output. Due to the narrow barrier flight clearance they should be avoided if a high-speed screw is to be designed. Another disadvantage lies in the long flushing time when changing the product or color because the barrier flight impedes the flushing of the solids channel.
- *What L/D ratio is desired?* The estimate for extrusion lines is 30D and 23D for injection molding plasticating units. Due to the efficient energy input in the melt, lines for the processing of high-viscosity or easy-melting materials are shorter to avoid overheating (e.g., HDPE pipe extrusion lines or HDPE blown-film lines). Lines for low-viscosity or slowly melting materials may have a length of up to 33D (e.g., PET, low-viscosity PP). Frequently, the L/D ratio is already determined because a screw is to be designed for an existing extruder.

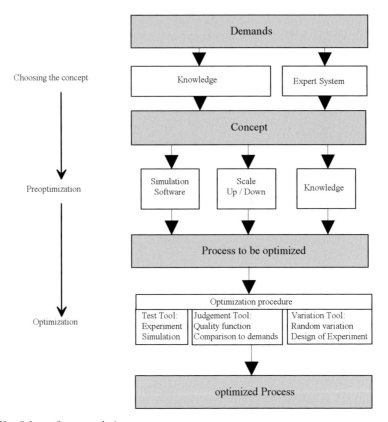

Figure 3.73   Scheme for screw design

After these questions are answered, a screw concept should have been defined. These decisions can also be made by means of an expert system and the corresponding software if the rules above (or others) are stored accordingly in computers. So far, a commercial software is not known.

In the next step the basic geometry of the screw concept (flight depth, channel depth, zone length) must be determined. There are different ways to proceed:

- The basic geometry is determined on the basis of the user's experience.
- The basic geometry is yielded by basing it on a similar satisfactory screw of the same type, which has been used for another product. This method bears only a low risk because the results are confirmed by experiments. If the underlying screw does not have the desired diameter or the desired length, a recomputation by means of scale up/down laws is necessary.
- The basic geometry is determined interactively by the user, using mathematical and physical models (e.g., a simulation software). This proceeding is not based on experience or experiments and can therefore generally be used in any case. Unfortunately, it also bears the highest risk because the modeling quality has a crucial influence on the result. A possible systematic proceeding is presented in Chapters 3.3.4.2 and 3.3.4.3.

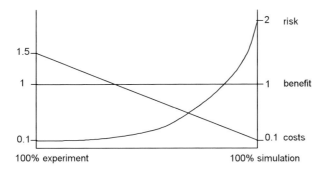

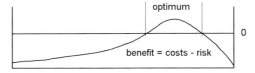

Figure 3.74   Cost and risk

After the basic geometry is determined, a detailed optimization must be done by examining the influence of small changes in the geometry on the process behavior. The aim is to find the global optimum, which can be approached only step by step. Procedure either by means of experiments or by means of a simulation software can be applied (Fig. 3.74).

- An experimental procedure means the production and test run of the basic geometry to deduct a new, better, and slightly different screw geometry from the process behavior. This can be repeated as often as required (i.e., as time and money allow) until a satisfactory result is obtained.
- When using a simulation software [11–14], the procedure is the same, but an immensely high amount of money and time is saved because the screw fabrication is not necessary and the test run is executed on the computer. However, this advantage is associated with a high risk because there are no experimental results.

- The ideal way is to combine the two methods by operating the screw experimentally and by making simulations. By analyzing the two results the changes of the screw geometry for the next experiment can be determined by means of simulation computations. Thus, a comparison is made between the experiment and the simulation, which will lead to a useful result.

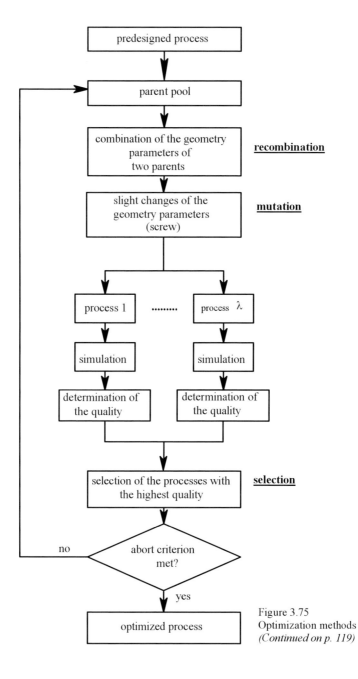

Figure 3.75
Optimization methods
(Continued on p. 119)

The detailed optimization presented here can be supported by software where the usual systematic optimization methods are applied. Fig. 3.75 shows the optimization via a statistical design and a statistical evaluation via regression functions, which is also useful for the experimental proceeding. The evolution strategy that is presented here uses random geometry variations and is appropriate for the computerized optimization.

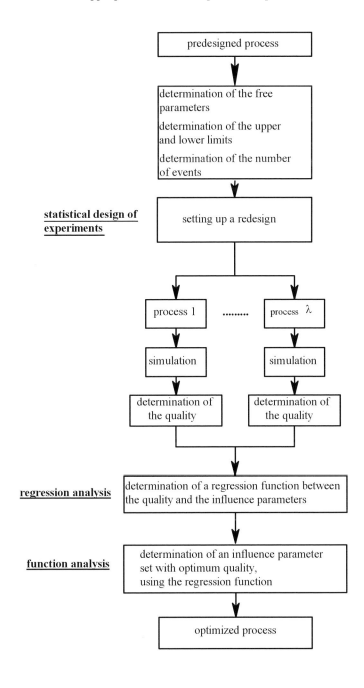

3.3.4.2   Definition of an Appropriate Specific Throughput

The crucial factor for the screw design is the specific throughput, i.e., the quotient from throughput and speed. Figs. 3.76 and 3.77 show simulation results of a 120 mm screw with a constant throughput and different screw speeds, i.e., a different specific throughput. The melting profile shows an increasing length of the melt zone with an increasing channel depth. Fig. 3.77 shows the decrease of the melt temperature with an increasing specific throughput. The basic idea of the design strategy is to choose a specific throughput that is high enough to achieve just the necessary melt quality. This proceeding leads to an optimal low melt temperature.

First, the chosen screw concept must be provided with estimated values for the zone length, the channel depth, and the flight depth. Then, the melting profile with a preset nominal throughput is determined for different screw speeds by means of simulation computations. The melting length (and thus the melt quality) improves with an increasing speed because of the constant preset throughput so that the point of the "correct" specific throughput can be selected. Figs. 3.78–3.80 show the corresponding diagrams for this point in different applications. The single function zones can be adjusted correspondingly (cf. Chapter 3.3.4.3) so that the desired specific throughput is actually attained, to obtain balance between the melting and conveying rate.

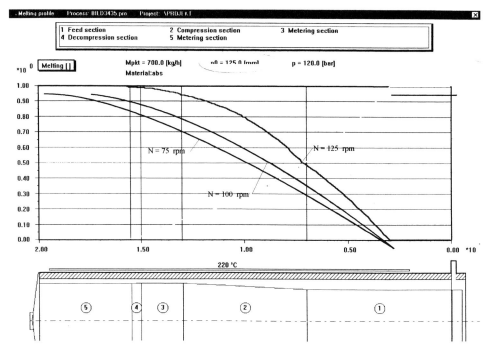

Figure 3.76   Melting profile when varying the specific throughput

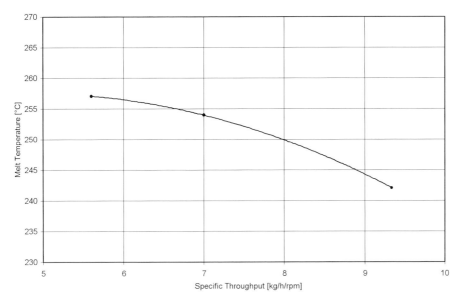

Figure 3.77   Melting profile when varying the specific throughput

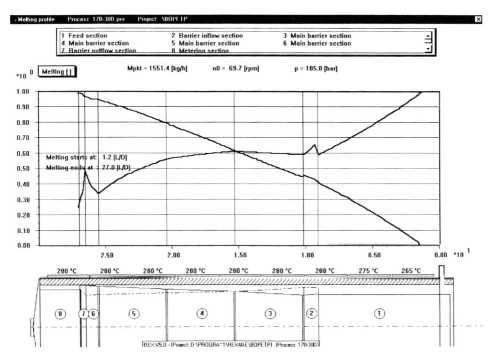

Figure 3.78   Melting profile in barrier screws

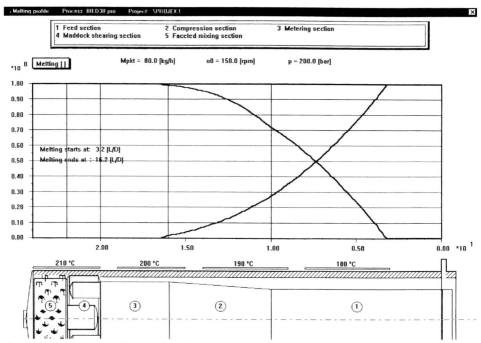

Figure 3.79   Melting profile for high-quality products (e.g., thin films)

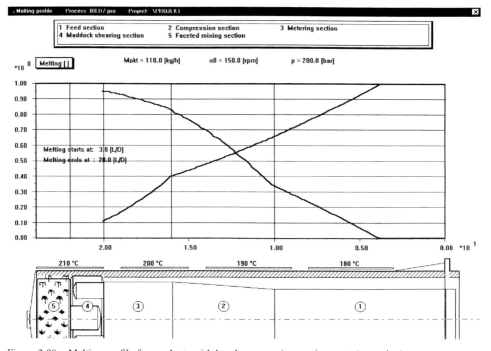

Figure 3.80   Melting profile for products with low homogeneity requirements (e.g., pipe)

3.3.4.3   Optimization of the Single Function Zones

3.3.4.3.1   Feeding Zone

The number of flights, the channel depth, and the flight depth must be determined to convey the desired specific throughput. The crucial material variable is the bulk density. The conveying behavior is described by the conveying equation [15, 16]:

$$\pi_{\dot{m}} = \frac{\dot{m}}{\rho W H i v_{0z}} = \frac{\sin(\alpha)}{\cos(\varphi)\sin(\alpha+\varphi)} \qquad \text{with}$$

$$\cos(\alpha) - K\,\sin(\alpha) = M_1 + M_2 \ln\!\left(\frac{p}{p_0}\right) \tag{3.32}$$

$$M_1 = \frac{\mu_S}{\mu_Z} 0.5\, \frac{2HEi}{t-ie}\left(K\,\tan(\varphi)+E\right)+\frac{\mu_S}{\mu_Z} C\,\cos(\varphi)\left(K\,\tan(\varphi)+C\right) \quad K = E\,\tan(\varphi)$$

$$M_2 = \frac{HE\,\cos(\varphi)\sin(\varphi)}{\mu_Z L}\left(K\,\tan(\varphi)+E\right) \quad E = \frac{D-H}{D} \quad C = \frac{D-2H}{D}$$

Fig. 3.81 shows the principle dependency of the dimensionless throughput on the friction coefficient ratio. In usual smooth barrel extruders, an estimated dimensionless throughput of 0.5 can be assumed with a friction value ratio of 1; in grooved-barrel extruders this value lies between 0.8 (friction value ratio of 3, no material flow in the grooves for powder) and 1.1 (mold-clamping transport in the channel and transport in the grooves). In two-flighted screws this value goes down to about 0.2, according to the screw size for smooth-barrel extruders. In Fig. 3.81 the relation between throughput and friction coefficient ratio is plotted because the granulate transport should be stable and independent of the back pressure.

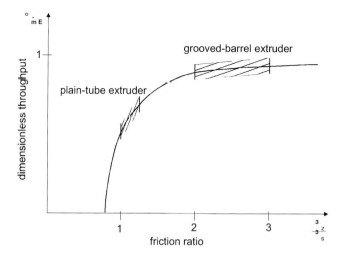

Figure 3.81   Dependency of the dimensionless throughput on the friction coefficient ratio

Although these Fig.s are operating mode-specific, the following general design guidelines can be given:

- Smooth-barrel extruders should be provided with a single-flighted feeding section; otherwise, the impeding screw surface would become too large compared with the driving barrel surface.
- The pitch in the feeding section should lie between 0.8D and 1D to realize a transport that is rather independent of the back pressure.
- The channel depth should be determined so that the desired specific throughput is obtained at a dimensionless throughput of 0.5 (plain-tube smooth barrel) and 0.8 to 1.1 (grooved-barrel).
- In smooth-barrel extruders, the feeding section extends into the melting section.
- In grooved-barrel extruders, the shallow-flighted feeding section should reach the end of the grooves.

3.3.4.3.2   Melting Zone

An optimization of the melting zones means the determination of the compression ratio, the compression length, and the beginning of the compression. The profile to strive for is illustrated in Fig. 3.82 (left). The following conditions should be met [17]:

- At its end, the deep-flighted feeding section should comprise up to about 30% of molten material so that no solid pellets are pressed to the wall of the compression zone [18].
- The compression ratio should at least correspond with that of the melt density and the bulk density; i.e., it should at least be about 1:1.5.
- The compression ratio is too high or the compression length is too short if a constant ratio of the solids bed width can be stated (see Fig. 3.82). In this operating mode the material at the barrel wall melts as fast as the channel depth decreases in the flow direction so that the produced melt is accelerated too much. As a consequence tensile stress is transferred to the solids bed, which could break [19].

3.3.4.3.3   Barrier Zone

Due to the many degrees of freedom, the optimal design of the barrier zone is quite complex. The following aspects should be considered.

3.3.4.3.3.1 Pitches, Flight and Channel Width

- To realize high melting rates the surface at the barrel wall should be maximized. This means that the solids channel has to remain wide and the melt channel has to be narrowly constructed. This can be achieved by parallel flights that provide a constant channel width within the barrier zone. Usually, the ratio between the width of the solids bed and of the melt channel lies between 1.5:1 and 2:1.
- The movement of the material from the solids channel to the melt channel is taken into account by the fact that the flow cross section in the solids channel decreases continuously, while the flow cross section increases continuously in the melt channel, Together with a required solids channel of an optimal width and a narrow melt channel

this leads to a continuously decreasing channel depth in the solids channel and an increasing depth in the melt channel.

- The solids bed should not be subject to deformations during the transport from the feeding section to the solids channel. This can be realized by an identical channel width in the feeding section and in the solids channel.

Fig. 3.83 shows the handling and cross sections of a barrier screw that has been designed according to the mentioned aspects. The pitches in the feeding section and in the barrier zone show a direct reciprocal relationship. Table 3.10 suggests useful pitch combinations when using the usual values of D/10 for the main flight width and D/15 for the barrier flight width.

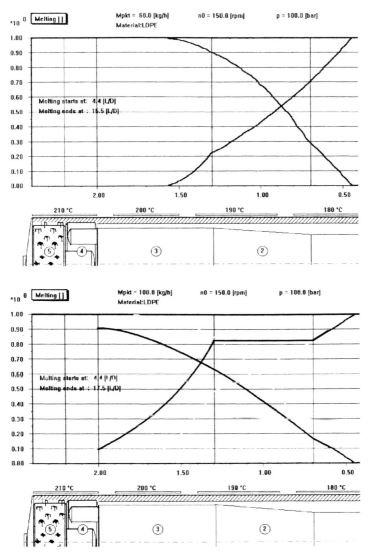

Figure 3.82   Melting profile in transition zones

Table 3.10   Pitches for Barrier Zones

Channel width ratio $W_S:W_M$	Pitch of the feeding section 0.8D	Pitch of the feeding section 0.9D	Pitch of the feeding section 1D
1:1	1.70	2.000	2.3
1.5:1	1.40	1.600	1.8
2:1	1.25	1.425	1.6

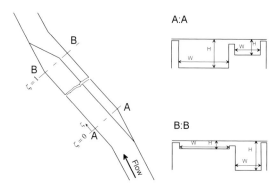

Fig. 3.83   Barrier screw

3.3.4.3.3.2 Zone Length, Feed, and Outlet Design

- The barrier zone should not begin unless about 25% of the solids has melted. Only then is there enough melt to flush the melt channel with a sufficient flow from the beginning to avoid deposits. This can be realized by making sure that at the beginning of the barrier zone the melt at the thrust face of the flight can pass easily and without any resistance into the melt channel via the barrier flight. This can be assured by appropriate fabrication.
- The length of the barrier zone should be adjusted to the melting length so that 100% of melt is available at the end of the solids channel.
- The end of the solids channel has to be designed so that no unmelted granulate may enter the metering zone. This means that the clearance of the barrier flight must never be exceeded. The design of the solids channel outlet has to assure that there are no "dead areas" in the shade of the barrier flight.

3.3.4.3.3.3 Channel Depth

To determine the channel depth, certain criteria have to be specified that concern the flight depth H_S of the solids channel and H_M of the melt channel. Furthermore, there are criteria that are applicable to the cross-sectional areas of the melt channel $A_M = H_M W_M$ and of the solids channel $A_S = H_S W_S$.

- At the beginning of the barrier zone the entire available flow area should be approximately equal to the cross section of the feeding zone:

$$A_{S0} + A_{M0} \approx W_F H_F \qquad (3.3.3)$$

At the end of the barrier zone the entire available flow area should be approximately equal to the cross section of the following metering zone:

$$A_{S1} + A_{M1} \approx W_{Met} H_{Met} \qquad (3.3.4)$$

- The material should pass the barrier flight uniformly over the length of the barrier zone so that the melt stream distribution is linear. This should be supported by a corresponding design of the flight depths where the melt channel's part of the cross-sectional area increase linearly from the starting value to the terminal value:

$$\frac{H_M W_M}{H_S W_S + H_M W_M} \approx \frac{A_{M0}}{A_{M0} + A_{S0}} + \left(\frac{A_{M1}}{A_{M1} + A_{S1}} - \frac{A_{M0}}{A_{M0} + A_{S0}} \right) \zeta \quad 0 < \zeta < 1 \qquad (3.3.5)$$

- The material should be compressed uniformly over the barrier zone so that the total cross-sectional area decreases linearly from the starting value to the terminal value just like in a classic compression zone:

$$\frac{H_S W_S + H_M W_M}{A_{S1} + A_{M1}} \approx K - (K-1)\zeta \quad 0 < \zeta < 1 \qquad (3.3.6)$$

with

$$K = \frac{A_{M0} + A_{S0}}{A_{M1} + A_{S1}} \text{ (compression ratio)} \qquad (3.3.7)$$

These criteria can only be fulfilled approximately if the barrier zone is designed with parallel flights and a linear channel depth profile. Table 3.11 shows approximate values for the flight depth profile considering the mentioned simplifications.

Table 3.11   Flight Depth Profile in Barrier Screws

Geometrical value	Solids channel	Melt channel
Channel depth at beginning	H_F	0
Channel depth at the end	0	$H_M(\zeta = 1) = \dfrac{B}{K} H_F$
Channel depth profile	$\dfrac{H_S}{H_F} = (1-\zeta)\left(1 - \dfrac{K-1}{K}\zeta\right)$	$\dfrac{H_M}{H_M(\zeta=1)} = K\zeta - (K-1)\zeta^2$

ζ is the dimensionless coordinate in the barrier section, $0 < \zeta < 1$;
K is the compression ratio;
B is the channel width ratio W_S/W_M
- linear decrease of the total cross-section area (compression)
- linear increase of the cross-section area share of the melt channel

3.3.4.3.4 Metering Zone

Generally, the metering zone is designed like a normal screw channel. It both homogenizes the completely melted throughput and builds the pressure in front of the die. It should show a constant transport rate to achieve a high pulsation damping. The design is relatively simple and satisfies the following criteria:

- The zone shall contribute to the pressure buildup according to its length:

$$\Delta p_{Metering} = p_{Die}\, \frac{L_{Metering}}{L_{Total}} \tag{3.3.8}$$

- The length of the zone is mostly deducted from the total length of the screw, using torpedoes and mixing sections. After the end of the melting, about 3D should be left.
- In most cases the pitch is simply continued from the compression zone or from the barrier zone. It should lie between 1D and 1.2D; use of too small a pitch creates too high a melt temperature.

By means of these conditions, the channel depth of the metering zone can be estimated via the melt transport Eq. [20]:

$$\pi_{\dot{m}} = \alpha_1 - \alpha_2 \pi_p \tag{3.3.9}$$

$$\pi_{\dot{m}} = \frac{\dot{m}}{0.5\rho WH\pi DN\cos(\varphi)} \quad \pi_p = \frac{\Delta p H^2 \sin(\varphi)}{6 n^{0.94} \eta(\dot{\gamma},T)\pi DN\cos(\varphi)L}$$

$$\alpha_1, \alpha_2 \text{ in } [20]$$

The coefficients α_1 and α_2 can be set to 1 for simplification.

3.3.4.3.5 Spiral Torpedo Section

The design of a spiral torpedo section is illustrated representatively for all torpedoes because this variant offers many advantages compared with other systems and thus is the most common variant. Fig. 3.84 shows the geometry of such an element.

For the design a pressure-neutral transport behavior has to be aspired to; i.e., the element should not consume pressure. Principally, this is achieved by the pitch through which the material in the grooves is dragged to the screw tip by the screw rotation. The choice of an appropriate pair of values has not yet been examined systematically; the necessary mathematical Eq.s for a pressure-neutral design can be found in Potente and Stenzel [21]. Table 3.12 contains the example of a geometry that leads to satisfying results with a multitude of materials.

3.3.4.3.6   Faceted Mixing Torpedo

The same applies to the design of faceted mixing torpedoes (Fig. 3.85) [22]. In contrast to other geometries of mixing sections, the following geometry principle prevents deposits at the teeth as well as nonabraded areas at the barrel liner. This has led to a predominance of the faceted mixing torpedo. There shall be a circular flow between the unsymmetrical facets that cause a temperature balance by an exchange of material between the regions that are close to the barrel wall and those that are close to the screw root. It should be a pressure-neutral design so that there is enough space left for the stream between the teeth. A common example is illustrated in Table 3.13.

For reasons of production costs the channel depth of the faceted mixing torpedo is mostly kept constant. A channel depth that can be modified over the length of the mixing section will have a positive effect on the mixing because this causes strain flows. A continuous shallowing of the mixing section or a constriction in the middle of the torpedo can be imagined.

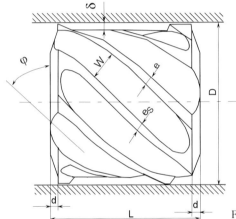

Figure 3.84   Spiral torpedo

Table 3.12   Standard Spiral Torpedo

Diameter D	100 mm
Length L	2D
Shearing fligth width e	3 mm
Fligth width e_s	6.32 mm
Pitch t (φ)	7D (65.82°)
Number of grooves i	5
Radial clearance δ	0.8 mm
Width of grooves W (semicircle)	21 mm

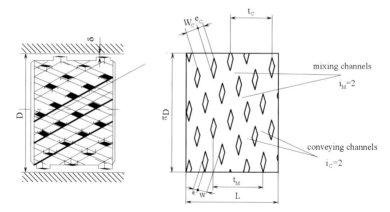

Figure 3.85   Faceted mixing torpedo

Table 3.13   Faceted Mixing Torpedo

Diameter	100 mm
Length	2D
Number of conveying channels	3
Number of mixing channels	4
Pitch of conveying channel	1D
Pitch of mixing channel	1.5D
Fligth width	10 mm
Channel depth	150% of the metering zone's depth

3.3.4.4   Series Design

The design strategy that has been presented so far uses mathematical and physical models and is intended for the optimization and/or new design of screws. However, for almost all extrusion products, machines of different sizes or output are required on the market. Their screws must satisfy identical quality requirements under the same boundary conditions.

By means of the model theory [23–26], the diameters of the screw geometries and thus the throughput can be varied. It is presumed that a screw that is computed correctly according to modeling theories produces the same melt quality as the underlying basic geometry if the input variables are identical. This process allows the geometrical determination of similar screws over the range of machine diameters that is required on the market.

The model theory has been developed for a multitude of boundary conditions. Here, only a strongly simplified variant is presented. The following restrictions are useful for the design of a new series:

- The melt temperature at the end of the screw shall be invariant.
- The pressure at the screw tip shall be invariant.
- The pitch shall be invariant.
- The drive power share and the temperature control share of the entire enthalpy increase in the material remains constant.
- The ratio between the drag flow and the pressure flow in the screw channel remains constant.
- The subdivision of the screw in single-function zones is maintained.
- The temperature control of the barrel wall is maintained.

Considering these boundary conditions, there are still three degrees of freedom: the screw speed, the screw length, and the channel depth; these are related via the constancy of the above-mentioned factors and the material data:

$$\frac{H}{H_0} = \left(\frac{D}{D_0}\right)^{\psi} \quad \frac{N}{N_0} = \left(\frac{D}{D_0}\right)^{-x} \quad \frac{L/D}{(L/D)_0} = \left(\frac{D}{D_0}\right)^{\omega} \tag{3.3.10}$$

The relations between these factors are illustrated in Fig. 3.86, the concrete position of the functions according to Table 3.14 being dependent on the flow curve of the material and the choice of the length exponent ω. The crucial Eq.s are summarized in Table 3.14.

Table 3.14   Model Law Equations

1: Constant heat flux at the barrel wall		$\psi = x + \omega$
2: Constant temperature difference between polymer and barrel wall		$\psi = \dfrac{x + \omega}{2}$
3: Constant ratio between shear energy input and throughput		$\psi = 1 - \dfrac{n}{n+1} x + \dfrac{\omega}{n+1}$
Intersection lines 1 and 3	$\psi = \dfrac{1 + n + \omega + n\omega}{2n+1}$	$x = \dfrac{1 + n - \omega n}{2n+1}$
Intersection lines 2 and 3	$\psi = \dfrac{1 + n + \omega + n\omega}{3n+1}$	$x = \dfrac{2n + 2 + \omega - \omega n}{3n+1}$

Concerning the efficiency of the equipment, the melt throughput that can be achieved is of great importance. It is based on the model law

$$\frac{\dot{m}}{\dot{m}_0} = \left(\frac{D}{D_0}\right)^{2+\psi-x} \tag{3.3.11}$$

and shall increase as high as possible with the diameter by economic reasons. The following strategies are usually applied for the final determination of the model transfer point:

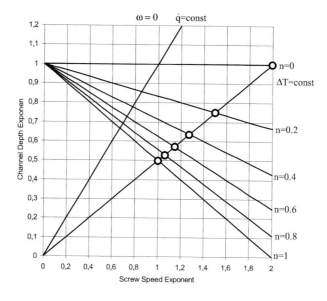

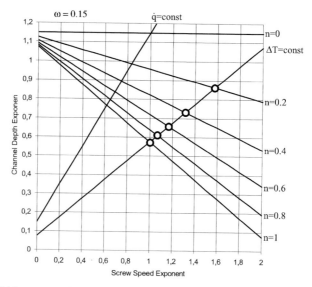

Figure 3.86   Model law exponents

- Under the boundary condition of a constant L/D ratio, the throughput exponent is considerably less than 2. Despite this, the model transfer is still done at the intersection point shown in Fig. 3.86. The correlation between the throughput exponent and the power law factor of the flow curve is demonstrated in Fig. 3.87.
- To increase the efficiency of the machine, the model is transferred at the point of a line with the same dissipation with a higher throughput. This may even reach the indicated line with constant heat flow. The boundary condition of a constant temperature control of the barrel wall is, however, abandoned, so that the melt quality changes. Fig. 3.88 shows the melt profiles of such a series. The continuous deterioration of the melt profile with an increasing diameter can be recognized.
- To achieve the desired high throughput exponent with the same melt quality, the boundary condition of a constant L/D ratio is abandoned [8].

In reality, a combination of these ideas will be favorable. So the scale-up can be done on a point of equal dissipation in the middle, betweeen the points of similar temperature difference and similar heat flux with a smooth increase in L/D ratio to have reasonable output rates. The result could be the conversion of a 90 mm-27D basic machine with 300 kg/h to a machine of 180 mm-30D for a throughput of 1060 kg/h, e.g., when taking the material data of a PP homopolymer ($n = 0.4$).

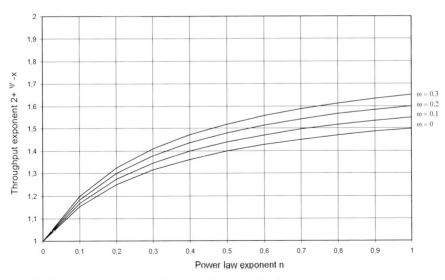

Figure 3.87   Throughput exponent as function of the power law factor

3.3.5   Example: Extruder ⌀90mm for 300 kg/h PET IV 0.8

As an example for the above-mentioned screw design procedure, we will discuss the design of an ⌀90mm screw for 300 kg/h PET (IV 0.8) for extrusion of thick film (500 µm unstretched film for thermoforming).

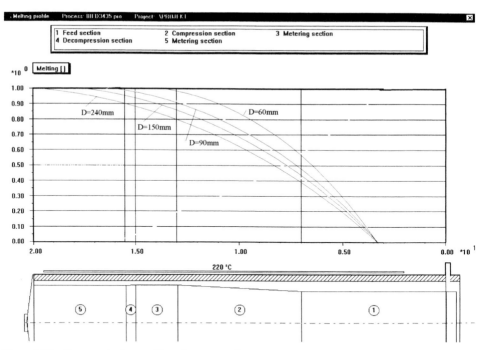

Figure 3.88   Homogenization profiles of a screw series

Looking for the suitable screw concept, the following decisions are necessary:

- The extruder shall have a smooth barrel to avoid wear problems in the feeding section cause by the high strength of the pellets. A second reason for using a smooth barrel is the necessity of a deep feeding section to deal with the edge trim. Using a grooved barrel will cause an undesirable correlation between the share of virgin material and the throughput.
- The extruder shall not have a degassing system but shall be equipped with an upstream drying unit.
- Due to the high viscosity of the resin and the thick-walled product, there is no reason for using a barrier screw or a shearing element.
- To ensure a good thermal homogeneity in the coathanger die the screw shall be equipped with a mixing element at the screw tip
- For a good melting behaviour and for a melt temperature as cold as possible we want to adjust independently the metering zone channel depth and the compression ratio. So we have to introduce a decompression after the melting regime
- Taking into account all of these points, the resulting screw has six sections:

For the predesign, the screw is square-pitched and the channel depths are chosen in a medium range.

The following step is the choice of the screw speed and the resulting specific throughput. The following Fig. shows the results of the REX-simulation of the melting profile. The calculation was done with a preset throughput of 300 kg/h with different screw speeds. For a screw speed of 80 rpm an acceptable behaviour is obtained. Lower screw speeds lead to insufficient homogeneity. Higher screw speeds will result in too high temperatures. So for the predesign a specific throughput of 3.75 (kg/h)/rpm is suitable.

The next procedure is the adjustment of the channel depth of each screw section to the desired throughput. Taking into account the solids and melt conveying analysis by throughput calculation with REX, the screw has to be designed with deeper channels than the estimated values

The simulation of this screw leads to the desired throughput with 80 rpm and a good pressure profile (Fig. 3.93). The pressure maximum at the end of the compression section is not desired but unavoidable. The melting profile (Fig. 3.92) looks like the desired one (Fig. 3.90 with 80 rpm).

The last step in the screw design is the fine-tuning of the geometry by varying the channel depths, the pitches and the section lengths. This can be done using an optimization method or by hand. Unfortunately the combination of simulation software with optimization strategy is not available up to now. So the following result is obtained by a „handmade" optimization. The screw has an improved section length distribution and small changes in the channel depth

The melting profile in Fig. 3.95 has small improvements compared to Fig. 3.92. Looking on the pressure profile (Fig. 3.96 vs. Fig. 3.93), the pressure maximum is avoided so that an almost constant pressure gradient is obtained.

The last step should be the experimental validation of this screw.

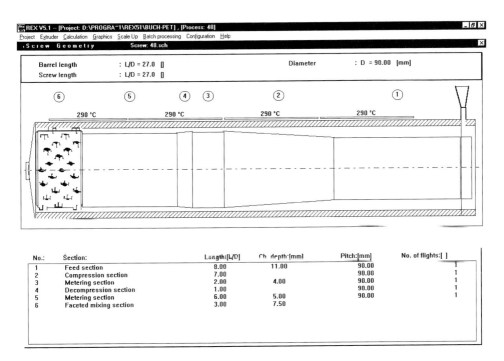

Figure 3.89   Screw after choice of concept

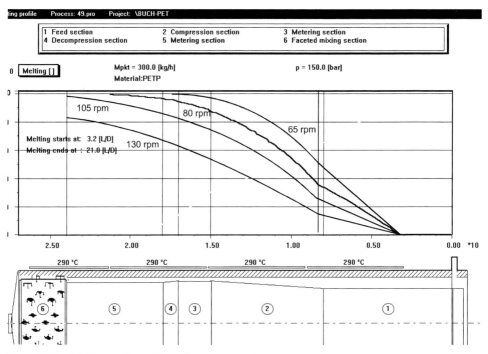

Figure 3.90   Melting profiles with varying screw speed

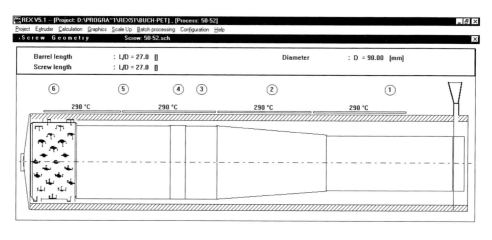

No.:	Section:	Length:[L/D]	Ch. depth:[mm]	Pitch:[mm]	No. of flights:[ ]
1	Feed section	8.00	15.00	90.00	1
2	Compression section	7.00		90.00	1
3	Metering section	2.00	5.00	90.00	1
4	Decompression section	1.00		90.00	1
5	Metering section	6.00	5.50	90.00	1
6	Faceted mixing section	3.00	8.00		

Figure 3.91   Screw after adjusting channel depths to the conveying requirement

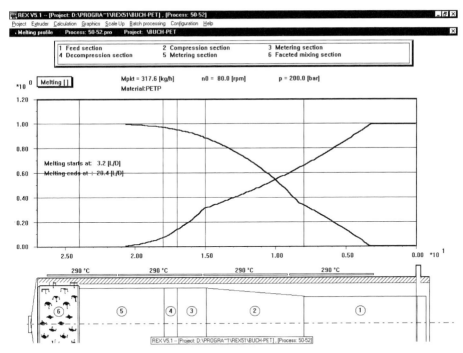

Figure 3.92   Melting profile of the preoptimized screw

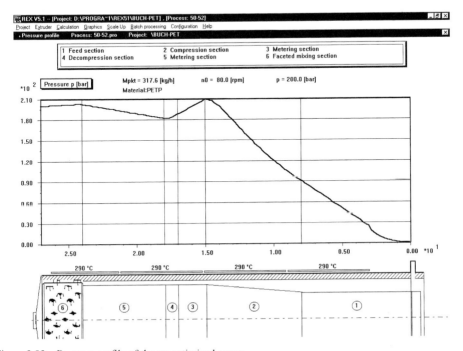

Figure 3.93   Pressure profile of the preoptimized screw

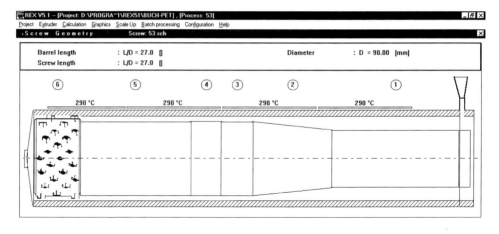

Figure 3.94   Screw after fine-tuning

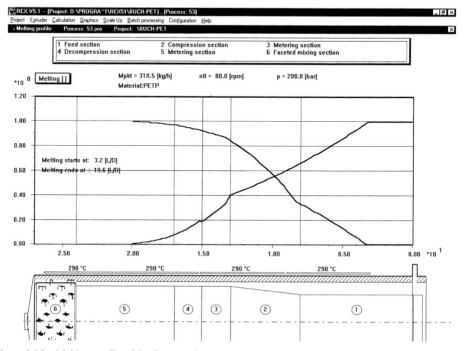

Figure 3.95   Melting profile of the fine-tuned screw

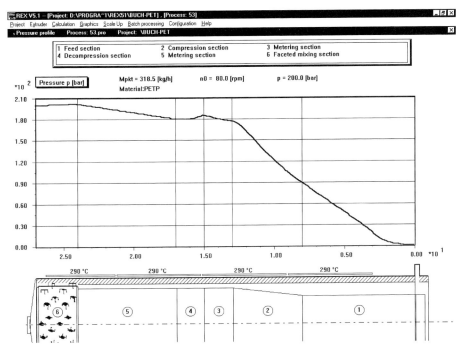

Figure 3.96   Pressure profile of the fine-tuned screw

References

1. Potente, H., *Kunststoffe* (1988) 78, p. 355
2. Langhorst, H., *Entwicklung von Schnecken für Hochleistungsextruder.* (1989) Dr. Ing. Dissertation, RWTH Aachen
3. Bürkle, E., Bauer, M., Würtele, M., *Kunststoffe* (1997) 87, p. 1272
4. Kruder, G.A., Calland, W.N., *SPE ANTEC Tech. Papers* (1990) 36, p. 74
5. Plumley, T., Spalding, M., Dooley, J., Hyun, K.S., *SPE ANTEC Tech. Papers* (1994) 40, p. 332
6. Wang, C., Manas-Zloczower, I., *Polym. Eng. Sci.* (1994) 34, p. 1224
7. Christiano, J., *SPE ANTEC Tech. Papers* (1991) 37, p. 183
8. Wortberg, J., Michels, R., *Innovative Entwicklungen in der Einschneckenextrusion – Übersicht über die Extruder- und Schneckenentwicklung – Möglichkeiten und Grenzen der Hochleistungsextrusion – Einschnek-kenextruder: Grundlagen und Systemoptimierung* (1997) VDI-K, Baden-Baden
9. Hensen, F., Imping, W., *Kunststoffe* (1990) 80, p. 673
10. Potente, H., Jungemann, J., Zelleröhr, M., *Plastics Special* (1997) 10, p. 16
11. Sebastian, D., *Applications of Extrusion Simulation,* Stevens Institute of Technology
12. Potente, H., Jungemann, J., *Plastics Special* (1996) 9, p. 4
13. INJEXX-PC and EXTRUD-PC, Scientific Process and Research Inc., New Jersey, U.S.A.
14. Agur, E.E., Vlachopoulos, J., *Polym. Eng. Sci.* (1982) 22, p. 1084
15. Schneider, K., *Der Fördervorgang in der Einzugszone eines Extruders.* (1968) Ph. D. Thesis, RWTH Aachen
16. Potente, H., Schöppner, V., Int. Polym. Process. 1995) 10, p. 289
17. Potente, H., Stenzel, H., *Int. Polym. Process.* (1991) 6, p. 297
18. Paller, G., Schwab, M., *Kunststoffe* (1996) 6, p. 792
19. Spalding, M., Hyun, K.S., Hughes, K.R., *SPE ANTEC Tech. papers* (1996) 42, p. 191
20. Effen, N., *Theoretische und experimentelle Untersuchungen zur rechnergestützten Auslegung und Optimierung von Spritzgießplastifiziereinheiten.* (1996) Dr.-Ing. Dissertation, Universität-GH Paderborn

21. Potente, H., Stenzel, H., *Kunststoffe – German Plastics* (1991) 18, p. 26
22. Potente, H., Kunststofftechnisches Seminar: Rechnergestützte Extruderauslegung, Universität-GH Paderborn (1992)
23. Carley, J.F., McKelvey, J., *Ind. Eng. Chem.* (1953) 45, p. 989
24. Maddock, B., *SPE J.* 11 (1959) 11, p. 983
25. Potente, H., *Rheol. Acta* (1996) 17, p. 406
26. Potente, H., *Kunststoffe* (1990) 80, pp. 80, 206

3.4   Machine Design and Construction

Klaus Schäfer, Georg Stausberg, and Friedel Dickmeiss

The single screw extruder is the most important type of extruder used in the polymer industry. Its key advantages are relatively low cost, straightforward design, ruggedness and reliabilty, and a favorable performance/cost ratio [1]. Because single screw extruders can be used for processing nearly all types of polymers, a wide range of requirements have to be covered by the design of these machines. Depending on the polymer, the required throughput, and the conditions defined by the specific process (acceptable melt temperature, back pressure, etc.), the correct size of the extruder (diameter, length) as well as the correct design has to be choosen. Beside the extruder size, the design is determined by the heating/cooling system, the layout of the drive, the layout of the feed section, and special features like venting ports. Before discussing the machine design in detail, let us have a look at the most common extruder sizes.

3.4.1   Extruder Size

The extruder size is mainly characterized by the nominal diameter of the extruder, which is equal to the inner diameter of the extruder barrel. Table 3.15 gives an overview of the most common sizes in combination with typical throughput ranges. The throughput range shown in Table 3.15 does not consider polymer or the length of the extruder. The maximum values shown in the table can be reached only with some polymers under special process conditions.

There are also extruders on the market with diameters different to the most common ones shown in Table 3.15. Extruders with these diameters (35, 40, 70, 75, 100, 105, 130, 135, 160, 175, 180, and 225 mm, etc.) are designed by some machine producers for their special applications. There are also single screw extruders available designed with diameters larger than 350 mm. Only a few of these large extruders are used in a direct process like film, fiber, profile, or sheet extrusion. Most of the single screw extruders with diameters greater than 350 mm are used in compounding lines.

Beside the diameter, the nominal length of the extruder is the second parameter determing the size of the machine. The nominal length is expressed as a ratio between length and diameter (L/D). For standard extruders, L/D ratios between 20 and 30 are normally used, with a most common range of 24 to 26. For special applications the L/D ratio can be as low as 10 or as high as 40.

Table 3.15   Standard Diameters and Typical Throughput Ranges for Single Screw Extruders

U.S. standard diameter (inch)	European standard diameter (mm)	Throughput range (kg/h)
¾	20	0.5–20
1	25	1–30
	30	5–50
1½		7–80
	45	20–120
2	50	25–150
2½	60	50–250
3½	90	100–600
4½		200–900
	120	300–1000
6	150	400–1500
8	200	600–2600
10	250	1000–4000
12	300	1500–6000
14	350	2000–8000

When designing an extruder, the first step is to choose the right size of the machine, depending on throughput and process requirements. After the size has been fixed the second step is to design and to choose the right components, for example, drive motor, gearbox, and heating system.

3.4.2   Components of a Single Screw Extruder

A common design example of a horizontal, single screw extruder is shown in Fig. 3.97. The extruder components are explained in the following sections.

3.4.2.1   Frame

The extruder is mounted on a welded steel construction to achieve the required extrusion height. The frame has to carry the weight of the gearbox, the feed housing, and the barrel. Additionally, the rigidity of the construction has to be sufficient to absorb the torque of the drive unit. The frame is built in a modular design principle to take into account specific errection conditions of the machine. Depending on the fixed point of the extruder, the frame can be fixed on the floor or mounted on a rail system, ball units, or self-aligning, double-row ball-bearings. Fig. 3.98 shows the principle design of a ball-bearing system. The extruder rest on ball casters, which, in turn, slide on polished steel plates to absorb the thermal expansion in two directions. If the extrusion unit is fixed at the measuring head or the die, the thermal expansion has to be compensated by a rearward movement of the extrusion unit. This is possible by rail systems or ball units. If the extruder frame is fixed on the ground the thermal expansion of the extruder barrel has to take place in the forward direction.

The barrel centre has to remain almost in the ideal position. Therefore, the vertical thermal expansion of the support should be low. Barrel supports fixed outside the hot area and insulation parts between barrel and support avoid a large heat transfer to the

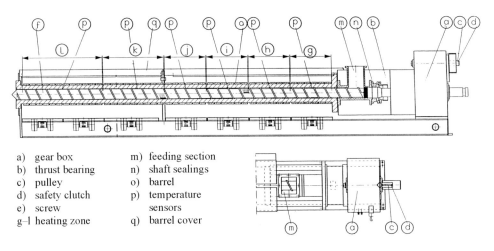

a) gear box
b) thrust bearing
c) pulley
d) safety clutch
e) screw
g–l heating zone
m) feeding section
n) shaft sealings
o) barrel
p) temperature
   sensors
q) barrel cover

Figure 3.97   Single screw extruder

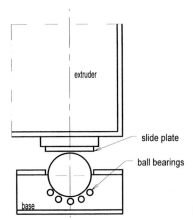

Figure 3.98   Ball bearing

frame. A open design of the frame improves the access and thus simpifies maintenance of the extruder.

3.4.2.2   Extruder Drive

The extruder screw is rotated by a reduction gear driven by a AC or DC motor drive system. The drive should meet following technical requirements:

- Adjustable speed over a wide range
- Constant speed independent of the load
- Adequate, constant torque
- High efficiency
- Durability and long operating life
- Low electrical and mechanical maintenance requirements
- Low noise level

The speed of the motors should be adjustable in a range 1:10 up to 1:30. The most common extruder drives are the AC induction motors and the DC shunt wound motors. Hydraulic drive systems are common in the field of injection molding extruders. A motor technology has several advantages: simplicity, ruggedness, no commutators and brushes, low maintenance, and compact construction [1].

To choose the right motor for an extruder, the required mechanical drive power, screw speed, and torque at the screw shaft must be known. The required energy to melt and transport the polymer can be calculated by the energy balance equation:

$$P + \dot{Q} = \dot{m}\Delta h + p\dot{V}$$ (3.4.1)

Where
P   =   motor power at screw shaft
$\dot{Q}$ =   heating power
$\dot{m}\,\Delta h$ =   enthalpy increase of the polymer
$p\dot{V}$ =   pump power

Most extruder drives have a "constant torque characteristic." This means that the nominal torque obtainable from the motor remains constant about the range of the screw speed. Thus, the power is direct constant with the speed. The relationship between torque and power is given by

$$P = \frac{2\pi}{60}MN$$ (3.4.2)

Where
P =   motor power (kW)
M =   motor torque (Nm)
N =   motor speed (1/min)

Table 3.16   Shows Common Drive Layouts of 24D and 30D Extruders

Screw diameter (mm)	45	60	75	90	105	120	150
Torque / standard (Nm)	3150	3150	5000	7,500	11,000	16,000	25,000
Torque / high (Nm)	3150	5000	7500	11,000	16,000	25,000	35,000
Motor / standard capacity (kW)	16	35	49	70	92	117	136
Motor / high capacity (kW)	30	49	70	117	136	136	193

3.4.2.3   Transmission and Clutch

The extruder can have a direct connection between the drive and the reducer by gears or with a belt or chain transmission. Drive systems up to 250 kW motor power are often equipped with a V-belt drive. The belt transmission have a sufficiently efficiency, reduce shock loads, and provides a optimum smoothness in a mechanical drive. It also increases the flexibility because it allows a relativly simple and cheap changeover to another reduction ratio by changing the belt pulleys. Additionaly, the belts allow a flexible arrangement of the motor in relation to the extruder. Disadvantages compared to a direct coupling are the efficiency loss and the maintenance costs.

The direct coupling between drive motor and gearbox by an axial meshed coupling is required if high drive power has to be transmitted. The efficiency is high but there is no flexibility of changing the reduction ratio or the position of the motor.

With small screw diameters up to approximately 75 mm it make sense to install a over-torque protection to avoid a damage of the screw or gear by a too powerfull drive torque. This can be achieved by a mechanical safety clutch at the gear inlet shaft or by electrical limitation of the motor current in the control cabinet.

3.4.2.4   Gear Unit

The gear unit must deliver the motor torque to the screw. The gearbox is the most expensive part of the extruder. Most extruder manufacturers supply their extruders with a range of standard gearbox size [2]. The installed torque and the reduction ratio of the transmission must be adapted to the process requirements. The mechanical power consumption is largely influenced by the design of the extruder screw. Typical reduction ratios range largely from 15:1 to 30:1. The installed maximum screw speed depends on the process, the screw diameter, and screw design. As rule of thumb small extruders up to 75 mm screw diameter have a maximum screw speed of 100 to 400 rpm; lager extruders have a maximum rated screw speeds of 100 up to 200 rpm. To avoid a damage of the gearbox by overload, the gear power should be equal or higher as the installed motor power [3]. The most frequently used reducer type is the spur gear reducer in a two- or three-step configuration. Helical spur gears or the V-shaped tooth design keep the operating noise on a low level and increase the efficiency and working life. The efficiency of these gears is about 98% at full load and 96% at low load. The transmissions use splash lubrications with a large oil reservoir or a forced lubrication by a oil pump system. The heat, generated by energy dissipation (2 to 4% of the transmitted power) is exhausted by surface convection or by a heat exchanger integrated in the gearbox. For extreme loads or high ambient temperatures a water cooling systems are used to control the oil temperature.

The axial force of the screw, generated by the melt pressure, is adsorbed by a thrust bearing. Axial thrust bearings positioned outside the gear unit avoid a undesirable axial effect on the transmission stages. The axial force of the screw F is calculated from the maximum die head pressure p and the cross-section area of the screw:

$$F = \frac{\pi D^2 P}{4} \tag{3.4.3}$$

where
F = axial force (N)
D = screw diameter (mm)
p = pressure in front of the screw (bar)

The correct layout of the high loaded thrust bearings is important for a long bearing life. The life of the bearing under normal load should be at least 100,000 to reach a life of more than 10 years. Generally, a splash lubrication is sufficient for the lubrication of the thrust bearing. Fig. 3.99 shows an example of a thrustbearing assembling in front of the gearbox housing.

The connection between screw shaft and gearbox usually consists of a key shaft, which allows an easy dismounting of the screw. The screw can be pushed out by a ejector device that passes through the hollow drive shaft.

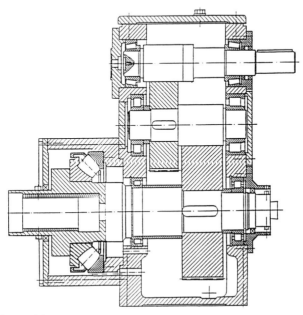

Figure 3.99   Gear box and thrust bearing

3.4.2.5   Feeding Section

The feeding section is positioned between the gearbox and the extruder barrel. To avoid an unwanted heat flow from the hot barrel to the feed throat, thermal insulation separation or a air gap is normal. One-piece barrel constructions with integrated feed does not offer this possibility. A water-cooling channel inside the feed housing reduces the heat transfer to the polymer chips and prevent the chips from sticking to the throat. Polymer sticking inside the throat causes feed-conveying problems. Temperature fluctuation in the feeding area should be avoided because the wall temperature influences the friction coefficient and the friction coefficient determines the feeding behaviour. Pressure and conveying fluctuation are often caused by this effect. The size of the opening must be designed to ensure a proper filling of the first screw channels. Circular and square inlet with approximately 1,5D to 2D length and 0,6D to 0,9D width usually reach good results.

Various feed throat designs are illustrated in Fig. 3.100. Figs. 3100a and b show a normal design. The eccentric opening of Fig. 3.100b avoid a backflow of chips from the screw channel, increasinng the filling level and stabilising the chip conveying. Fig. 3.100c shows a undercut feed with a feeding pocket, which is usually used for feeding molten polymers. The problrmwith this design is, if it is used for feeding polymer pellets, a high pressure can be created between pellets and the housing.

Between the feedhousing and the gearbox a stuffing box and a shaft sealing for protective gas seal or vacuum seal are installed. These seals the extrusion system from the surrounding air to avoid a degradation of the molten polymer during the extrusion process by humidity or oxygen. Fig. 3.101 shows the cross section of a feed housing with stuffing box, shaft sealing, feed-housing water-cooling channel, and connection for protective gas supply.

A) central, with = D   B) eccentrical   C) eccentrical with feed pocket

Figure 3.100   Feed throat design

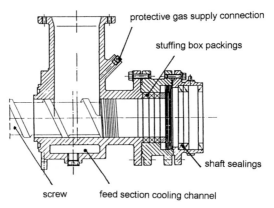

Figure 3.101   Feed section

Extruders equipped with grooved barrel sections have a special design with additional water-cooling channels and grooved bushings. A design comparison between smooth and grooved feed sections is given in Fig. 3.102. The grooved bushings are made of highly wear-resistant material to avoid high wear by the high stress between the polymer and the groove surfaces. Some grooving systems require special metal carbide feed inserts or bushings. The common shapes of the axial or spiral grooves are shown in Fig. 3.103. The shape of the groves has a decisive influence on the processing properties and the wear of the bushing.

The optimal shape of the grooves has to be determined experimentally [4]. Corresponding to experimental experiences the number of grooves is approxionated by

$$\frac{D}{10} \le n < \left(\frac{D}{10} + 2\right) \tag{3.4.4}$$

Many european extruders and blow molders uses the grooved barrel as standard to reach higher output rates.

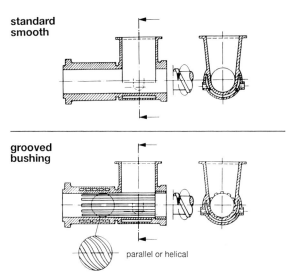

Figure 3.102   Feed section design

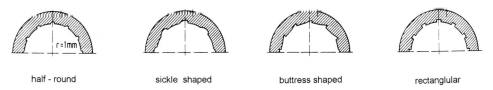

half - round   sickle shaped   buttress shaped   rectanglular

Figure 3.103   Design of grooves for grooved feeding

3.4.2.6   Hopper

The extruder hopper feeds the polymer into the extruder. Usually, the material, in form of pellets or powder, flows by gravity from the hopper into the extruder throat. The hopper has cylindrical-conical, conical, square, or rectangular shape. Common hopper volumes are 50 L up to 250 L depending on the extruder size. The load on the extruder should be kept low to avoid deformations of the barrel and consequent jamming the screw. Heavy-duty hoppers should be mounted on a adequately support or otherwise independently mounted.

The angle of the side walls of the hopper depends on the internal friction coefficient of the bulk material. Generally, the hopper angle should be lager than the angle of the internal friction to avoid the risk of bridging inside the hopper. Sometimes it is necessary to install agitors or feeding screws inside the hopper to ensure a steady flow into the extruder throat. To feed bulk materials with bulk densities below approximately 0,2 kg/m³, a force feeding is generally necessary. The bottom of the hopper can be closed by a slide. If hot chips are conveyed to the hopper the hopperwalls should be insulated to avoid a loss of energy. If the extruder operates with inert gas or vacuum the hopper must be equipped with additional sealings and slides to prevent leakage.

3.4.2.7   Barrel

Standard barrels are constructed of high-temperature-grade, gas-nitrated steel or with a bimetallic lining. The current trend is to use bimetallic barrels. Bimetallic barrels are centrifugal cast, with an abrasion-resistant liner material inside an alloy steel outer shell. The depth of the liner is about 1,5 to 2,0 mm, whereas total nitriding depth is about 0,4 mm. In general, wear and corrosion resistance of nitrided barrels is not as good as that of bimetallic barrels. Gas-nitrated barrels usually do not last as long as bimetallic barrels. Gas-nitrated barrels should be avoided where chemical resistance is an important factor. An additional drawback of the nitride surface is that once the thin nitrided layer is worn away, wear will increase rapidly. Some suppliers offer the option to install a wear bushing of 1,5D to 3D length in the feed of the barrel. Fig. 3.104 shows a Common wear bushing assembly. The bushing can be replaced to give the barrel added life. The correct fit between bushing and barrel is very important to ensure a good heat transfer.

Barrel

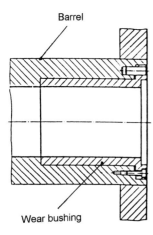

Wear bushing   Figure 3.104   Wear bushing

Extrusion barrels usually have an *L/D* ratio of 24:1 up to 36:1. The most common *L/D* ratios are 24:1 and 30:1. The extruder barrel is usually designed to withstand pressures of more than 500 bar in axial and radial direction. Barrels of more than 6 m length generally are flanged together of two or three parts because the manufacturing length is limited. Barrels have an integral feed opening or fastened to a separate feed throat housing. At the front and the rear, there are large-diameter flanges with bolt circles to attach the barrel to the feed throat and the melt pipe system. The free end of the barrel of is normally supported by a barrel support frame. The barrel is equipped with wells to install thermocouples in every heating zone. The distance between the inner barrel surface and the temperature sensor tip should be as small as possible to minimise the measurement errors. The maximum hole is determined by the pressure strength calculation of the remaining barrel wall thickness. In general, the sensor is spring loaded to ensure a good contact to the barrel.

3.4.2.8   Barrel Heating and Cooling Equipment

3.4.2.8.1   Heating

Heating of the extruder barrel is necessary during the startup of the machine and usually also during the operation. Heating contributes to the overall power requirement of the process. Electrical heating , fluid heating, or steam heating can provide the heating energy. Electrical heating is the most common type of heating and has almost totally replaced the fluid and steam heating. Electrical heating systems have decisive advantages compared to fluid or steam heating:

- Low cost
- Easy to maintain
- Low safety hazard
- Large temperature range
- Easy to control
- High efficiency
- Low required construction space
- Easy energy transfer
- Clean

The most common electrical heating elements are to electrical resistant heaters. Generally, the barrel heater bands can be divided in to mica, ceramic, and aluminium heaters. Mica heater bands are now being replaced by ceramic heater bands or cast aluminium heaters because they more reliable, with longer life and highers power density. Ceramic heater bands reach a high power density and temperatures. They usually consist of two semicircular shelves that are bolted together around the barrel. The third common type are the cast aluminum heaters. The heating element are cast in aluminum parts corresponding to the required shape. Cast heaters have a good heat transfer, are very reliable, and have a long operating life.

To reach a good heat transfer and long lifetime the heater must have a good contact to the barrel over the entire contact area. To ensure the contact, clamping or spring element press the heaters on the surface. Table 3.17 presents the operating temperatures and power density of the different elecrical heating elements.

Table 3.17   Operatig Temperatures and Power Density of Electrical Heating Elements

	Mica heater band	Ceramic heater band	Aluminum cast heater
Maximum temperature °C	~ 500	~ 750	~ 400
Power density (W/cm²)	2,5– 3,5	4,5– 6,5	4,5– 5,5

The heaters are arranged in three to ten heating zones. Fig. 3.105 shows a barrel equipped with cast aluminum heaters. The amount of the heating zones increases with increasing barrel length. Each zone is controlled independently to maintain a flexibility over the barrel temperature profile.

Figure 3.105   Barrel with aluminium cast heaters

Oil or steam heating requires a double-walled extruder barrel and an oil or steam heating boiler. The advantage of the fluid heating is a very even temperature over the heating area. The disadvantages are

- Safety hazards
- Additional space requirement for the equipment
- Increased installation and operating expenses
- Increased maintenance cost
- Chance of leakage
- Corrosion
- Increased heat losses
- Separate oil-heating-circulation necessary for every heating zone.

Steam is a good heat energy transfer fluid because of its high specific heat. It also allow a very even temperature over the heating area. The disadvantages are similar to disadvantages of the oil heating; additionally, the system has to be pressure coded. Oil and steam heating are rarely used on polymer extruders today.

3.4.2.8.2   Extruder Cooling

On most extruders, a barrel cooling is installed. There are two methods of barrel cooling: air cooling and fluid cooling. Forced air cooling is usually used. Fig. 3.106 shows the layout of an air cooling system according Barmag AG. A uniform cooling effect in the axial and radial direction of the barrel is achieved by controlled distribution of the cooling air. The air volume conveyed by the radial blower is distributed around the barrel by air guiding shields. The exhaust air leaves the cover at the top or is collected in an air channel. Each zone is separated, so the cooling air flow cannot influence neighboring zones. To reach a high cooling capacity, an air speed of more than 15 m/s is advisable.

The external cooling surface of the barrel or the heaters is increased by cooling ribs. At aluminum-cast heaters the ribs are direct integrated on the surfaces. The second method of increasing the heat transfer area is placing ribbed spacers between the heater bands. Small extruders sometimes do not require a forced cooling by blowers, because their heat losses provide enough cooling capacity due to the natural convection. Air cooling is a gentle type of cooling because the heat transfer rates of air is small. This is an advantage because the change of temperature occurs gradually and simplify the control of the cooling.

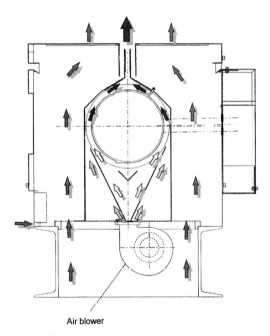

Air blower

Figure 3.106   Barrel with air cooling system

Usually the air cooling provides a sufficient cooling capacity in extrusion. If a high cooling capacity is required a fluid cooling system is necessary. Water cooling is the common system. Because of the high heat transfer rate, the water cooling rapidly change the temperatures. Therefore, the control of water cooling is more difficult compared to air cooling. Water cooling can be controlled by proportional valves or pulsed valves. Generally, the temperature controlling accuracy should be less than ±2 °C because barrel temperature fluctuation can influence pressure and output. The installed cooling capacity – as well as the installed heating capacity – has to be choosen in a range that allows an acceptable temperature control in each heating zone. Fluid cooling systems are relative expensive because of the costly control equipment and pipe system made of corrosion-resistant material. High-speed compounders and melting extruders with screw diameters of more than 200 mm are frequently equipped with a fluid cooling.

3.4.2.9   Screw

The extruder screw is heart of the machine. The screw is generally a custom-designed part that matches the resins being processed. The details of the screw layout are discussed in Chapter 3.3.

The screws are not subjected to a high bending force because they are always run inside a strong rigid barrel. The clearance between screw and barrel is small. Generally, the ratio of the clearance to screw diameter is around 0,0005 to 0,002 [5]. The gap between screw and barrel is small to prevent gelation of the melt or cracking. Leakage flow, caused by the clearance between barrel and flight land also reduce the melting efficiency. Therefore, extruder screws must be manufactured to very tight tolerances.

The critical strength requirement is resistance to torque. The maximum torque that can be applied to a screw can be calculated by the following formula:

$$M = \frac{9550P}{n} \tag{3.4.5}$$

where
M = torque in (Nm)
P = power of the drive (in kW)
N = screw speed (in rpm)

Unfortunately, the weakest area of a screw, the feed section, is the portion subject to the highest torque. The information about the highest torque can be used to determine the maximum channel depth in the feed section. Screw breaking is a problem in small diameters 25- to 45-mm screws because the depth required by the feed zone serverely weaken the root. To avoid the risk of screw breakage some processors prefer to operate a large-diameter extruder at slower speed rather than a smaller diameter extruder at high speed.

Alloy steel is by far the most common screw material. Other important common screw materials are stainless and tool steels. The choose of material is depending on factors as

- Yield strength
- Hardness
- Wear resistance
- Corrosion resistance
- Ease of machining
- Cost of material

Usually wear resistance is improved by hardening the screw surface. A very hard surface can be obtained by a nitriding process. Ion nitriding is currently preferred because it superior to the gas nitriding process. Ion nitriding is more expensive but give less distortion because of the lower processing temperatures. In case of high wear, the wear surface, primarily the top of the flights, which are the wear surface, can be protected by welding on special resistant alloys. The most popular of these alloys are Stellite and Colmonoy, but many other materials have been developed in recent years. Sometimes screws are chrome plated or nickel plated. In most cases the plating is used to improve the corrosion resistance. This option is questionable with polymer that can release acids during processing.

Screws are of one-piece construction or assembled of different parts. Additional mixing tips at the front of the screw usually threaded or bolted in place. This allows easier manufacturing of the mixer and offer the opportunity to change the mixing tip. Changing the mixing tip allows the shearing and mixing levels to be customized to the polymer and process conditions.

To incorporate screw cooling , screws may be cored internally, but the core should not extend very far beyond the feed throat (approximately four to five turns) [6]. In general, a screw heating or cooling is not necessary. An intensive cooling of the screw in the feeding section makes sense if there is the risk of premature melting of the polymer, which can result in an inconsistent feeding of the granules. Cooling the screw by water or oil is difficult because of the rotation of the screw. Rotary devices and cooper tubing have to be installed inside cavity in the screw. The flight width should be 0,08 up to 0,12D. Very narrow flights increase the leakage flow of polymer and risk a break of the flight because of the loading from the shear stress forces and differential pressure between leading and trailing side of the flights. Very wide flights increase the risk of a local overheating of the polymer. Typically screw channels cross section have a box profile with slightly rounded edges. The screw tip has round or cone design to avoid a stagnation of the polymer in front of the screw.

The main design features of the extruder screw are

- Screw length
- Zone distribution, zone length
- Flight depth, compression ratio
- Number of flights
- Flight pitch
- Mixing component/shearing component arrangement
- Mixing component/shearing component geometry (pin mixer, rhomboid mixer, grooved mixer, shearing surface mixer, maddock mixer)

Depending on the process and the required throughput the diameter of the screws varies from 19 mm for lab extruders up to 350 mm for melting extruders. Fig. 3.107 shows a melting extruder screw of 350 mm diameter and 30D lenght with a rhomboid mixing tip.

Figure 3.107   350 mm, 30D melting extruder screw

3.4.3   Special Designs

3.4.3.1   Spinning Extruder

Spinning extruders are used for the production of man-made fibers. Most of the polymers used for this application (except polymers for industrial yarn) are characterized by a comparatively low viscosity. Therefore, these polymers (mainly PETP, PA, and PP) can be extruded with a minimum of energy dissipation, which gives less importance to an effective cooling system for spinning extruders. Beside an excellent melt homogenity (fibers with diameters down to 10 μm, winding speeds up to 8000 m/min), energy efficiency is most important for spinning extruders [8]. All spinning lines need controlled climate conditions (ambient air temperature, ambient air humidity) to produce fibers with the required qualities at high winding speeds. Energy dissipation from the extruder to the environment has to be handled by the air conditioning system, which results in increased production cost.

To reduce these energy losses spinning extruders are designed not only with a protective cover around the barrel, but with an effective insulation (Fig. 3.108). Cooling fans are installed only in these heating zones, where it is absolutely necessary for the process. With these measures, energy losses can be reduced by more than 40%, which can be up to 6 kW, depending on the extruder size.

3.4.3.2   Adiabatic Extruder

The idea of the adiabatic extruder is to use all the dissipated energy for melting of the polymer. The heating system is used only for the startup period and for influencing the melt temperature. All the energy needed for melting is mechanical energy transferred by the screw (Fig. 3.109).

Adiabatic extruders are very effective for processing Polyolefins like Polypropylene or Polyethylene. Due to the physical properties of polypropylene (i.e., comparatively low heat conductivity, high specific enthalpy), extruders have to be designed with L/D ratios of 30 or more to have sufficient residence time inside the extruder for melting and homogenizing of

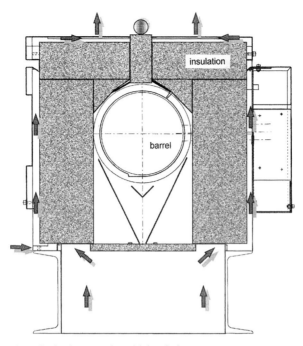

Figure 3.108   Crosection of spinning extruder with insulation

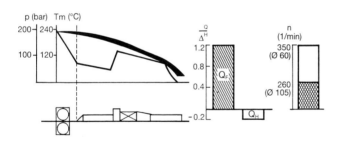

Δ T $\leq$ + 2 °C
T min = 240 °C (PP)
Δ p/pm $\leq$ + 0,75 %; pm > 100 bar

n: Screw Speed
Q_H Heat Content by Heating
Q_F Heat Content by Friction
T: Melt Temperature
pm: Melt Pressure

Δ H: Increase of melt enthalpy
        in the extruder

Figure 3.109   Operating conditions of adiabatic extruder

the polymer. With the adiabatic extruder, melting will be done more effective [9] and therefore high throughput ranges can be reached even with small extruders.

From the design point of view adiabatic extruders are characterized by high-capacity drives with high screw speeds. Due to special screw designs most of the adiabatic extruders are installed in cascade systems, either in combination with a gear pump or in combination with a metering extruder. In these combinations, the extruder has only the task of melting and homogenizing of the polymer. Metering will be done by the pump or by a second extruder.

3.4.3.3   Vented Extruder

Vented extruders are used for processing polymers with high contents of monomers, oligomers, or other materials boiling at temperatures below the melt temperature.

They are characterized by additional openings (vent ports) along the extruder barrel (Fig. 3.110). Volatiles can escape during the extrusion process.

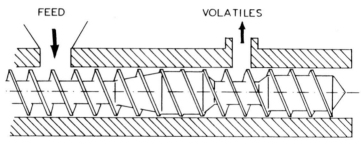

Figure 3.110   Vented Extruder [1]

Most important for vented extruders is the screw, which is designed as a two-stage or multi-stage screw. In the first stage feeding and heating of the polymer are achieved. Depending on the process, the first vent port can be positioned before the polymer starts melting to degas volatiles from the surface of the granules (for example, moisture). In other cases, its position is in an area where the polymer is completely molten to degas (for example, mono-mers).

The second stage of the screw starts at the first vent port. In case of a multistage screw, each venting port characterizes the transition from one stage to the next. In each stage the polymer is compressed. At the vent port it is decompressed so that the volume/surface ratio is decreased. Under these conditions, volatiles can escape from the melt. The number of stages depends on the required degree of devolatilization. When designing the screw, it is important to get the right balance between the different stages. The feed capacity of the following stage should always be higher than the capacity of the stage before; otherwise, there will be melt flow in the vent port. The last stage should be designed to allow constant feed of the polymer at constant pressure to the downstream process. Depending on the number and on the design of the stages, the length of vented extruders can be extended to 40–50 L/D.

3.4.3.4   Metering Extruder

Metering extruders are also known as liquid feed or hot melt extruders. The polymer fed to that machine is already molten at the extruder inlet. The task of the metering extruder is to homogenize the melt and to feed it with excellent temperature and pressure uniformity to the downstream process. The function of the metering extruder is comparable to the function of the metering section of a standard extruder, with the difference that the pressure profile is adjustable and that the melt temperature can be influenced over a wide range. Depending on the process, metering extruders can even be used for cooling the melt (Fig. 3.111).

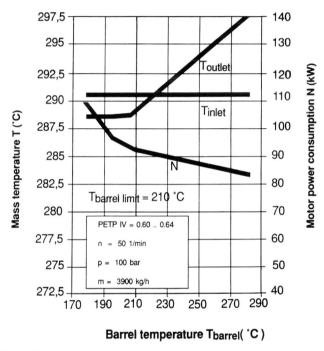

Figure 3.111   Effect of the barrel temperature on the mass temperature

With regards to the design of a metering extruder the most important part is the melt sealing between the feed housing and the gearbox (Fig. 3.112). Inlet pressures can be as high as 200 bar. To avoid an unacceptable meltflow along the screw shaft, an effective sealing has to be installed. Sealing can be done by freezing of the polymer or by using stuffing boxes.

Beside the sealing metering extruders are characterized by comparatively small drives and for some applications by a fluid heating system. The small drive results from the low energy requirement due to the fact that no energy is needed for melting the polymer. Fluid heating systems are used to get a uniform melt temperature; the metering extruder is used like a heat exchanger in such a case. Typical sizes for metering extruders range from 90 mm diameter to 400 mm diameter, with an *L/D* ratio between 15 and 35. Metering extruders are used in tandem extruders or below polycondensation plants.

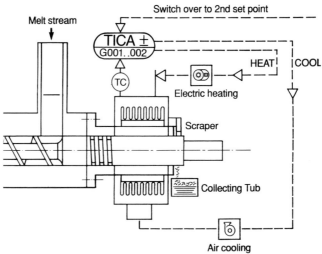

Figure 3.112   Melt sealing for metering extruder

3.4.3.5   Tandem Extruder

The combination of a melting extruder and a metering extruder is called tandem or cascade extruder. The idea of that combination is the separation of the process steps melting, homogenization, and metering (Fig. 3.113). By using such an extrusion system optimum melt qualities with regards to temperature and throughput uniformity can be achieved, especially at high throughput rates [10]. The melting extruder is a standard single screw extruder, optimized just for melting of the polymer. For polyolefins it can be designed as adiabatic extruder. Because it is not required to homogenize the polymer, the melting extruder can be shorter than an standard extruders. L/D ratios between 21 and 30 are common. The metering extruder is described in the previous section. Depending on the application, a melt filter can be installed between the melting and metering extruders to give a further improvement in melt quality. Tandem extruders are mainly used for processing PETP and PP in extrusion lines producing biaxally oriented films. Another application is the extrusion of foamed PS sheets and films.

3.4.3.6   Vertical Extruder

Depending on space requirements or special process conditions, single screw extruders can be designed in a vertical alignment (Fig. 3.114). In case it has to process polymers that are delivered as powder or paste, the vertical extruder is used as a "front-end-drive" machine, where the feed section of the screw is extended to the hopper. Drive and gearbox are connected to outlet side of the extruder. These machines require a melt sealing similar to the metering extruder. Beside the different feeding behavior, there is no difference between a vertical and a horizontal extruder with regards to the process.

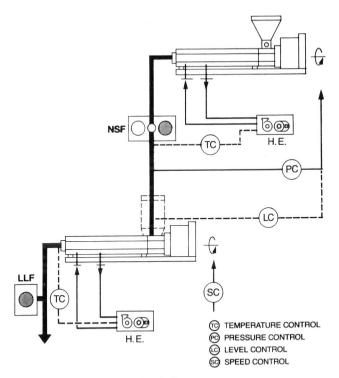

NSF

TC

H.E.

PC

LC

LLF

TC

H.E.

SC

(TC) TEMPERATURE CONTROL
(PC) PRESSURE CONTROL
(LC) LEVEL CONTROL
(SC) SPEED CONTROL

Figure 3.113   Tandem extruder with integrated melt filters

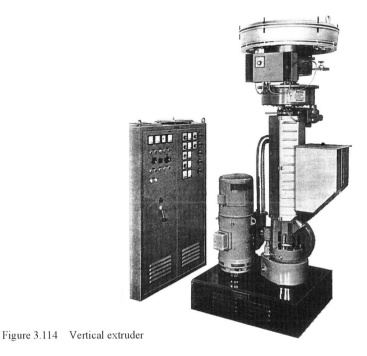

Figure 3.114   Vertical extruder

3.4.3.7   Extruder with Conical Feed Zone

Another special design for solving feeding problems is the extruder with conical feed zone (Fig. 3.115). This machine is mainly used for processing materials with low bulk density, for example, ground up film or fiber waste. The extruder is characterized by an enlarged screw diameter in the feed section below the hopper, which results in an increased volume of the screw channel. In the transmission zone, the screw diameter is reduced. The nominal diameter of the extruder is the diameter of the metering section, which is mainly responsible for the throughput of the machine. The ratio between the diameter of the feed section and the diameter of the metering section is determined by the ratio between the bulk density and the melt density of the extruded material and by the compression ratio that is necessary for melting that polymer. Screw cooling in the feed section or grooves in the barrel can improve the feed behavior of an extruder with a conical feed zone.

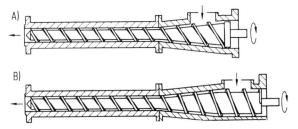

Figure 3.115   Extruder with conical feed zone [4]

References

1. Rauwendaal, C., *Polymer Extrusion* (1986) Carl Hanser Verlag, Munich
2. Barmag AG., *Technical Information EX 75/2,* Remscheid
3. Michaeli,W., *Umdruck Kunststoffverarbeitung II: Extrudertechnik* (1988) Institut für Kunststoffverarbeitung, Aachen
4. Hensen, F., Knappe, W., Potente, H. *Handbuch der Kunststofftechnik,* Band I: Grundlagen (1989) Carl Hanser Verlag, Munich
5. Schroeter, B., *Der Extruder im Extrusionsprozeß* (1989) VDI-Verlag, Düsseldorf
6. SPIREX-Corporation, *Plasticating Components Technology* (1992), booklet
7. Kohan, M.I., *Nylon Plastics Handbook* (1995) Carl Hanser Verlag, Munich
8. Schäfer, K. *Components for Efficient Production of the Highest Yarn Qualities* (1996) World Congress PET 96, Zürich
9. Langhorst, H., Dissertation an der RWTH Aachen (1988)
10. Hensen, F., Imping, W., Stausberg, G., *Int. Polym. Process* (1992) 7, 20

3.5   Measurement and Open-Loop and Closed-Loop Control Engineering

Hans Recker and Gerd Wiegand

The function of measurement devices and open-loop and closed-loop control systems is to ensure compliance with defined quality requirements, placed on the product or on the process itself, by monitoring these during production. Quality-relevant data are thus not only acquired and documented during production processes for purposes of providing evidence of a specific production quality, but are also further processed in downstream open-loop or closed-loop control systems in a large number of cases.

Tables 3.18 and 3.19 provide an overview of the sensors employed on extrusion plants [35]. Since new systems are constantly being developed and others withdrawn from the market, these tables cannot provide an exhaustive list.

Table 3.18   Methods Used for Measuring Process Variables

Process parameter	Sensor	Measuring principle
Temperature	Thermocouple	Seebeck effect
	Resistance thermometer	Change in resistance due to temperature
	Infrared thermometer	Planck's intensity distribution
	Ultrasonic emitter/receiver	Time measurement
	Liquid-filled thermometer	Thermal expansion
Pressure	Strain gauge sensor	Change in resistance due to strain
	Piezoelectric sensor	Piezo effect
	Ultrasonic transmitter/receiver	Time measurement
Rpm/speed	Tachometer/pulse speed sensor	Induction
Throughput	Gravimetric sensor	Weighing
	Volumetric sensor	Volume determination
Power	Current and voltage measuring units	Induction
Torque	Strain gauge	Change in resistance due to strain

Table 3.19   Methods Used for Measuring Product Properties

Product property	Sensor	Measuring principle
Thickness	Position transducer	Induction
	Radiometry sensor	Absorption
	Infrared sensor	Absorption
	Laser sensor/CCD sensor	Triangulation
	High-frequency sensor	Electrical attenuation

Table 3.19   Continued

Product property	Sensor	Measuring principle
	Capacitor sensor	Geometry-dependence of capacity
	Ultrasonic transmitter/receiver	Time measurement
Gloss	CCD sensor	Intensity measurement
Contours	Position transducer	Induction
	CCD sensor	Image processing
Material defects	Radiometry sensor	Image processing
Color	CCD sensor	Planck's intensity distribution
Roughness	Strain gauge	Change in resistance due to strain
Tensile strength/ elongation at break	Strain gauge	Change in resistance due to strain
Inherent stresses in transparent products	CCD sensor	Image processing

Table 3.18 shows the methods used for measuring process variables. Table 3.19 shows the methods used for measuring product properties. A number of key process parameters and product properties shown here will be looked at in greater detail in what follows, together with the possibilities that exist for their open-loop and closed-loop control.

3.5.1   Temperature Measurement and Closed-Loop Control

3.5.1.1   Wall Temperature

The temperature of the barrel wall is recorded via plug-in sensor types that are equipped with a spring bayonet connection; in most cases, use is made of iron/copper-nickel thermo-couples. The sensor element should be positioned as close to the melt channel as possible to avoid high system-conditioned deviations between the wall temperature and the melt temperature and also to minimize the heat dissipation error via the sensor element (Fig. 3.116). Fig. 3.117 shows the effect of the sensor hole depth [1]. The temperature displayed here deviates from the actual wall temperature by up to 10 °C as a function of the immersion depth. If the influence of draughts is observed, then an additional velocity-dependent error of 10 °C is observed.

3.5.1.2   Melt Temperature

Melt temperatures are recorded virtually exclusively using thermocouples. The different designs represent compromises between the call for sufficient mechanical stability on the one hand and a satisfactory measuring accuracy (heat dissipation and response rate) on the other. Fig. 3.118 shows the standard designs that are employed [1]. Under production conditions, use is generally made of sensor elements that are installed flush with the wall (Fig. 3.118a). These record the melt temperature close to the wall (apart from potential heat dissipation errors). All the other designs shown are basically more suitable for recording melt temperature, but, because of their mechanical sensitivity, they are generally used only on laboratory plants.

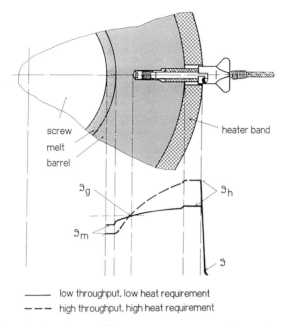

Fig. 3.116   Section through a barrel zone with a temperature sensor and temperature profile

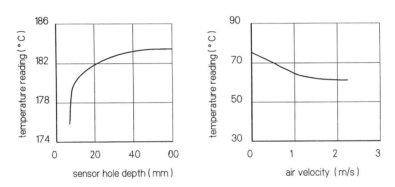

Fig. 3.117   Heat dissipation error with temperature sensors

In contrast to Fig. 3.118a, the sensor elements that are immersed in the melt can be installed only in the space between the die and the screw tip. The velocity and temperature profiles that prevail here are a function of the extruder operating conditions. The designs of Figs. 3.118d and 3.118e are suitable for measuring this temperature profile, while the design of Fig. 3.118c can be used to record a representative mean value for the temperature, providing that the sensor is immersed to a depth of roughly one-third of the channel diameter [1–5].

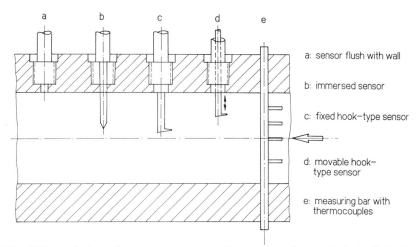

Fig. 3.118   Different designs of a temperature sensor [1]: (a) sensor flush with the wall, (b) immersed sensor, (c) fixed-location hook-type probe, (d) movable hook-type probe, (e) measuring bar with several temperature probes

There are two further methods for measuring melt temperature, which have the major advantage of not requiring mechanical components that protrude into the flow channel.

While indirect melt temperature measurement on the basis of an ultrasound travel-time method was unable to become permanently established [6, 7], a method in which the heat radiation of the melt, which is a function of wavelength, is conducted to an infrared (IR) detector by appropriate optical fibers is in increasingly widespread use [8, 9].

Unless an ideal black radiator is involved, the measured radiation value must be corrected with the emission factor of the radiator. Plastic melts that are located in a space enclosed on all sides, such as a die, behave like virtually ideal black radiators, since the radiation components that are reflected off the surrounding walls and the transmitted components return to the IR detector once again. For this reason, emission coefficients of 0.99 are observed in dies for plastics that have an emission coefficient of 0.9 in air [10].

3.5.1.3   Surface Temperature

Surface temperatures can be suitably measured by contact-free measuring methods and, to a certain extent, by contact methods. The contact methods involve a temperature sensor being brought directly into contact with the object to be measured. In the case of moving objects, use is made of a wheel or a grinder with a low coefficient of friction to this end. The advantages of this form of measurement include

- The low cost of the sensors and the evaluation electronics
- The material-independence of the measured values

The contact methods do, however, have distinct drawbacks:

- A heat dissipation error always results from the contact with the sensor element. The smaller the heat capacity of the object that is to be measured (e.g., film), the greater the expected error will be.

- A mark is left on the product at the point of measurement if the measurement is conducted at below the solidification temperature.

By contrast to the IR measurement in dies described in Section 3.5.1.2, the emission of the object being measured has a considerable effect on the measurement result when surface temperatures are measured by infrared radiation and this has to be taken into account. There are thus certain drawbacks involved in this method:

- The emission coefficient must be known.
- The measuring equipment is highly elaborate.

The advantages of the method are its contact-free recording of the temperature, without any heat conduction, and its rapid response rate, which is not affected by the mass of the object being measured.

3.5.1.4   Closed-Loop Temperature Control

Extrusion plants have become increasingly automated over the past few years – a development that started with closed-loop temperature control systems for the different barrel sections. Apart from controlling the wall temperatures of the barrel sections, controllers are employed for

- Temperature control of the extruder barrel
- Temperature control of dies and tools
- Temperature control of cooling media, etc.

Fig. 3.119 shows the basic layout of a closed-loop control circuit for controlling the wall temperature of a barrel zone. The temperature is recorded with a thermocouple of the type described in Section 3.5.1.1 and constitutes the actual value for the control circuit. This actual value is compared with a setpoint value, and a controller output signal is generated as a function of this comparison.

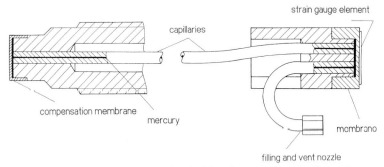

Fig. 3.119   Temperature control circuit for an individual barrel zone (schematic diagram)

A distinction is drawn between controllers with standardized, continuous-action outputs and those with discontinuous-action outputs as a function of the type of output involved. Nowadays, control algorithms are only rarely still implemented in analog circuits and, instead, are implemented as digital circuits in a microcomputer. This has made it possible for these digital controllers to incorporate considerably more tasks than just the pure control task. They thus also provide interfaces to higher-ranking control rooms. Their reactions to a wide range of error states can be programmed in advance, thereby avoiding production stoppages. Microcomputers of this type are available ready-integrated in a single component, with analog inputs for the temperature sensors and analogue outputs for the actuators, as well as digital inputs and outputs for the recording and controlling of different states and communication interfaces. Controllers have a feedback loop that is aligned to the controlled system. The time response of this feedback loop has a considerable influence on the control-action result.

Fig. 3.120 shows the control response of an analog two-position controller to a change in the setpoint value (feedback response) [11]. Fig. 3.120a shows the response of a

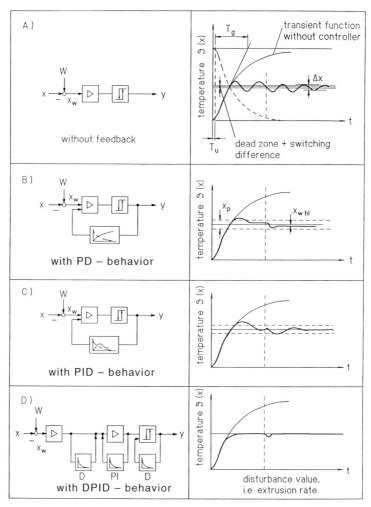

Fig. 3.120   Comparison of controller circuits and results for temperature control systems [2], where Tg is the compensation time, Tu the delay time, xwbl the system deviation that remains, and xp the proportional range

controller without any feedback. The setpoint value has an oscillation superimposed on it which is a function of the input/output time. Figs. 3.120b and d show the control-action results of standard controllers employed today (irrespective of whether an analog or digital controller is involved). If the feedback has proportional differential (PD) action (Fig. 3.120b), then a lasting deviation will result between the setpoint value and the actual value of the wall temperature. A controller with DPID action (additional integral/differential part) in its feedback, by contrast, will operate without any permanent system deviation.

On the standard commercial controllers used for plastics processing machines, the constants that define the feedback characteristics are aligned to the behavior of the controlled system and are frequently permanently set to give a satisfactory control response to both changes in the setpoint value and any process malfunctions that may occur. In special cases, such as if the control response at a fixed operating point needs to be optimized for malfunctions, it can be wise to reset the controller parameters. Tried-and-tested methods and guidelines for setting controller parameters may be found in [12–18].

Today, digital controllers are also available that automatically adapt their constants to the prevailing operating state. A learning cycle is triggered, either during startup or by the user, during which the optimum controller parameters for the system are established. In this way, if there is a change in the setpoint, the controller can work with a set of parameters tailored to an optimum response to a setpoint change and, once the new setpoint has been attained, it can work with a parameter set optimized for malfunctions [19, 20].

3.5.2   Measurement and Closed-Loop Control of Pressure

3.5.2.1   Melt Pressure

The pressures that prevail during extrusion act for long periods of time and are subject to only a low rate of change. Different methods can be suitably applied to measure these pressures [5, 21].

The chief type of pressure sensors employed today are those based on strain gauges made up of a metal grid that is evaporation coated onto a carrier film [10]. The high temperatures that occur and the sensitivity of the strain gauge to heat means that the section of the sensor that is in contact with the melt is spatially separated from the measuring element (Fig. 3.121).

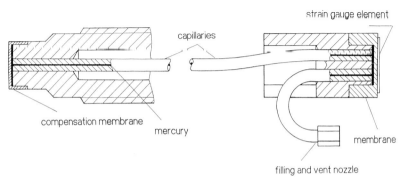

Fig. 3.121   Principle behind a pressure sensor incorporating a strain gauge

The sensor thus comprises two chambers, both of which are sealed by a membrane. The two chambers are joined together via a capillary that is filled with mercury. The melt pressure can then be transferred to the measuring membrane without any friction. During steady-state operation, the temperature that develops on the measuring membrane is a function of the room temperature and the heat transfer conditions that result from radiation and conduction. This temperature will be in the region of the ambient temperature and thus considerably lower than the actual melt temperature.

Strain gauge melt pressure sensors are available that have a thermocouple incorporated in the direct vicinity of the compensation membrane. This means that pressure and temperature measurements can be conducted at one and the same location [10]. The use of pressure sensors containing mercury is not permitted in the extrusion of packaging material or semifinished products for medical applications. In this case, use is made of pneumatic pressure sensors or sensors with a compensation membrane linked to the measuring membrane by means of a ram or a piston.

For some time now, melt pressure sensors have been available for use in extrusion that have a membrane in wear-resistant, monocrystalline aluminum oxide (Saphir). Silicon piezo resistances are applied directly to the rear of this membrane, which is in contact with the melt. The pressure sensors do not contain any elements for transmitting pressure or force, such as capillaries, rods, or pistons, and thus possess considerably shorter response times. Versions with integrated temperature sensors are available as an option in this case, too [22].

3.5.2.2   Open-Loop Pressure Control

In extrusion, the melt pressure is measured primarily for purposes of monitoring extruder operation. Pressure changes point to deviations from setpoint values in the area of feed, plastization, or homogenization. When producing semifinished products that are subject to stringent requirements in terms of melt purity and homogeneity, it is standard practice to install melt filters between the extruder barrel and the die. These filters gradually clog up during running production, causing an increase in the melt pressure. With the aid of melt pressure sensors upstream and downstream of the screen unit, it is possible to establish the degree of soiling from the pressure drop at the filter. The filter can then be changed in good time and production monitoring or, in critical cases, a shutdown in the event of malfunctions can be achieved.

3.5.2.3   Closed-Loop Pressure Control

Increasing use is being made of melt pumps as an aid to process control. In the case of coextrusion, in particular, a constant material throughput is of decisive importance for achieving a uniform thickness in the individual extruded layers. Melt throughput fluctuations or fluctuations in the pressure buildup inside the extruder have the effect of impairing quality. These fluctuations can be smoothed out by melt pumps inserted between the extruder and the die. To ensure that constant melt pressures build up at the inflow into the pump, at not too high a level, the extruders are fitted with a closed-loop pressure control. The melt throughput necessary for extrusion is set via the speed of the melt pump. The pressure/speed control system on the extruder then ensures that the necessary quantity of melt is available at the set pressure. The makeup of a melt gear pump is shown in Fig. 3.122 .

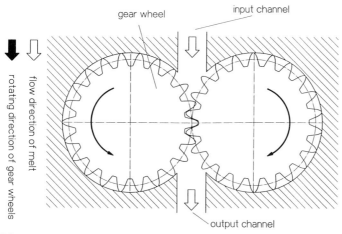

Fig. 3.122   Melt pump

3.5.2.4   Closed-Loop Thickness Control with the Aid of Pressure and Temperature Measurements

Flat film and pipe extrusion can be taken as an example for the indirect control of a process parameter. Here, the local thickness distribution at right angles to the direction of takeoff is calculated and controlled via pressure and temperature measurements in the die, on the basis of an adaptive process model [23]. The process model is verified by a downstream thickness measurement unit and corrected where necessary. The advantage of this approach is that the actual values for pressure and temperature are recorded immediately prior to the die gap and the control can thus be implemented with virtually no dead time.

3.5.3   Measuring Specific Product Properties

Specific properties are monitored and documented during the running process as a function of the use to which the extruded product is to be put. Apart from the specified dimensions, the surface finish is a key parameter for almost all extruded products. This is generally recorded on a continuous basis through gloss measurements and measurements of the surface structure.

3.5.3.1   Dimensioning

A large number of extrudates are of a simple geometry (film, panels, pipes), which can be recorded with a one-dimensional measurement. Recording more complex geometries (such as window profiles) is more complicated, since there are more individual dimensions to be checked and maintained within a specified tolerance. In such cases, one-dimensional measurements are still frequently employed at characteristic or critical points. Although the technology is available for measuring complex geometries.

Apart from purely mechanical control and measurement units and also measuring gauges for conducting random sample checks, use is made of electromechanical methods involving contact and contactfree optical, electrical, and radiometric methods for the continuous recording of dimensions. Preference should basically be given to contact-free methods, since the contact methods can be applied only when the extrudate has cooled to below its solidification temperature [24–27]. This is a particular drawback if the measurement signal is to be used as the actual value for controlling the takeoff rate, since control systems with a pronounced dead time result here in control engineering terms.

Contact-free radiometric thickness measurement is in widespread use for panel, film, and pipe extrusion. Here, the absorption or scatter of radioactive isotopic radiation is determined as it passes through the product. The measured value correlates approximately with the thickness at a specified density. This is a one-dimensional measurement; the radiator is generally passed over or around the product and the mean value for a single passage obtained [28].

Infrared thickness measurement technology is gaining increasing importance in film extrusion. One particular advantage of IR technology is that it does not need the elaborate safety precautions required for radiometric measurement, and it can also be used to determine the individual layer thickness of composite films [24]. Optical thickness measurement units are used for the determination of diameters especially in conductor, cable, and pipe extrusion.

Fig. 3.123 shows the principle of diameter measurement using a laser scanner. A motor, which holds a mirror, is controlled by a quartz oscillator. The mirror conducts the laser beam over the object to be measured in a parallel configuration. During this time, which is proportional to the diameter, the photoelectric cell positioned behind the object being measured is in the shade. The diameter is then determined from the ratio of the light/dark times or from the mean radiation intensity over an individual measurement period. Similar measurement techniques are employed for measuring the thickness of films and foamed panels [27].

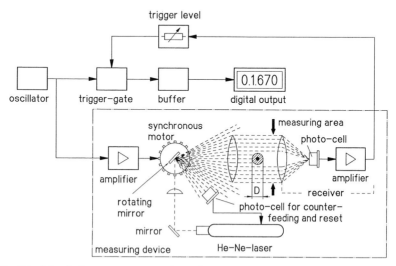

Fig. 3.123   Principle of a laser scanner measuring unit

3.5.3.2   Surface Defect Recognition on Panels and Films

Fig. 3.124 shows a surface inspection system that is employed in panel and film extrusion [29]. The light from the laser is conducted to a lens by an optical system and shines on a rapidly rotating, polyhedral reflecting wheel. As the reflecting wheel rotates, it deflects the laser beam onto a flat mirror, and from there into a concave mirror. The axially parallel beams reflected by the concave mirror are focused on the material web by a cylindrical lens. Fiber-optic rods are used as receivers for the reflection and transmission, and their intensity or modification is then assessed.

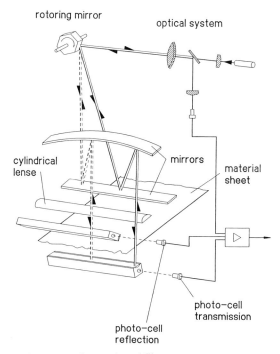

Fig. 3.124   Surface inspection system for panels and film

3.5.3.3   Surface Inspection of Profiles

Although it rarely happens that there are surface defects on profiles, such as dimensional deviations, waviness, cracks, encapsulated foreign bodies, scratches, pores, and sink marks, these nonetheless account for approximately 80% of overall waste.

The light section method shown in Fig. 3.125 can be suitably employed for the surface inspection of extruded profiles [30]. A semiconductor laser projects a thin line of light onto the profile at a defined angle by means of a special optical system. A video camera records the reflected line of light, which is distorted as a function of the topography of the object being inspected. The shape of the surface is then calculated on the basis of this pattern and the underlying geometrical constraints. Nowadays, video cameras for applications of this type are equipped with CCD sensors, which are arranged in either a linear or two-dimensional configuration. CCD sensors are light-sensitive semiconductor elements that convert incident light into a voltage and charge up a capacitor. The stored charge is output serially on a line-by-line basis in the form of a voltage signal that complies with the video standard. Since this measurement principle only permits the detection of topographical changes on the surface, it cannot be used to determine gloss differences or color changes in the same way as other systems [30].

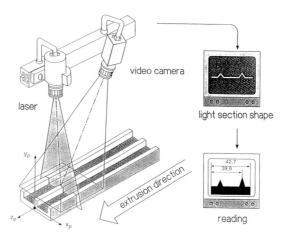

Fig. 3.125   Surface testing with the light section method

3.5.4   Process Control on the Extruder

In the simplest case, an extruder control system comprises a heating/cooling facility for the screw barrel and a variable speed drive. On state-of-the-art extruder control systems, both the temperatures and the extruder speed can be set and monitored in a central control room. Alongside these, parameters are displayed that are not subject to control but are required for purposes of assessing the current state of the process. These include the pressure prior to the screw tip and the prevailing drive power.

Records of the properties that have a bearing on quality are gaining increasing importance. These often have to be recorded and documented on a continuous basis, without any gaps. A further task to be assumed by the extruder control system is the automatic protection of the plant and the process [31]. It has to be guaranteed that the plant cannot be damaged by what are obviously operating errors or by pronounced deviations in the controlled variables. A large number of different approaches are possible, and just a few will be set out here by way of example:

- The screw drive must be locked during the extruder heating phase.
- If the drive consumes too much power over a prolonged period of time, then the drive must be switched off.
- In the event of slight differences between the target temperature and the actual temperature, a warning must be issued. In the event of major differences, the plant must be stopped.

- If the melt pressure prior to the screw tip is too high, the plant must be stopped or an automatic screen changer triggered, where available.

The latest control systems permit the storage and verification of complex error states, such as those that can be detected through the interlinking of a wide range of signals. These extruder control systems also supply formulations for individual products. The machine operator does not then need to reestablish the settings, such as the temperature, each time, but can have recourse to previously determined data. Since there is a very large quantity of data to be stored, electronic storage media, such as magnetic tapes or electronically erasable, programmable read-only memories (EEPROMs), can suitably be used. These facilities for reading stored data and manipulating it from the control room also harbor dangers, however. It is essential for hierarchically based access rights to be defined and implemented. These requirements are now essentially fulfilled by two differently structured automation concepts [32].

3.5.4.1   Automation System Structures

Centralized and networked system structures are widespread today [33]. In the decentralized system structure previously employed, all the components operated independently of each other and it was not possible to link information supplied by different components.

3.5.4.1.1   Centralized System Structure

When it comes to automation systems, extrusion plants are relatively small units. This makes it possible for all the process states to be recorded by a central computer, which then triggers the appropriate reactions [34]. A concept of this type made sense at the time when computer hardware was still relatively expensive. The advantages of this concept are that

- The costs of the hardware are low.
- Data management is easy, since all the data are available at a centralized point.

Set against this comes the crucial drawback that production will be closed down in its entirety in the event of computer failure or a software error. The current low cost of computer hardware, coupled with the high risk of a potential production stoppage, means that networked structures are generally employed today.

3.5.4.1.2   Networked Automation Systems

The different tasks are assumed by individual systems here; these systems are then linked together via interfaces for purposes of data exchange. The advantages of this structure are that

- Even in the event of individual components failing, it is highly likely that the plant will be able to continue operating.
- The entire configuration is flexible and can thus be readily modified and aligned to different products.

The negative features are the higher outlay required on data communication and the expense involved in the adaptation of potentially dissimilar interfaces [32].

3.5.4.2   Material Metering

The material costs dictate the costs of extruded products to a considerable extent. Measures that will reduce the amount of material required thus frequently justify high investment costs.

Material metering units are used in a wide variety of applications. These are employed for

- Determining and adjusting the melt throughput
- Achieving specific mix ratios on-line, such as for the addition of pigments or for mixtures of regrind and virgin material
- The material feed, if extrusion is performed with a partially filled screw
- Coextrusion.

Apart from metering screws and vibrating conveyors, gravimetric systems are also available for recording melt throughput. A gravimetric system will be presented by way of an example (Fig. 3.126). A hopper weighing unit is filled with material. The mass throughput in the extruder is established from the weight reduction. When the hopper is almost empty, more material is poured in. The takeoff rate is controlled in such a way that the per meter weight of the extruded product remains identical, despite fluctuating throughputs. If the other process parameters are also stable, then the operator can adjust the machine to the minimum meter weight, leaving a slight safety margin, and thus save on material.

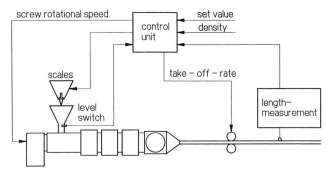

Fig. 3.126   Gravimetric system for measuring melt throughput

3.5.4.3   Closed-Loop Startup Control

The startup of a plant is a highly complex process. More recent control systems thus incorporate startup aids to assist the machine operator. Semiautomatic startup aids repeatedly ask the machine operator to confirm that a specific startup step has been completed. Intervention is possible and, indeed, necessary at all times. Individual plant components can also be operated manually, independently of the higher-ranking control computer.

An automatic startup control can be implemented only on an automated plant. The control computer must have access to all the adjustable variables. With automatic startup aids, it is possible for the startup process to be programmed in the control computer during one or more learning cycles. When the startup process is called up, the individual control steps that the machine operator would normally perform are run through by the control system.

Systems of this type are supported by the sensors on the extrusion plant. True closed-loop control of the startup process is then made possible in this way. By measuring the current melt throughput, the melt pressure downstream of the die, and the takeoff force, the system is able to establish whether a specific state has already been attained during the start-up process before triggering the next step. A key problem in startup control systems is the program sequence, which is determined by the product and the material. The system has to be reprogrammed for each different product and material. An experienced machine operator can thus frequently set a stable and sufficiently good operating point more flexibly and more rapidly. With the advent of new sensors and automatic optimization algorithms, considerably more complex and refined systems can be expected to be available here in future.

3.5.4.4   Process Control for Selected Extrusion Methods

3.5.4.4.1   Tubular Film Plants

Tubular film plants are generally consistently equipped with comprehensive open-loop and closed-loop control facilities [35]. Fig. 3.127 shows a plant of this type with its individual components. Control facilities are available, for instance, for

- Film thickness
- Film width
- The winder

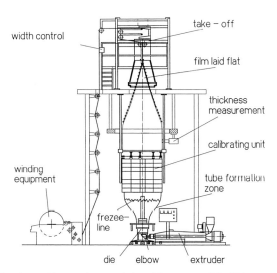

Fig. 3.127   Tubular film plant with open-loop and closed-loop control facilities

The film thickness is controlled by recording the thickness distribution around the circumference of the tube. A reversing measuring heat records the profile. Radioactive β radiators, infrared measuring units, and capacitive thickness measuring units are suitable measuring methods. Automatic dies, on which the width of the die gap can be adjusted or the die wall temperature varied, are employed for the control action in the event of deviations.

Recording the film thickness over the tube diameter and time offers the additional possibility of in-line quality documentation, as is now specified for a large number of products. The film width can be measured with the aid of light barriers once the film has been laid flat. It can then be adjusted via the blowing ratio and the diameter of the calibration basket.

The third component that is to be presented here is the closed-loop control of the takeoff and the winder. Constant longitudinal stretching is achieved via the measurement of the takeoff force. The winder must also be equipped with facilities for an automatic roll change.

All the components together form an interlinked and interacting multiparameter control system. One of the biggest problems encountered here is the extensive dead time that occurs between the control action and the change in the measured value. Experience has shown that control systems with a prolonged dead time are particularly difficult to design for stable operation.

3.5.4.4.2   Flat Film Plants

The task of open-loop and closed-loop process control systems on flat film plants is once again to ensure that the film is produced as economically as possible. This means that a constant, high quality is to be attained with a lower material input. At the same time, the plant is to be easier to operate. Closed control circuits are thus implemented, which fulfil this task (cf. Section 3.5.2.4).

Achieving the correct setting of the die lip profile is a job that is highly time-consuming and thus very costly. It is possible for the film thickness distribution to be influenced via this profile. A thickness measuring device records the thickness profile and calculates the optimum die lip setting from this. The thickness measuring device uses different physical effects, as a function of its design, which are described above. The profile is then adjusted via actuators. Electrically operated expansion bolts or piezoelectric translation units can be used for this.

3.5.4.4.3   Pipe and Profile Plants

Fig. 3.128 shows all the components of a pipe or profile plant. All the open-loop and closed-loop control equipment provided on a modern plant is also included. It is easy to see that a large number of interlinked control facilities are available. It thus makes sense to link all the components together in a network for computer engineering purposes so that all the adjustable variables can be readily accessed. The chief advantage of a fully networked system of this type is its flexibility when there is a change of product. In many cases, a change of product will necessitate a change in the plant configuration. New components are added, while others are no longer needed.

A solution is provided here by systems that are able to withstand the harsh operating climate and that permit the integration of different makes of peripheral equipment via standardized connections and transmission protocols.

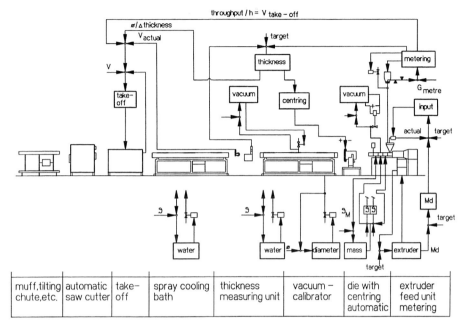

muff,tilting chute,etc.	automatic saw cutter	take-off	spray cooling bath	thickness measuring unit	vacuum – calibrator	die with centring automatic	extruder feed unit metering

Fig. 3.128   Pipe and profile plant with open-loop and closed-loop control facilities

References

1. Görmar, H., Meßwertaufnehmer und Sensoren. In: *Automatisierung in der Kunststoffverarbeitung* (1986) Carl Hanser Verlag, Munich, pp. 37–65
2. Leeuwen van, J., *Polym. Eng. Sci.* (1967), 7, pp. 98–109
3. Scholl, K.H., Wichtige Verarbeitungsparameter und ihre meßtechnische Erfassung. In: *Der Extruder als Plastifiziereinheit* (1977) VDI-Verlag, Düsseldorf, pp. 177–194
4. Mennig, G., *Kunststofftechnik* (1969) 8, pp. 13–16
5. Wiegand, G., *Prozeßautomatisierung beim Extrudieren und Spritzgießen* (1979) Carl Hanser Verlag, Munich
6. Petrik, M., *Kontinuierliche Überwachung von Kunststoffschmelzen im Extrusionsprozeß durch Ultraschallmessungen* (1975) Dissertation, RWTH Aachen
7. Lüscher, H.G., *Kunststoff, Plastics* (1976) 23, pp. 16–23
8. Intrierei, A.J., *Messen und Prüfen* (1978), 12, pp. 849–859
9. Galskoy, A., Wang, K. K., *Plastics Eng.* (1978) 34 (11), pp. 42–45
10. Company documentation from Dynisco Geräte GmbH, Heilbronn
11. Allerdisse, W., Überprüfen und Angleichen von Thermofühlern, Thermofühlerleitungen und Reglern an gegebene Regelstrecken. In: *Messen an Extrusionsanlagen* (1978) VDI-Verlag, Düsseldorf, pp 105–128
12. Meissner, M., *Regelungstechnische Untersuchungen an Kunststoff-Extrudern* (1971) Dissertation, RWTH Aachen
13. Pressler, G., *Regelungstechnik, B.I.-Hochschultaschenbuch 63/63a* (1967) Hochschultaschenbuchverlag, Munich
14. Ziegler, J.G., Nichols, N.B., *Trans. ASME* (1942) 64, p. 759
15. Nyquist, H., *Bell Systems Technol. J.* (1932), pp. 126–147
16. Oppelt, W., *Chem. Ing. Tech.* (1951), 23, pp. 190–193
17. Oppelt, W., *Kleines Handbuch der Regelungstechnik* (1956) 2. Auflage, Verlag Chemie, Weinheim
18. Takahashi, Y., Chen, C.S., Anslander, D.M., *Reglungstechnik und Prozeßdatenverarbeitung* (1971) 19, pp. 237–244

19. Company documentation from Omron Electronics GmbH, Düsseldorf

20. Company documentation from M.K. Juchheim GmbH & Co, Fulda

21. Menges, G., Recker, H., *Messen, Steuern, Regeln in der Kunststoffverarbeitung* (1976) IKV Aachen, Carl Hanser Verlag, Munich

22. Company documentation from Schlaepfer AG, Winterthur

23. Michaeli, W., *Zur Analyse des Flachfolien- und Tafelextrusionsprozeßes* (1975) Dissertation, RWTH Aachen

24. Upmeier, H., *Plastverarbeiter* (1982) 33, pp. 267–270

25. Scholl, K.H., *Plastverarbeiter* (1982) 33, pp. 278–283

26. Bierwolf, H., Berührungslose Dickenmessung an Extrudaten. In: *Messen an Extrusionsanlagen* (1978) VDI-Verlag, Düsseldorf

27. Modern Plastics Int. (1979) 9 (9), pp. 106–109

28. Company documentation from Betacontrol GmbH, Freudenberg

29. Droscha, H., *Kunststoffberater* (1979) 24, pp. 145–147

30. Behrens, M., Schäfer, B., Schmitz, G., Verbesserung der Qualität extrudierter Produkte. Report at the 17th Plastics Technology Colloquium (1994) Aachen, pp. 29–58

31. Bolder, G., Bergweiler, E., Breil, J., Rechnergesteuerte Extrusion – Automatisierungskonzepte des IKV Aachen. Report at the 11th Plastics Technology Colloquium (1982) Aachen, pp. 112–137

32. Spix, L.S., *Automatisierung von Kunststoffextrusionsanlagen – Sensorfehlererkennung, Diagnose, Datenreduktion* (1994) Dissertation, RWTH Aachen

33. Isermann, R., *Digitale Regelsysteme, Grundlagen Deterministischer Regelungen* (1998) Springer-Verlag, Berlin

34. Fürchtenicht, H., Kollmann, M., Kollmann, M., *Regelungstechnische Praxis* (1983) 25 (4) pp. 136–140

35. Recker, H., Spix, L.S., Meß-, Steuer- und Regeleinrichtungen, In: *Kunststoffmaschinenführer* (1992) Carl Hanser Verlag, Munich, 3. Auflage, pp. 1123–1180

4 Technology of Single Screw Extrusion with Reciprocating Screws

4.1   Screw Injection Molding

Martin Würtele and Erwin Bürkle

4.1.1   Particularities Regarding the Transition from the Continuous to the Discontinuous Method of Operation

Besides extrusion, injection molding is considered to be the most important process for the production of moldings from thermoplastics. Injection molding machines are nowa equipped with screw-plunger plasticizing units exclusively, apart from exceptional cases. The screw-plunger units not only plasticize and meter the material, they also inject it (Fig. 4.1). In contrast to the continuous mode of operation with extrusion lines, the injection molding process therefore represents a discontinuous method (Fig. 4.2).

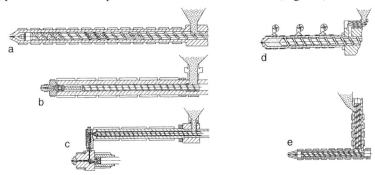

Figure 4.1   Usual plasticizing systems for injection molding machines: (a) conventional plasticizing and conveying screw (*L*=26D), (b) conventional system with feed of positive conveying action and special screw; (c) preplasticizing screw for metering and injection cylinder, (d) dynameltorsystem for metered feeding into the plasticizing unit, (e) twin-screw plasticizing system

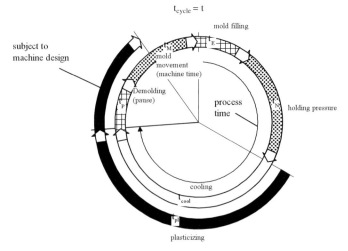

Figure 4.2   Time blocks of an injection molding cycle – borderline conditions for the plasticizing time [1]

For economical reasons and in the interest of rational processing, standard injection molding machines are equipped with three-zone, general-purpose screws, as a rule. They are capable of processing almost all conventional thermoplastics satisfactorily [2]. Similar to extrusion, the screws are divided into three zones, designed for individual tasks:

- Feed zone
- Transition section (compression zone) d
- Discharge section (metering zone)

It is usual for the screw tip to be fitted with a check valve, which prevents the melt from passing back into the screw's antechamber during the injection and holding pressure phase [3].

Injection molding screws are shorter than extrusion screws on the whole. Due to the cyclic and therefore discontinuous operation of the injection molding process, there is an additional energy exchange of pure heat conductivity, when the screw stops rotating. This benefits the melting process. Modern standard screws have an effective length of 20D to 23D, with the feed zone taking up approximately half the screw length. The compression and metering zones are of about identical length, screw pitch amounts to 1D, and the flight depth ratio between the feed and metering zones is between 2 and 3 [4].

With the example of standard conditions, Fig. 4.3 shows for an injection molding machine series, which process- and machine-related individual times can be expected as a function of the shot weight. From this, the ratio of plasticizing time $t_{pl.}$ to idle period t_{idle} can be deduced.

In view of the much improved mold and machine technology as well as article design, it has been possible to shorten cooling, movement, and cycle times drastically over the past

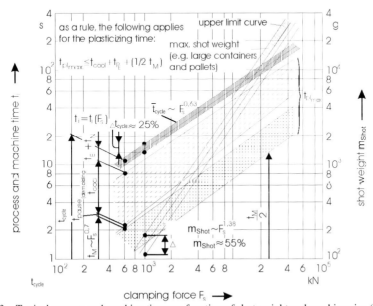

Figure 4.3   Typical process and machine times as function of shot weight and machine size (standard conditions, scatter ranges are shown with shot weight and cycle time)

20 years. At very short cycle times and therefore high melt throughputs, as can be found in the packaging industry, in particular, with the production of bulk containers and pallets ($t_{cycle}/t_{pl.} \rightarrow 1$), the plasticizing rate can become the restrictive variable for an injection molding machine's capacity. The result of this has been that the maximum possible plasticizing rates of injection units also had to be increased. The following possibilities for achieving this are available:

- Increasing the screw speeds
- Development of plasticizing units with grooved barrels [2]
- Machine concepts that allow plasticizing during the clamping unit's machine times (see Fig. 4.2)
- Improvement of the screw geometry, with the aim of increasing the specific plasticizing rate (plasticizing rate per screw revolution (g/R) is a constant that characterizes the geometry and/or the system)

Whereas extruders fitted with grooved feed systems had become state of the art, grooved infeeds were installed by injection molding machinery manufacturers in special cases only if high output rates were demanded (Fig. 4.4). This applied to difficult to feed molding materials mostly. The requirement for high torques here caused problems. This was due, in particular, to the cyclic startup process, which with plasticizing units for injection molding can result in considerable difficulties [2]. However, there was to be no deterioration in melt-quality as a result of the increases in melt output rates. That, in conjunction with the increasing demands for direct coloring (mechanical homogeneity) and the compounding of plastics formulations with the injection molding machine, as well as the particular demands of new raw materials (greater shear and residence time sensitivity), increasingly confines the range of activity for today's general-purpose screws. For such special cases, longer screws from 23D to 27D, partially even equipped with additional shear and mixing elements, are employed. These are known as high-performance screws and the elements as barrier concepts.

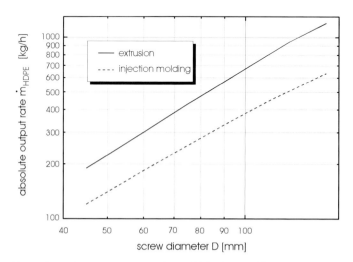

Figure 4.4   Comparison of output rates: extrusion and injection molding

With injection molding machines, plasticizing units have thus been assigned the most important task within the process: They must carry out material- and operating-point-compatible processing across a wide range of application at all times [1]. Similar to extrusion, the most important aims of the plasticizing process that are affected by the screw geometry are

- Feed performance, which is repeatable and ensures consistent conveying
- Intensive, but gentle melting at as low a material degradation as possible
- Good melt homogeneity, with regard to both thermal and visual/mechanical properties
- Optimum, feedback-controllable melt temperature
- An adequate plasticizing rate

Due to the discontinuous mode of operation, the system is also subject to the most different stress characteristics, such as

- Internal pressures of up to about 3500 bar (injection pressures)
- Operating temperatures of up to 450 °C
- Load changes at a frequency ≤ 1 Hz
- Sliding speeds of up to 1.8 m/s
- Wear caused by adhesion, abrasion and corrosion
- A wide range of materials, such as
    - Unreinforced thermoplastics (e.g., HDPE, PP)
    - Reinforced/filled plastics (e.g., PA + GF, ABS/PC + FS)
    - Transparent plastics (e.g., PC, PMMA, SAN)
    - High-temperature plastics (e.g., PEEK, PPS, PFA)
    - Polymer blends (e.g., PBT/PC elastomer-modified, PP/EPDM)

4.1.1.1   Feed Zone

The most important task for an injection molding screw will always be its capacity for processing all injection-moldable plastics. But as practice has proved, the application of a general purpose screw is dominated by the serviceability of its feed zone. Naetsch [5] first pointed out its importance, and work carried out at the IKV (Institut für Kunststoffverarbeitung, Aachen/Germany), in particular, emphasized the great importance of feed performance and its influence on the subsequent melting process as well as pressure buildup and melt output rate. In this connection, the work specifically carried out on injection molding machines by Elbe [2] must here be mentioned.

As with extrusion, so the most important demands made of this zone with the injection molding process are:

- Pick-up, conveyance, and compression of the plastics raw material
- That the conveying characteristic must be repeatable at any time
- That the various molding materials must not affect the conveying characteristic significantly
- That the filling rate of the screw channels must be as high as possible
- That the unit must operate as free as possible of conveying resistances and other influencing factors

Because of the discontinuous mode of operation and the axial displacement of the screw, it is not possible to transpose directly onto injection molding machines the results gained by extrusion. For an analysis of the processes in the feed zone of an injection molding machine, one has to differentiate between the feed performance itself and the onward conveying of the molding materials entering the screw channels. So far, this is identical to extrusion [6]. With this, the feed performance can be differentiated even further between the free-flowing process in the area of the hopper discharge (inlet connection) and the screw-channel filling by material (Fig. 4.5).

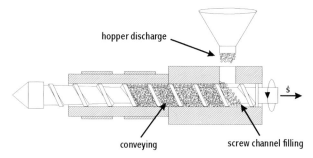

Figure 4.5   Division of the processes in the feed zone [3]

A prerequisite for problem-free operation of the plasticizing units is the uniform flow of the material _ granulate, chips, powder _ from hopper into screw channel. During free-flow, a certain flow pressure $p_{T,a}$ occurs at the hopper discharge. This must always be higher than the counterpressure p_1 in the screw channel below the hopper. According to [7], the flow pressure is about 13% lower than the vertical static pressure $p_{v,a}$.

Solids conveying is essentially determined by the feed-zone geometry (screw, cylinder) and the material properties. In stationary operation, material output is calculated by the continuity equation consisting of the free annular cross section A_{Ring}, the axial speed v_a, and the apparent density ρ_s (see Fig. 4.6):

$$\dot{m}_F = A_{Ring} v_a \rho_S f \tag{4.1}$$

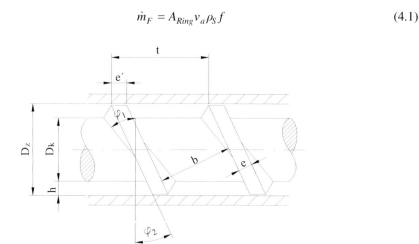

Figure 4.6   Schematic diagram of a screw $\left(\varphi_1 = \varphi_Z \; ; \varphi_2 = \varphi_K \right)$

Video recordings by Effen [3] have proved that an Archimedean conveying principle has to be assumed here Fig. 4.7). This particularly concerns the material that trickled down during the injection phase, because in this case the screw channels were filled only incompletely. If it is not possible to count on a completely filled compartment volume, $f < 1$ results as the volume of filling. This means that frictional influences can be ignored, so that the axial speed v_a results directly from the screw speed n_{red} and the pitch t:

$$v_a = v_{flight} = tn_{red} = \tan(\varphi_Z) D_Z \, \pi n_{red} \tag{4.2}$$

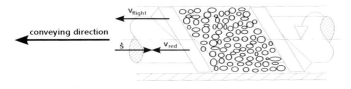

Figure 4.7
Archimedian conveying

Here the reduced speed takes into consideration

$$n_{red} = \frac{\pi D_Z n_0 \tan(\varphi_Z) - \dot{s}}{\pi D_Z \tan(\varphi_Z)} \tag{4.3}$$

the influence on the screw return speed $\dot{s}$ (Fig. 4.8).

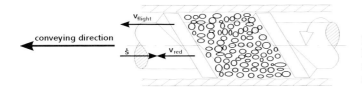

Figure 4.8
Relative movement of screw to cylinder during injection molding

The screw return speed is calculated through the cylinder_s cross-sectional area from the proportionality between screw return speed $\dot{s}$ and the melt volume $\dot{V}$ conveyed into the screw's antechamber.

$$\dot{s} = \frac{4\dot{V}}{\pi D_Z^2} \tag{4.4}$$

Thus, at reduced speed n_{red}, an extrusion screw is capable of conveying just as much solid material, as an injection molding screw at speed n_o and a screw return speed $\dot{s}$. The solids throughput is therefore calculated to [3]

$$\dot{m}_F = \rho_S \, \pi D_Z \, n_{red} \, \tan(\varphi_Z) \left[\frac{\pi}{4} \left(D_Z^2 - D_K^2 \right) - \frac{ieh}{\tan(\overline{\varphi})} \right] f \tag{4.5}$$

According to [8], the apparent density determined by DIN [9] has to be corrected, because of fringe influences resulting from the existing flight depth h, the channel width b, channel length z, as well as the particle size d and the apparent density $\rho_{S,\infty}$, determined as per [9].

Calculation of solids conveying by the Archimedean theorem proves difficult, as it is not possible to make any accurate statements regarding the apparent density ρ_S and the filling rate f. The apparent density ρ_S is a function of the material height filling the screw channel, and thus, in turn, of the filling rate f.

Partial filling of the screw channel under the material hopper may occur through two phenomena. The first possibility is that the hopper output rate $\dot{m}_{hopper}$ is insufficient for filling the screw channel volume under the hopper completely within the time available. From a defined output rate $\dot{m}_{partfill.}$ or a defined rotational speed $n_{partfill.}$ onward, the screw channels are thus partially filled. This output is subject to the hopper and particle geometry. If the screw speed is increased over and above that setting, the hopper discharge cannot keep up with the amount of material the screw can convey. The output rate therefore approaches a limiting value $\dot{m}_{limit}$, corresponding to the hopper's output rate $\dot{m}_{hopper}$ (Fig. 4.9). It is thus no longer possible to increase the material output rate at higher screw speeds.

A second equation states that with a sufficient material flow from the hopper there must exist a screw speed $n_{partfill.}$, at which the filling cross section b is being reduced to such an extent that there is insufficient time for the solids to fill the screw compartment completely. There is also only partial filling of the screw channels under the hopper. This screw speed is subject to the screw and particle geometry. The screw speed and thus the feed zone's output rate can be increased only until the reduction of the feed cross section allows less material into the screw channels than the screw can convey. At a critical screw speed $n_{crit.}$, the output rate therefore approaches a maximum, the limiting value $\dot{m}_{limit}$ (Fig. 4.10). If the screw speed is increased further still, the filling cross section is reduced even more, the material output rate drops, and in the extreme case it may even reach a stage at which particles no longer have sufficient time for dropping fully into the screw compartment.

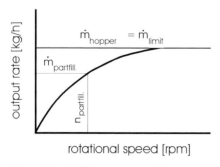

Figure 4.9   Solids output rate as function of the hopper's output rate

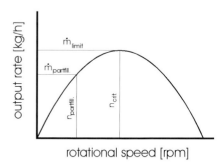

Figure 4.10   Solids output rate as function of the screw's rotational speed

At complete filling of the screw channels, i.e., filling rate $f = 1$, the granulate is being transported due to the frictional forces between the granulate particles and the cylinder wall and/or the screw. In this case, the granulate's speed vector has a component in the axial as well as the radial direction (Fig. 4.11). Besides, with the nonstationary injection molding operation the screw retraction movement will have to be taken into consideration when examining speed vectors [10]. Due to the backward screw movement in axial direction, the relative speed between solids bed and cylinder changes. Therefore, there is a lower solids output rate with the injection molding mode than with a stationary operation. This has been confirmed by Effen [3] through experimental comparison of specific feed-zone output rates under extrusion and injection molding conditions.

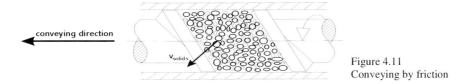

Figure 4.11
Conveying by friction

All known conveying models show a clear influence of friction-value fluctuations on the calculation result. Figure 4.12 shows that the solids conveying angle to Schneider [11], for instance, increases at a cylinder friction value, which grows relative to the screw's friction value. Increasing friction forces on the cylinder result in a greater driving force acting on the solids bed. To achieve a reproducible, consistent feed and conveying performance, it must be ensured that the driving friction forces in the cylinder are greater, than the retarding friction forces of the screw surface.

With counterpressureless conveying and equal friction coefficients on cylinder and screw, the models of Hyun-Spalding [12] and Campbell–Dontula [13] are indifferent to changes in the number of screw channels. The Schneider model [11], on the other hand, reacts sensitively to that. With this model, conveying collapses at high-friction coefficients on two- or three-start screws. Therefore, trials involving single-, twin-, or triple-start screws are highly suitable for checking the validity of the various models [14].

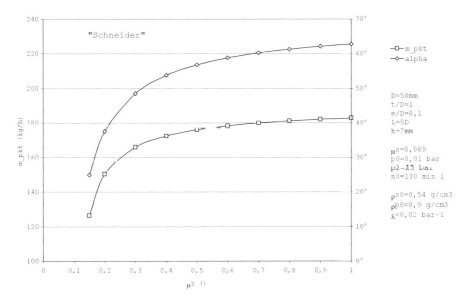

Figure 4.12   Solids conveying angle according to Schneider [11]

4.1.1.2   Transition Section

The most important task of an injection molding machine's plasticizing unit is the material conversion from the solid state into melt. The overall "plasticizing unit" system is primarily judged by its melting capacity [15]. The transition performance has a significant influence on the melt homogeneity in the screw's antechamber and therefore on the article quality. For extrusion screws, numerous visual quality checks were carried out to obtain greater knowledge of the transition process. Essential results were achieved by Maddock [16], Klenk [17], and Lindt [18]. The Maddock model has largely asserted itself with the processing of wall-adhering plastics melts. It will therefore form the basis of our subsequent examinations.

To analyze the transition process taking place in an injection molding machine's plasticizing unit, it is expedient to look at the whole course of the process during injection molding, which can be divided into the following sections:

- Metering process
- Injection process
- Screw stoppage time

There have been a few publications on the subject of material melting in injection molding machines up to now. Changes in the solids bed profile during the injection cycle have been investigated by Donovan [19]. He also developed a model for estimating the melting and pressure profiles along the screw's axis, as well as the melt temperature [20]. A quite accurate analysis of the idle time's influence on the solids bed profile has been established by Lipshitz et al. [21].

4.1.1.2.1   Melting Process During the Metering Period

The melting process during the metering phase can be regarded as a stationary process, which one can fall back on for the analytical formula by Tadmor and Klein [22]. The axial screw displacement must also be taken into consideration. The whole of the transition zone is shown as schematic diagram in plan view as well as cross section in Fig. 4.13 with the aid of a screw channel, opened out as layflat. The theoretical explanation is given in Chapter 5.

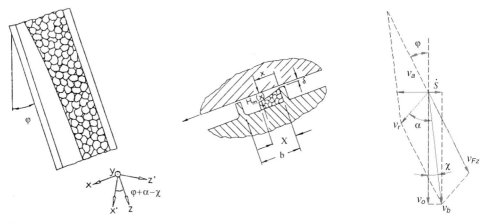

Figure 4.13   Schematic diagram of the transition zone, showing the speed vectors

4.1.1.2.2   Melting Process During Injection

During injection, the screw moves forward in axial direction. This relative movement of the screw also dissipates mechanical energy as heat. This causes a film of melt to form on the cylinder wall, which is being scraped off by the active flight flank through the screw's axial movement. The melting mechanism is thus identical with the melting process during the metering phase. If the following assumptions are made

- The solids bed is rigid.
- No rotary momentum is transmitted to the solids bed by the active screw flight.
- The solids bed is being pushed against the cylinder wall.

Then only the relative movement v_r resulting from the peripheral speed v_b has to be replaced by the axial relative speed v_{ra} of the screw. The screw's relative speed results from the metering stroke s_{DH} up to the changeover to holding pressure s_{UN}, divided by the injection time t_E as

$$v_{ra} = \frac{s_{DH} - s_{UN}}{t_E} \tag{4.6}$$

Because of the screw's axial movement, the distance between screw flights in axial direction has to replace channel width b:

$$b_a = \frac{\pi D \sin(\varphi) - ie}{i \cos \varphi} = \frac{t}{i} - e \tag{4.7}$$

This allows the average melt-film thickness and/or the melting speed during metering to be calculated, analog to the melting process. It is necessary to check however, whether with very fast injection speeds the assumptions made still apply. It is feasible, for instance, that a fast axial movement loosens the solids bed because of the material's inertia, so that contact with the cylinder wall is lost. At least, no new unmelted material is being pushed against the cylinder wall.

4.1.1.2.3   Screw Idle Time

During the screw's idle phase, melting depends entirely on pure heat conductivity between cylinder wall and solids bed. Melting on screw root and flights is being neglected [15]. A detailed presentation of the calculations will not be given at this point, but a description will be given of the complete melting process in injection molding machine plasticizing systems, linked to individual calculation models. In this case, although the melting process during the injection phase will not be taken into consideration.

By way of introduction, the process is illustrated by the melting profiles shown in Fig. 4.14. The material is introduced at A, and subsequently turns into a compact solids bed. It stays as pure solids conveying up to the start of the first heating zone (B). Complete heat insulation between flange area and first heating zone is assumed as an ideal situation. The melting process starts at B, melt fluidization is apparent at C. Following the start of melt fluidization, melting is calculated analog to the stationary extrusion process. On reaching position D, screw rotation stops and the idling phase starts. The distance between A and D

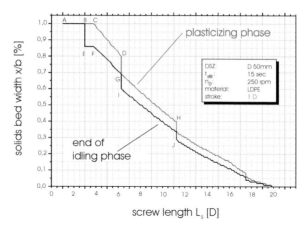

Figure 4.14   Standardized solids bed width as function of the screw length for a stroke of 1D

along the plasticizing cylinder results from the article weight m_{shot}, the screw geometry and the apparent density ρ_S of the material in the screw channel. In [15], the connection

$$\Delta z = \frac{m_{shot}}{\rho_S \, bh} \tag{4.8}$$

is shown for estimating the stroke Δz traveled by a solids element in the screw channel. Following the plasticizing phase, melting is determined through the pure heat conductivity between cylinder wall and solids bed. The dimensionless solids bed content at G is the starting point for the next plasticizing phase during screw rotation. The line connecting G to H shows the calculated solids content according to the model for melting during the metering phase. By including the changing screw–channel geometry, position H is determined by Equation 4.8. After the plasticizing phase, the solids content is determined once more, based on the subsequent screw idling phase (I–J). That calculation process is continued to the same pattern until there are no solids left in the screw channel or the checkring valve has been reached.

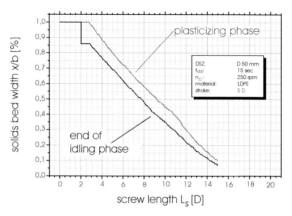

Figure 4.15   Standardized solids bed width as function of the screw length for a stroke of 5D

This extreme case is shown in Fig. 4.15. Due to the large screw stroke, the melt volume in the screw channel is smaller than required by the article weight, and for the end of the screw a solids content of about 10% is calculated, which will be conveyed into the screw's antechamber. Analog to extrusion, the curve for the plasticizing phase is determined by taking into consideration injection molding's additional speed components. At low article weights, the whole length of the screw can be utilized for plasticizing. Should the melt volume required for the article be smaller than that in the screw channel, a considerably more favorable melting profile can be achieved in the forward screw position.

4.1.1.3   Metering Zone

The metering zone is the area of theoretically pure melt conveying. It is also known as compensating zone, because this is where homogenization takes place in classic 3-zone screws. The metering zone acts as a kind of melt pump and thus is output-determining with screws of a more flexible conveying design. That is the reason why most of the model laws are based on basic equations for the metering zone's conveying effect.

To transpose the flat plate model – well known from extrusion – onto injection molding screws, it will have to be extended by the speed vectors for the axial screw movement.

Because an added particularity of an injection molding screw is its axial displacement during the plasticizing phase, it is feasible for the rotating screw to move back in such a way that an element of melt does seem stationary, when viewed relative to the cylinder. Therefore, the melt stream should also be considered relative to the cylinder at this point, because it corresponds to the actual output rate. With reference to [10], this reads as follows for Newtonian melts:

$$\dot{V} = \frac{1}{2} ibhv_{1z} - \frac{ibh^3}{12\eta} \frac{dp}{dz} - \frac{ibh}{\sin \varphi} \dot{s} \qquad (4.9)$$

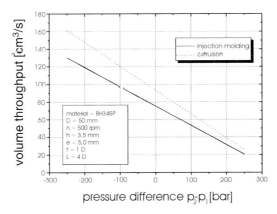

Figure 4.16   Comparison of melt streams for extrusion and injection molding

Fig. 4.16 shows the volume throughput under constant conditions, with and without axial screw movement. At constant pressure difference, the melt stream is determined by the metering zone. The curve profile of both characteristics is similar, but with injection molding outputs is lower, due to the screw's return movement. This is discussed in theoretical detail in Chapter 5.

4.1.1.4   Mixing Zone

At high output rates, melting will only be completed well inside the metering zone. This shortens the material's residence time in the metering zone to such an extent that remaining granulate pellets can no longer be fully melted by the heat of the surrounding melt. There is a risk of unmelted particles getting into the mold. Besides, the mixing effect of conventional three-zone screws is relatively low and depends heavily on such process variables as back pressure, rotational speed, and temperature, for instance. To achieve a homogeneity that is sufficiently independent of the process, these screws are additionally equipped with shear and mixing elements (see Fig. 4.17).

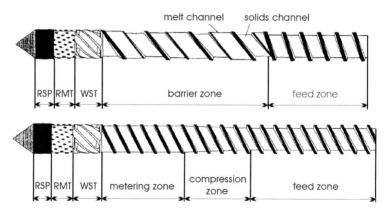

Figure 4.17   Screw designs showing shear and mixing elements

The present multitude of shear and mixing elements available is most confusing, not least because of the patent situation. A detailed examination of various mixing and shear elements is to be found in [23] and [24], where simulation and experiment are also compared with each other. Pointers have also been compiled on the design of various mixing and shear elements.

Mixing elements are fitted to the ends of screws for dispersion purposes. Mixing elements achieve an improvement in the melt's thermal and mechanical homogeneity by distributive mixing; i.e., the melt is being divided and reunited many times over.

With conventional three-zone screws, the shear elements are arranged in the metering zone at about two-thirds of the screw length. Shear elements mixing actions are primarily dispersive, i.e., they divide particles through the shearing of adjacent layers.

These days, Maddock's shear element is the one most often applied in the field. It consists of dead-end grooves in axial direction, with openings alternating in and against the direction of flow. In each case, pairs of grooves are linked by a shear bar, whose clearance gap is larger than that of the opposite blocking bar.

The spiral shear element (see Fig. 4.18), in which the grooves have been cut at an angle of $\varphi < 90°$ to the screw's axis, offers an improvement. The Maddock shear element can thus be considered as being on the borderline. The advantage over the Maddock shear element consists in the possibility for arranging the grooves, so that they assist the material flow. This reduces pressure losses caused by the shear element, resulting in a lower melt temperature and thus an improved mixing effect. However, the width of the shear gap must clearly differ from the play of the blocking bar, otherwise the pressure buildup performance will be adversely affected [25].

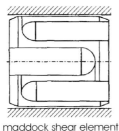 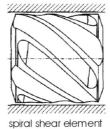

maddock shear element   spiral shear element

Figure 4.18   Shear elements

4.1.2   Size Transposition

Model theory is an important instrument for screw development, in most cases employing small diameters, where new screws are concerned. The thus established ideal geometry (model screw) must then be "projected" for larger diameters (main design) of higher output rate, i.e., identical energetic properties, as well as physical flow characteristics. Papers by Schenkel [26] and Fischer [27] are the foundation stones of model theory. At the IKV, Potente's model theory for injection molding machines [28] has been developed further by Menges, Paar, and Schmelzer [29, 30]. It has also been extended by Bürkle [1].

4.1.3   Considering the Plasticizing Unit as a Whole

In contrast to continuous extrusion, the melt on screw-plunger plasticizing units is not processed under stationary conditions during cycling. During the plasticizing process, the effective screw length is being reduced, until the set plasticizing stroke is completed. Particularly with longer metering strokes, this affects the thermal and mechanical homogeneity of the melt in the screw's antechamber, right up to the unmelted material. The thermal homogeneity of the compounded melt volume for producing the article is directly related to the temperature development in the screw channel. Axial and radial temperature differences are inevitably generated by the melting of the cold solids and the discontinuous displacements of the screw during plasticizing, as well as the screw's idling phase [15]. One will also have to take a metering-stroke-dependent torque into consideration. With the subsequent injection of the melt into the cavity, the screw moves axially forward, thus allowing material to enter the screw channels of the feed zone now becoming exposed.

4.1.3.1   Feed Performance

The feed performance presents a decisive criterion for an injection molding machine's achievable output rate and maximum possible metering stroke. Poor feed and conveying performance can cause air entrapments, so that the output rate is well below the expected value and/or may also be subject to severe fluctuations. If twin- or multistart screws are to be employed, it will have to be clarified, from which screw diameter onwards such a design or feed performance makes sense.

4.1.3.2   Air Entrapment

As part of the process with injection molding, the screw moves forward axially during injection, so that material can trickle into the screw channels becoming exposed. As a rule, the time available during the injection phase is not long enough for filling the screw channels completely. Air is being sucked in, which – if it cannot escape – detracts from the article quality in the form of bubbles or streaks. Air entrapments cause problems, moreover, with the unwelcome degradation of heat sensitive plastics, as well as rising component wear rates, due to aggressive reaction products.

Partial filling causes solids islands and voids in the screw channel, as shown schematically in Fig. 4.19 for the situation at the start of injection. The air-filled voids migrate in the direction of the screw tip during plasticizing. Whether air gets into the screw antechamber depends on the length of the metering stroke, the solids bed friction, and the speeds of the

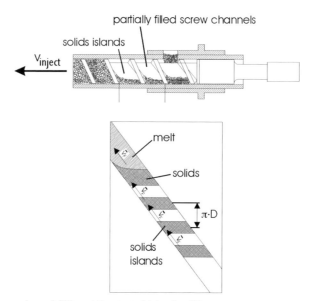

Figure 4.19   Screw channel filling at the start of injection [3]

main bed v and the solids islands v_F. Air entrapments can be prevented only if the solids islands are of a higher speed than the main bed. The design of a screw geometry for large metering strokes must at least meet the specification

$$\frac{\bar{v}}{v_F} < 1 \qquad (4.10)$$

so that the air entrapments can be pushed out by the solids bed extending from the hopper.

4.1.3.3   Swept Volume

With various materials and at a metering stroke longer than 3D, research carried out by Effen [3] on "rifled screws" has shown air entrapments immediately behind the checkring valve. The utilized metering stroke should therefore not exceed 3D. Such limitation of the metering stroke also makes sense with regard to the melting process and thus the article quality. With very long metering strokes the effective length of the feed zone is restricted so severely that during transition of solids to the compression zone the volume of melt, or solids starting to melt, becomes too small. Subject to the material, the solids content can increase to such an extent, that the cross-sectional reduction in the compression zone compresses the solids so much that volumetric flow rate is reduced and under certain circumstances even stops altogether (see Fig. 4.20).

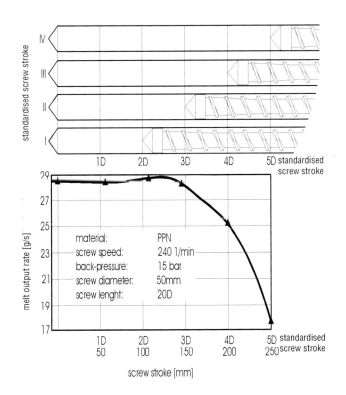

Figure 4.20
Melt output rate as a function of the set screw stroke

Generally speaking, from all that has been said, one can only arrive at the conclusion that maximum screw strokes ranging from 5D to 6.5D and mostly installed in injection molding machines (see Fig. 4.21) are not basically utilizable. Strokes exceeding 3D will have to be considered as critical transition zones rather, which should be exploited only in special cases. This can be rectified by increasing the effective screw length to $L_S/D = 23D - 26D$. The lower limit, on the other hand, is determined by the checkring valve's pick-up accuracy and shutoff speed (this determines the repeatability of the article quality), as well as the thermal stability of the plastics to be processed. From 7% to 12% of the maximum swept volume must be considered as the lower limit [31] (see Fig. 4.22). Stroke exploitation and the resultant dwell time are a significant criterion for the material's molecular rearrangement, affecting polymer blends and the high-temperature-resistant thermoplastics in particular.

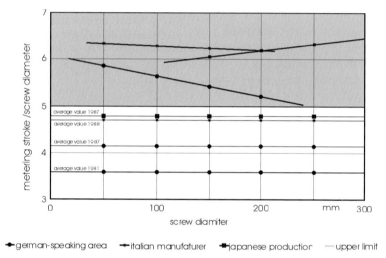

Figure 4.21   The ratio of metering stroke to screw diameter as a function of the screw diameter

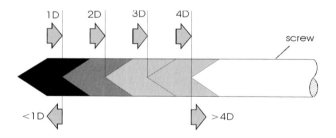

1D to 3D optimum area
3D to 4D in exceptions possible
<1D and >4D not recommandable

Figure 4.22   Useful and possible means of measurement for injection molding screw

4.1.3.4   Feed Problems

Effen [3] defines the dimensionless characteristic E for predicting feed problems as a function of the trickling-in material, the chosen geometry (diameter), and the rotational speed:

$$E = \frac{\dot{m}_{trickle-in} - \dot{m}_F}{\dot{m}_{trickle-in}} \tag{4.11}$$

Here $\dot{m}_{trickle-in}$ is the material flow trickling into the screw from the hopper [32]. The maximum solids output rate $\dot{m}_F$ is calculated with Eq. 4.5. Particular attention must here be paid to the fact that the material can trickle into the screw channels only, when the material-dependent minimum feed width b_{min} in accordance with [33] has been achieved between the hopperedge and screw flight (see Fig. 4.23). According to Effen, for double flighted screws it follows for characteristic E that this must be greater than 0.5, in any case, to prevent feed problems.

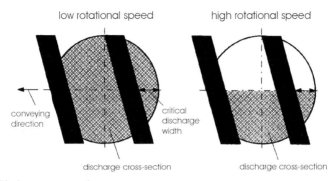

Figure 4.23   Discharge cross section

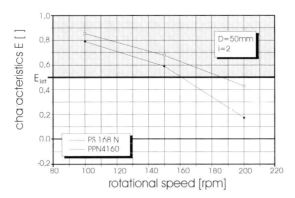

Figure 4.24   Characteristic E as function of the rotational speed for different materials [3]

Fig. 4.24 shows that feed problems become greater at higher speeds with double flighted screws of increasing diameter. Here it must also be taken into consideration that at increasing diameters identical peripheral velocities are achieved at lower rotational speeds, so that the influence of the second screw flight decreases with increasing diameter. The experimental research in [3] has proved for the presently practice-relevant peripheral speed range and the standard feed geometries employed that screws for polyolefin processing can be of the double flighted type from 50 mm diameter onward. For materials that are critical with regard to feed performance, such as PS, Effen [3] recommends the use of a double flighted screw from 70 mm diameter onward.

For screw designers this means that at diameters below 50 mm they must endeavor to design the plasticizing unit's feed geometry in such a way that problem-free trickle feeding of the plastics granulate is ensured, the solids output rate in the feed zone exceeds the melt output rate of the metering zone, and thus characteristic E can be increased. The characteristic E and therefore the limiting speed for the start of feed-opening reduction will migrate toward the higher speeds at the following structural modifications [3]:

- single flighted design of the feed zone
- Increase in channel width
- Moving the feed opening in direction of rotation, as shown in Fig. 4.25 (feed pocket)

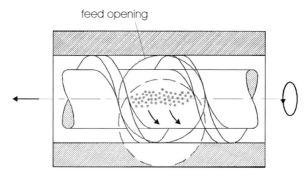

Figure 4.25   Moving the feed opening [3]

These measures ensure a higher melt output rate so that the „limiting speed" is increased. Due to the single flighted feed zone, the h/b ratio is lowered and therefore higher solids conveying performances are achieved. The influence of other structural feed-zone modifications are shown in Fig. 4.26 [34].

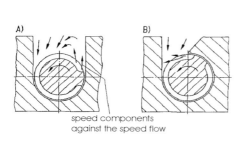

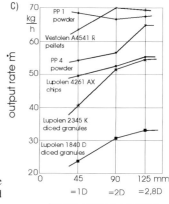

Figure 4.26   (A) Obstruction of the trickling-in process because of too large an opening; (B) ideal geometric design of the feed opening; (C) output rate as function of feed-opening length

By extending the feed opening in axial screw direction, it is possible to increase the output rate of single- and double flighted screws at low peripheral speed. This means that feed problems arising from feed zone reduction cannot be affected by an extension of the feed opening (Fig. 4.27), as the front screw channels below the extended feed throat are already filled with material, because of the screw's axial movement.

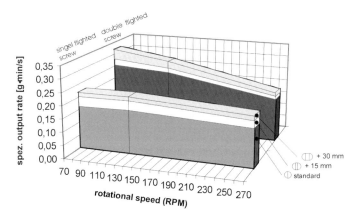

Figure 4.27   Output rate as function of the rotational speed for different feed throats [3]

Heatbalancing in the feed-throat area (flange heat balancing) is decisive for both feed performance and conveying consistency. Particular attention will therefore have to be paid to the tribological processes within the solids range of the feed zone. Because the coefficients of friction between granulate and cylinder wall are temperature-dependent (see Fig. 4.28), the temperature profile in this area must be adapted to the respective operating conditions and tribological characteristics. This would mean that the conveying performance of the feed zone decreases with increasing flange temperature. But since the metering zone determines the output rate with injection molding as a rule, and the solids output is generally greater than the melt throughput, these effects are of no consequence. Under certain circumstances, an inverse temperature profile can even have a positive effect on the plasticizing unit's output rate (see Fig. 4.29), subject to material and screw geometry. This effect remains active even when the cylinder temperatures of subsequent zones are reduced to prevent the melt temperature from rising $\left(T_M = const.\right)$.

The output rate is decisively influenced by the solids content and/or the solids progress along the length of the screw. If solids conveying is dominant, e.g., with fast cycling applications, lower plasticizing performances are achieved than during longer cycles (see Fig. 4.30). In the latter case, a larger volume of solids is being melted, so that melt conveying can be assumed. This makes it clear that for simulation of a plasticizing screw, a combination of solids and melt conveying is required in the area of the transition zone. However, this action for increasing the capacity has its restrictions. On the one hand, care must be taken that not too high a flange temperature is chosen, otherwise solids would start melting as early as within the first zones. This would result in the plasticizing unit "stalling." On the other hand, melt could be carried into the hopper area through the screw's axial movements at long strokes. This would result in "gumming up" the feed throat, and plasticizing performance would suffer.

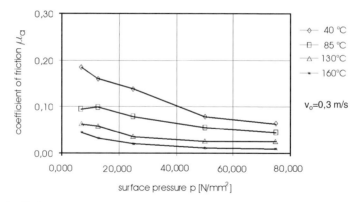

Figure 4.28   Coefficients of friction for Hostalen GF 7740 [35]

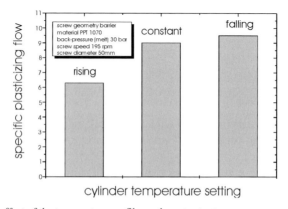

Figure 4.29   The effect of the temperature profile on the output rate

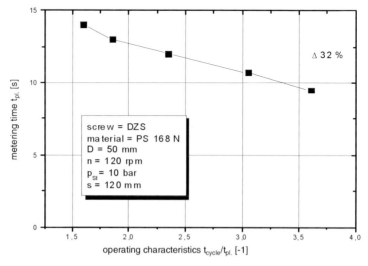

Figure 4.30   Metering time at varied operating characteristics

4.1.3.5   Melting Section

The time taken for the maximum plasticizing rate must not determine the cycle time, if at all possible. Plasticizing must be completed either within the cooling time or – with the appropriate machine layout – during the mold closing movement at the latest. Nowadays capacity increases of 25% to 40%, achieved through cycle times, are not uncommon. This has become possible in particular due to the improved technology applied to molds and machines, resulting in higher machine capacities (see Fig. 4.31). However, the metering time cannot be reduced arbitrarily, since the material must be sufficiently melted before it is injected into the mold to produce articles of the specified quality ("zero defect" production). The reasons will become obvious when considering the melting capacities required for the processing of amorphous and part-crystalline thermoplastics. The wide band-spread of these material groups' melting enthalpies already points to one of the problems encountered with the processing of part-crystalline thermoplastics at high melt throughputs as well as short cycle times, and the related short residence and metering times (see Fig. 4.32). Thus, the specific enthalpy of polyolefins is higher by a factor of 1.3 to 1.7 than with amorphous materials. Therefore, the following countermeasures must be carried out, if it is intended to operate exclusively within such "extreme areas":

- Longer screws ($L_S/D \geq 23$), for increasing the residence times, and/or
- Altering the screw geometry to intensify the melting process

With the aid of the simple Tadmor model for Newtonian melts, the essential influencing factors on the melting rate Γ (melting rate per time and area unit) of a screw are quickly established:

$$\Gamma = \Phi \sqrt{X} \qquad (4.12)$$

$$\Phi = \left\{ \frac{v_{bx}\, \rho_S \left[\underbrace{\lambda_M \left(T_Z - T_M\right)}_{conduction\, of\, heat} + \underbrace{\frac{1}{2}\, \eta_{SF}\, v_r^2}_{dissipation} \right]}{2 \left[\underbrace{c_F \left(T_M - T_F\right) + i}_{\substack{enthalpy + \\ phase\, transition}} \right]} \right\}^{\frac{1}{2}})$$

with

$$v_r = \sqrt{v_b^2 + v_{Fz}^2 - 2 v_b v_{Fz}\, \cos(\varphi - \chi)}$$

$$v_{Fz} = \frac{\dot{m}}{\rho_F\, bh}$$

It is demonstrated with the specific enthalpy (Fig. 4.32) and the enthalpy term in the denominator of Eq. 4.12, how different materials can occupy varying melting sections along the screw (see Fig. 4.33).

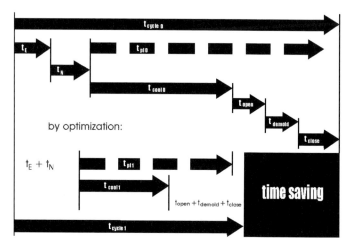

Figure 4.31   Cycle analysis

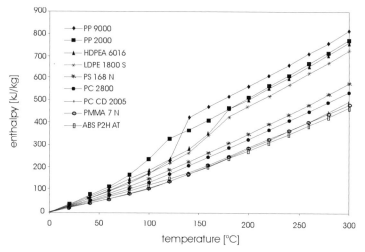

Figure 4.32   Specific enthalpy of diverse thermoplastics as function of the temperature

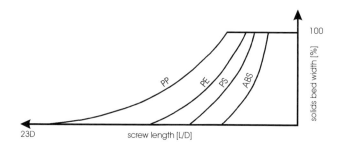

Figure 4.33   Melting trends of different plastics

An increase in heat conductivity terms can be achieved by putting up the cylinder temperature. This must be regarded as a restricted measure only, however, because the plastics melt can suffer thermal damage at higher temperatures. Raised melt temperatures also result in longer cooling times and thus an extension of the cycle time. Also, an increase in temperature must not inevitably lead to improved melting efficiency. As mentioned before, an energy input by dissipation is generally greater than that by heat conductivity. Although heat conductivity is improved by raising the cylinder temperature, the dissipation content decreases, as viscosity in the melt-film also reduces with increasing temperature. Melting performance drops once the decrease in dissipation is greater than the increase in heat conductivity.

An improvement in melting can be achieved by the following designs (see Fig. 4.34):

- Multistart screws
- Barrier screws

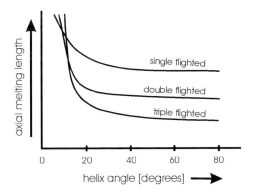

Figure 4.34   Effect of the number of flights on the melting length [23]

An increase in dissipation level and therefore the melting rate Γ may also be achieved by a faster relative speed v_r, e.g., by increasing the rotational speed n_0. However, if the cycle time is also reduced during a fast-cycling application, heat conductivity drops in accordance with

$$\dot{m}\Delta h - \pi D \left\{ \alpha_M L_S \left[T_Z(x) \quad T_M(x) \right] \right.$$

$$\leftarrow \text{convection}$$

$$+ \frac{\lambda_M}{\delta} L \left[T_Z(x) - T_M(x) \right] \left(\frac{t_{cycle}}{t_{pl.}} - 1 \right)$$

$$\leftarrow \text{heat conductivity}$$

$$+ \left(\frac{\pi}{60} \right)^2 \eta \, n_0^2 \frac{D^2}{h_2} L_S + \underbrace{\frac{\pi}{60} \mu p L_F D n_0}_{} \right\} \tag{4.13}$$

$$\underbrace{\phantom{+ \left(\frac{\pi}{60} \right)^2 \eta \, n_0^2 \frac{D^2}{h_2} L_S}}_{\rightarrow \text{dissipation}} \quad \rightarrow \text{fricton}$$

Maximum screw rpm is limited by the machine-installed drive control, as well as the permissible, material-related shear loading and/or shear deformation. The latter is the product of shear speed and residence time within the shear section. Impermissibly high shear deformations result in molecular changes within the material. The melt temperature here is a significant influencing factor. As substitute for a description of a practice-related working window, material-specific peripheral screw-speed limits may be taken as guide values for permissible rotational speeds of respective screw diameters (Fig. 4.35). Subject to screw geometry, such limiting zones may be utilized to a greater or lesser extent; it is self-evident that geometries of gentle melting performance allow for a larger working range.

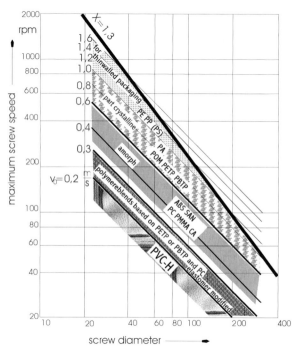

Figure 4.35   Limiting values for rotational speeds of screws [1]

The injection speed can have a negative influence on the length of the melting section. As discussed earlier, shear forces _ from melt film and melt pool _ acting on the solids bed, increase toward the end of the compression zone, due to a consistent decrease in channel depth. These shear forces, plus the invasion of the solids by the melt, result in a moment being exerted on the solids bed so that the latter is broken up quite soon.

At very high injection speeds, it is possible for the melt to be pushed through the leak gap from the melt pool of the thrust face of the screw flight to the trailing edge of the previous channel section. Thus, the solids bed is encapsulated by melt as soon as a melt-pool has been generated. At very high rotational speeds and the related high shear forces, the solids bed may even get broken up as early as at the end of the feed zone (see Fig. 4.36).

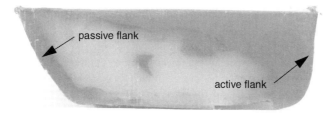

Figure 4.36   Microtome section from a dual start screw's feed zone end

4.1.3.6   Mixing Quality

4.1.3.6.1   *Homogeneity and Melt Temperature*

With the quality characteristic homogeneity one has to differentiate between thermal homogeneity, which expresses itself in the form of temperature differences, and the optical-mechanical homogeneity (mixing quality), which describes pigment or filler distribution. Both are subject to a series of influencing factors, such as

- Material characteristics
- Screw geometry, ratio of stroke and channel volume, as well as mixing and shear systems
- Screw speed and stroke
- Screw driving moment
- Back pressure
- Temperature profile on cylinder
- Proportion of plasticizing and cycle time

Experience has shown that the mixing effect of single screw machines is not all that good. If pigments are to be admixed in the injection molding machine directly, it is frequently

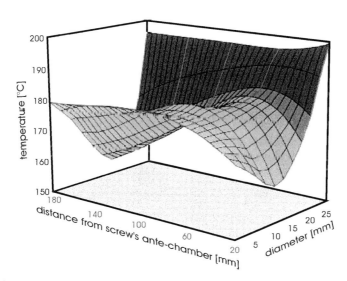

Figure 4.37   Temperature profile in the screw's antechamber; LD-PE, $T_Z = 200$ °C, $n_0 = 114$ rmp, $p_{Stau.} = 35$ bar, $D = 60$ mm

possible to see evidence of poor mixing quality in the article, such as color inhomogeneities, streaks, or clouding. It becomes even more difficult when material components of different viscosities have to be processed

Mixing quality deserves special attention, because there is hardly any chance for improving it in any way, once the plasticizing process is completed. Besides, mixing quality also represents a characteristic that is the most difficult to quantify.

4.1.3.6.2 Influence of Processing Parameters on the Mixing Quality

Fig. 4.38 demonstrates to what extent screw rotational speed and back pressure exert an influence on plasticizing rate and mixing quality. Mixing quality has been defined as optical homogeneity, determined by the visual assessment of pigment dispersion, and classified. An increase in back pressure results in improved mixing quality, but reduces the plasticizing rate.

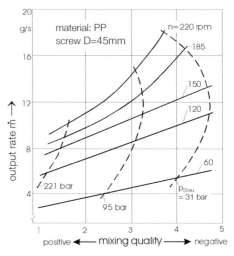

Figure 4.38   Curve diagram for the influence of back pressure, screw rotational speed, and plasticizing rate on the mixing quality

High screw speeds affect the mixing quality adversely. When production demands shorter plasticizing times – i.e., higher rotational speeds – the back pressure must also be increased to achieve comparable mixing quality. Similar effects on the thermal homogeneity can also be expected. If the back pressure is increased, the melt temperature differences in the screw's antechamber are reduced. At the same time, the melt temperature level as a whole is increased, however. Large melt temperature differences in the screw's antechamber, particularly in axial direction, again result from high screw speeds (see Fig. 4.39).

To get the melt temperature in the screw antechamber onto a more consistent level, the logical step is a progressive feedback control of back pressure and rotational speed over the plasticizing stroke (Fig. 4.40). At long cycle and short plasticizing times, the melt temperature can be influenced by the cylinder's temperature profile, where the cylinder temperature is conventionally feedback controlled by zones. To prevent large differences in melt

temperatures, care must always be taken when setting cylinder temperatures that temperature gradients selected between hopper and cylinder head or nozzle are not too great.
Temperatures in individual heating zones ought to be as uniform as possible and particularly – starting with the transition zone – should be close to the melt temperature. With fast-
cycling applications, where high plasticizing performances at high rotational speeds and
simultaneously low cycle times have to be achieved, closed-loop control of the melt
temperature by the cylinder temperatures is no longer possible.

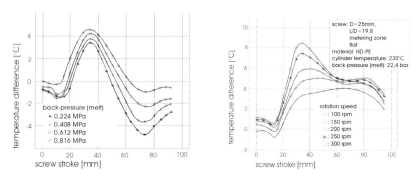

Figure 4.39   Axial temperature profile as a function of back pressure and rotational speed

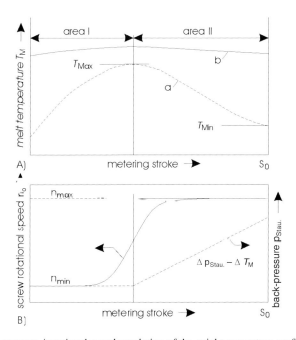

Figure 4.40   Back-pressure-/rotational speed regulation of the axial temperature profile in the screw antechamber

4.1.3.7   Residence Time

As increasingly tighter processing tolerances accompany the development of higher-performance polymer blends in particular, this will give cause to ever rising processing specifications. In practice, it is just those characteristic differences regarding the flow, solidification, and shrinkage properties that make processing them so much more difficult. It is the multiphase characteristic of these material systems and the chemical structure of their basic components (amorphous or part-crystalline molecular bonding) that determine the residence time, processing temperature, shear sensitivity (peripheral velocity), as well as the required energy supply for the melting process. All of this affects not only the screw geometry, but also the residence time and the energy to be supplied. Also, in this case reduction of the cycle time is particularly critical, because the minimum residence time must not be undercut, so that the required energy supply is available for the melting of individual material components.

The minimum and average residence time is composed of the proportions in the metering and idlingphase

$$t_V = t_{V,metering} + t_{V,idle} \tag{4.14}$$

The melt does not move in the screw channel during the idling phase, and the minimum/average residence times are identical to the idling time for the duration of that phase. The residence time during the idling period can be calculated from the cycle time, minus the metering time:

$$t_{V,idle} = t_{cycle} - t_{pl.} \tag{4.15}$$

The average residence time during the metering phase is calculated as

$$\bar{t} = \frac{m_{Shot}}{\dot{m}} = \frac{\rho V}{\dot{m}} \tag{4.16}$$

and is produced from the quotient of channel material and output rate. The minimum residence time results from the sum of the individual times in the zones

$$t_1 = t_{1M} + t_{1K} + t_{1E} \tag{4.17}$$

Fig 4.41 shows in exemplary fashion the calculated minimum residence time over the screw length. Up to the point of melt fluidization, the two residence times do not differ, as up to that position a solid flow is being assumed [15]. Since the calculated residence time is equal to the metering time, the idling time and the injection time are here added to the residence time. This process is continued until the screw tip is reached.

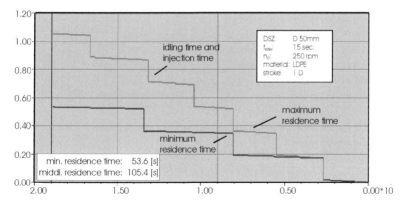

Figure 4.41   Calculated minimum residence time over the screw length

Nomenclature

A_{Ring}	free circular cross sectional area
b	channel width
b_{min}	critical discharge width
c_F	specific heat capacity of the solid material
D	screw diameter
DZS	metering screw
D_K	core diameter
D_Z	cylinder diameter
E	characteristic
e	flight width
f	killing rate
F_S	clamping force
h	channel depth
h_2	channel depth of metering zone
i	number of flights
L	length
L_F	length of solids bed
L_S	effective screw length
$\dot{m}_{trickle-in}$	output trickling into the screw channels
$\dot{m}_F$	solids output rate
$\dot{m}_{limit}$	maximum possible hopper output rate
$\dot{m}_{max.}$	maximum output rate
m_{Shot}	article (melt) weight
$\dot{m}_{partfill}$	hopper output rate, resulting in partial screw-channel filling
$\dot{m}_{hopper}$	hopper output rate
n_0, n	rotational speed (rpm)
$n_{crit.}$	critical rpm
$n_{red.}$	reduced rpm

$n_{partfill.}$	screw speed resulting in partial screw-channel filling
p	pressure
p_0	pressure on front-edge of hopper
p_1	pressure at start of feed zone
p_2	pressure at end of feed zone
$p_{Stau.}$	back-pressure
$p_{T,a}$	yield pressure
$p_{v,a}$	vertical quiescent pressure in a cone-shaped hopper
$\dot{s}$	screw return speed
s_{DH}	metering stroke
s_{UN}	changeover point to holding pressure
t	time
t	pitch
$\bar{t}$	average residence time
t_E	injection time
T_F	solids temperature
T_N	holding pressure time
t_{cool}	cooling time
t_M	melt temperature
t_{pl}	metering time
$t_{idle.}$	idling time
t_V	residence time
t_m	machine movement time
T_Z	heating zone temperature
$t_{cycl.}$	cycle time
$\dot{V}$	volumetric flow rate
v	speed
$\bar{v}$	average speed
v_a	axial speed
v_b	corrected peripheral velocity
v_F	solids speed (conveying speed)
$\dot{V}$	volume throughput
v_F/v_r	relative speed
v_{flight}	flight speed
X	solids bed width

Greek

Δh	enthalpy difference
Δp	pressure difference
ΔT	temperature difference
Φ	melting speed
α	solids conveying angle
χ	angle
δ	melt film thickness

η	viscosity
φ	pitch angle
$\bar{\varphi}$	average pitch angle
φ_K	pitch angle at the screw root
φ_Z	pitch angle at screw diameter
λ_M	melt-heat conductivity
μ_Z	cylinder friction coefficient
μ	fiction coefficient
π	constant
ρ_F	solids density
ρ_S	apparent density
$\rho_{S,\infty}$	apparent density at measuring unit to DIN 53468
Γ	melting rate per time and area unit

Indices

pl.	metering phase
E	injection
F	solids channel
M	melt
max.	maximal
min.	minimal
idle	idling phase
x, y, z	directions
M, 2	metering zone
E, 1	feed zone
K	compression zone

References

1. Bürkle, E., *Verbesserte Kenntnis des Plastifiziersystems an Spritzgießmaschinen* (1988) Dissertation, RWTH Aachen
2. Elbe, W., *Untersuchungen zum Plastifizierverhalten von Schneckenspritzgießmaschinen* (1973) Dissertation, RWTH Aachen
3. Effen, N., *Theoretische und experimentelle Untersuchungen zur rechnergestützten Auslegung und Optimierung von Spritzgießplastifiziereinheiten* (1996) Dissertation, UNI-GH Paderborn
4. Johannaber, F., *Injection Molding Machines* (1994) Carl Hanser Verlag, Munich
5. Naetsch, H., Kunststoff und Gummi (1968) 7 (5), pp. 161–164; 7 (6), pp. 207, 208
6. Hegele, R., *Untersuchungen zur Verarbeitung pulverförmiger Polyolefine auf Einschnekkenextrudern* (1972) Dissertation, RWTH Aachen
7. Langecker, G., (1977) Dr. Ing. Dissertation, RWTH Aachen
8. Schöppner, V., *Simulation der Plastifiziereinheit von Einschneckenextrudern* (1994) Dissertation UNI-GH Paderborn
9. DIN 53.466 (1984) Deutsches Institut für Normung e.V., Beuth Verlag, Berlin
10. Rauwendaal, Ch., Int. Polym. Process (1992) 7 (1), p. 26

11. Schneider, K., *Der Fördervorgang in der Einzugszone eines Extruders* (1969) Dissertation, RWTH Aachen

12. Hyun, K.S., Spalding, M.A., SPE ANTEC Tech. Papers (1997) 43, p. 211

13. Campbell, G.A., Dontula, N., *Int. Polym. Process* (1995) 10, pp. 30–34

14. Potente, H., Kunststoffe (1999) 89, p. 1

15. Schulte, H., *Grundlagen zur verfahrenstechnischen Auslegung von Spritzgießplastifiziereinheiten* (1990) Dissertation, UNI-GH Paderborn

16. Maddock, B.H., SPE J. (1959) 15, p. 383

17. Klenk, K.P., *Plastverarbeiter* (1970) 21, p. 819

18. Lindt, J.T., *Polym. Eng. Sci.* (1976) 16, p. 284

19. Donovan, R.C., *Polym. Eng. Sci.* (1974) 14 (2), p. 101

20. Donovan, R.C., Thomas, D.E., Leversen, L.D., *Polym. Eng. Sci.* (1971) 11 (5), p. 353

21. Lipshitz, S.D., Lavie, R., Tadmor, Z., *Polym. Eng. Sci.* (1974), 14 (8), p. 553

22. Tadmor, Z., Klein, I., *Engineering Principles of Plasticating Extrusion* (1970) Van Nostrand Reinhold Book Co., New York

23. Rauwendaal, Ch., *Polymer Mixing* (1998) Carl Hanser Verlag, Munich

24. Michaeli, W., Wolff, Th., M*ischen und Mischteile, Der Einschneckenextruder – Grundlagen und Systemoptimierung* (1997) VDI-Verlag

25. Stenzel, H., *Grundlagen zur verfahrenstechnischen Auslegung von Barriereschnecken in Glattrohr- und Nutbuchsenextrudern* (1992) Dissertation, UNI-GH Paderborn

26. Schenkel, G., *Kunststoff-Extrudertechnik* (1963) Carl Hanser Verlag, Munich, 2. Auflage

27. Fischer, P., *Auslegung von Einschneckenextrudern auf der Grundlage verfahrenstechnischer Kenndaten (Modelltheorie)* (1976) Dissertation, RWTH Aachen

28. Potente, H., *Auslegung von Schneckenmaschinen-Baureihen – Modellgesetze und ihre Anwendung: Kunststoffe-Fortschrittsberichte,* Band 6 (1981) Carl Hanser Verlag, Munich

29. Menges, G., Paar, M., Schmelzer, E., Kunststoffe (1984) 74

30. Schmelzer, E., *Modelltheorie für Spritzgießmaschinen* IKV, Aachen

31. Bürkle, E., Kunststoffe (1988) 79, p. 4

32. Tomas, J., *Modellierung des instationären Auslaufverhaltens von kohäsiven Schüttgütern aus Bunker* (1991) Chem. Techn. 43. Jg., Heft 8 (August 1991) Dissertation B, Bergakademie Freiberg

33. Molerus, O., *Schüttgutmechanik-Grundlagen und Anwendungen in der Verfahrenstechnik* (1985) Springer-Verlag, Berlin

34. Hensen, F., Knappe, W., Potente, H., *Handbuch der Kunststoff-Extrusionstechnik I Grundlagen* (1989) Carl Hanser Verlag, Munich

35. VDMA (Hrsg.), *Kenndaten für die Verarbeitung thermoplastischer Kunststoffe, Teil 3: Tribologie* (1983) Carl Hanser Verlag, Munich

4.2   Buss Kneader

James L. White

4.2.1   Introduction

The Kokneter or Kneader developed and manufactured by *Buss AG* (now Coperion Buss) has long played a role as an important continuous mixer in the polymer and food industries. The original machine was invented by List [1–4] in the mid 1940s and shortly thereafter commercialized [4] by *Buss AG* in Europe and by *Baker Perkins* in the Unites States. Today *Buss AG* is the primary manufacturer. The machine consists of a screw with slices sitting in a barrel containing rows of pins. The screw both rotates and axially oscillates during the operation of the machine. During these motions, the pins wipe the screw flights and clean them. The combination of flow through the interrupted screw flights and the oscillating action of the screw through the pin barrel gives a unique mixing action. The machine has a modularly constructed screw that sits in a barrel containing rows of pins. Additional ports for feeding or venting are located along the length of the machine.

It is our purpose in this section to review the development of the Kneader, and to summarize the literature on its flow and mixing mechanisms.

4.2.2   Machine Technology

4.2.2.1   Earlier Related Machines

The original Buss Kneader brought together several threads in the engineering literature of the 1930s and 1940s: (1) extruders with pins in their barrels, (2) extruders with reciprocating screws, and (3) extruders with self-wiping action. The concept of placing pins into the barrel of a screw extruder is an old idea dating at least to the turn of the century. A 1901 German patent by Casimir Wurster [5] describes a pin barrel Fleischschneidemaschine. This is a machine intended for a butcher, with the pins being used to knead the meat as it flows through its channel to a die, which would produce *Hackfleisch* (or ground beef). The screw flights contain slices through which the pins pass. A 1930 U.S. patent application by F. B. Anderson [6] of the *V.D. Anderson Company* of Cleveland, Ohio for a sausage-stuffing machine has a section containing cap screws of "knives" entering from the barrel and appropriate slices in the screw flights. The purpose of the cap screws or pins is to "agitate and knead" the material being processed to remove air from it.

Screw processing machines with reciprocating screws were first introduced for injection molding machines. A February 1939 French patent application by H.P.M. Quillery [7] describes a reciprocating screw injection molding machine for rubber compounds. In December 1943, H. Beck [8] of the *IG Farbenindustrie* filed a patent for a screw injection molding machine for thermoplastics. Beck was employed in the Ludwigshafen site of the *IG Farbenindustrie,* which had been before 1925 and would be again *BASF*.

The idea of developing screw extruders with self-wiping mechanisms dates to the turn of the century. Patents by the German, A. Wunsche [9] in 1901 and later by the Englishman, R. Easton in 1916 [10] and 1920 [11] describe self-wiping, co-rotating twin screw extruders and the advantage of self-wiping behavior. Easton writes of keeping screws clean and avoiding the collection of dirt. In the same period, fully intermeshing, counter-rotating machines were developed [12–15], which operated as positive displacement pumps. The steel surfaces of the screw flights of one screw were in close contact with the roots of the opposite screw. The purpose of this design was to allow the movement of contained volumes of the fluid along the length of the screw to the exit port.

4.2.2.2   Origins of the Kneader

It is often not realized that the period of 1939 to 1945 in central Europe was one of remarkable development in the polymer and associated processing machinery industries. This was the time when synthetic rubber and various thermoplastics were first being manufactured on a large scale. In 1939, *Lavorazione Materie Plastische,* an Italian company based in Turin, commercialized an intermeshing co-rotating twin screw for extrusion of thermosets and polyvinyl chloride. The machine was designed by Roberto Colombo, a principal of the firm, who filed and received a patent for it in 1939 [16]. He filed for a Swiss patent in 1941 [17] and a German patent in 1944 [18]. The company was able not only to market their machines in Italy, but also to sell machines to the *IG Farbenindustrie* in Germany. The *IG Farbenindustrie* applied the LMP Colombo machines to polymerization processes [19]. Self-wiping twin screw extruders were also developed in the early 1940s at the *IG Farbenindustrie's* Wolfen works by Walter Meskat and Rudolf Erdmenger [20–22]. They were applied to mixing and devolatilization. In 1941 the German steel and machinery manufacturer, *Friederich Krupp* announced the commercialization of an intermeshing, counter-rotating continuous mixer. The machine was known as the Knetwolf and was used for masticating synthetic rubber. Its inventor was Willi Ellermann [23] who filed a patent application in 1941 [24].

Heinz List [1–4], the inventor of the Kneader, was born in Germany and attended the Technical University of Munich where he received a Dipl Ing degree in Mechanical Engineering. He subsequently worked with the *IG Farbenindustrie* in Leverkusen, where he was involved with the design of mixing equipment. In 1945 List moved to Basel, Switzerland, where he invented the Kokneter or Kneader, a machine with an oscillating screw that rotates in a barrel with pins. The pins wipe and clean the screw flights. List's

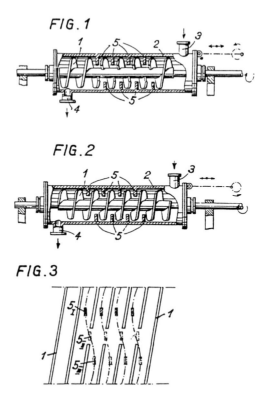

Fig. 4.42   List's 1945 patent drawing of the "Kokneter" [1]

mentality and thought processes resembled those of other central European engineers of his time, such as Roberto Colombo and Rudolf Erdmenger; all sought to develop self-wiping mixing and processing equipment for the new difficult and viscous materials of the polymer industry.

List [1–3] describes many different continuous mixers in his patent. The original patent drawing of the invention of List [1], which gave rise to the Kneader, shows a machine with an oscillating barrel containing pins with a rotating screw containing slices in it flights (Fig. 4.42). However, the machine that List actually had in mind is only briefly described. It was this machine he convinced the Swiss firm *Buss AG* of Pratteln to build. While not clearly described in the early patents and papers of List, the principles were, of course, clearly understood and are well illustrated in Fig. 4.43.

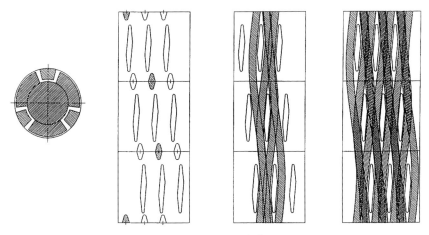

Fig. 4.43   Schematic showing movements of pins in "Kneader"

4.2.2.3   Development of the Kneader

The machine called a Kokneter in List's patent was licensed to *Buss AG* of Pratteln, Switzerland. North American rights were initially licensed to *Baker Perkins*. The major improvements in the machine in the years following List's invention were made by *Buss AG* engineers. In a 1961 patent application, Gubler [25] describes a modified Kneader, which is divided into a kneading/mixing section and a plasticizing section. The latter had deep pins in the barrel and the former had set back pins so as to reduce the level of mastication in this section. This machine is shown in Fig. 4.44.

In a 1962 German and 1963 American patent application, Sutter [26] of *Buss AG* described a modular Kneader design in which screwlike elements are fitted together to form a modular screw. This screw is combined with a barrel containing pins (Fig. 4.45). This represents the beginning of the modern modular Kneader.

In a 1965 Swiss and 1966 American patent application, Gresch [27] of *Buss AG* describes a machine of this type with barriers present in the barrel and vent ports. The barriers induce pressure drops and the development of starved regions under vent ports where an applied vacuum induces devolatization (Fig. 4.46).

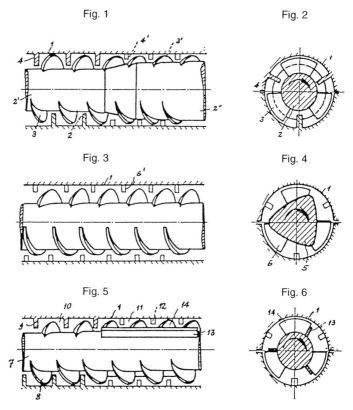

Fig. 4.44   Gubler's 1961 patent drawing showing a Kneader with mixing and kneading section with pins and plasticizing section without pins

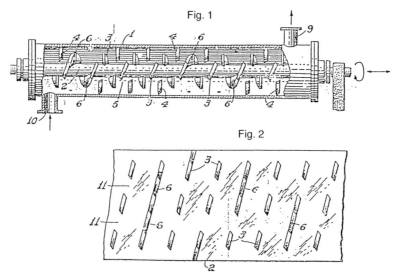

Fig. 4.45   Sutter's 1963 patent drawing showing a modular Kneader with screw reciprocating in a barrel

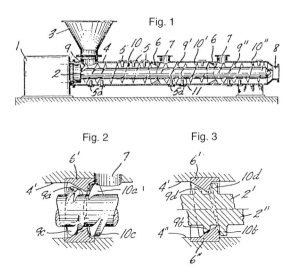

Fig. 4.46   Gresch's 1966 patent drawing showing a modular Kneader with barriers and vent ports in the barrel

In a 1967 Swiss and 1968 American patent application, Ruettener and Sutter [28] of Buss AG describe an oscillating screw machine with two vent ports that feeds into a starved cross-head screw extruder. One of the vent ports is integrated into the connection between the oscillating screw machine and the cross-head extruder (Fig. 4.47).

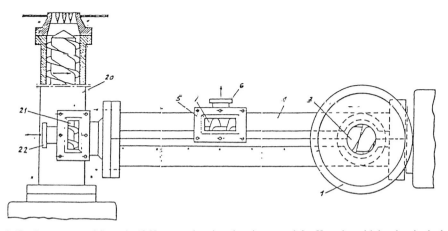

Fig. 4.47   Ruettener and Sutter's 1968 patent drawing showing a modular Kneader with barriers in the barrels, vent ports and a crosshead extruder

Various patents filed in this period address the problem of the oscillating screw leading to an oscillating output [29]. A 1959 U.S. patent application by Geier and Irving [29] of *Baker Perkins* describes a screw design with an elevated section following the section containing pins and partial flights, which is in turn followed by a right-handed screw and a left-handed screw. The intention is to damp out the effects of the oscillation. A 1962 Netherlands and 1963 U.S. patent application by Schuur [30] of *Shell Oil* describes a machine with an oscillating screw with pins that possesses two entry ports and has right-handed screw elements in front of one hopper and left-handed threads subsequent to the second. In a 1962 Swiss and 1963 U.S. patent application, List [31], and List and Ronner [32] of *Buss AG* describe a machine in which the screw rotates and the barrel is divided into two sections, one of which closer to the feed hopper has pins and oscillates to give self-wiping character. The mixed material flows out of the oscillating barrel through a barrier section into a stationary housing where flow oscillations are damped out.

4.2.2.4   Modern Kneader

The modern Buss Kneader is a continuous mixing machine with a modular screw construction in a barrel containing usually specially shaped pins and with three (or more) entry or devolatilization parts. It is available with screw/barrel diameter sizes of 46 mm up to 200 mm. Machines are available that are either oil or electrically heated. The Buss Kneader is usually combined with a crosshead screw extruder or a gear pump to eliminate the time-dependent flow associated with the oscillation of the screw.

The Kneader has three distinct types of screw modules (Fig. 4.48). The EZ element is a screw element with one screw slice and one row of pins. This is intended as a transporting element. The KE element is a screw element with three parallel screw slices and three parallel rows of pins. The ST element is a combination of a KE element and a barrier fitted into the barrel. These elements may, in large part, be placed in any order along the axis of the screw shaft. Hollowed pins are now used to position thermocouples along the screw axis or as feed parts for injecting liquids.

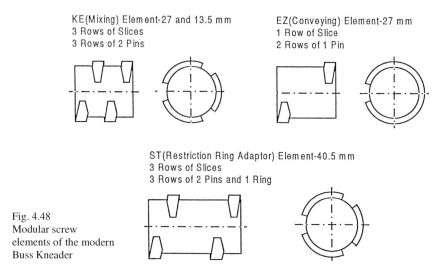

KE(Mixing) Element-27 and 13.5 mm
3 Rows of Slices
3 Rows of 2 Pins

EZ(Conveying) Element-27 mm
1 Row of Slice
2 Rows of 1 Pin

ST(Restriction Ring Adaptor) Element-40.5 mm
3 Rows of Slices
3 Rows of 2 Pins and 1 Ring

Fig. 4.48
Modular screw
elements of the modern
Buss Kneader

4.2.3   Basic Experimental Studies

There have been few experimental studies of flow in the Buss Kneader. Most early published investigations have been largely technological in character [33–44]. These papers included 1977/78 studies by Stade [36, 37] of *Buss AG* on producing glass fiber-reinforced thermoplastics. The first basic experimental study was published by Elemans and Meijer [45] in 1991. There have also been more recent studies by Lyu and White [46–48] and by Shon et al. [49, 50].

Elemans and Meijer [45] describe experiments carried out on a 46 mm Kneader with the screw fitted into a polymethyl methacrylate (Plexiglas) cylinder with small glass tubes that act as an open manometer attached to it. They describe flow experiments using a silicone oil (η = 1.0 Pas) and a parafinic oil (0.2 Pas) under isothermal conditions at room temperature. They measured throughput versus screw speed, throughput versus pressure generation, as well as locally filled lengths. With these measurements, they were able to construct screw characteristic curves that are plots of throughput Q versus pressure drop Δp at specific screw speeds for the two Newtonian oils in different Buss modular screw elements. Filled lengths were determined in different screw elements with different processing conditions. Lyu and White [46–48] described experimental studies with a polypropylene in a 46 mm Buss Kneader connected to a crosshead extruder. They measured fill factor profiles, temperature profiles, conditions of melting, and residence time distributions as a function of screw configuration and processing conditions. The ST elements with their restriction rings produced fully filled regions in the barrel immediately in back of them on the hopper side. These authors made the first study of melting in a Buss Kneader using polypropylene. The melting of the polypropylene was found to occur between the initial hopper and the first ST element. Lyu and White found that if the crosshead extruder was detached from the

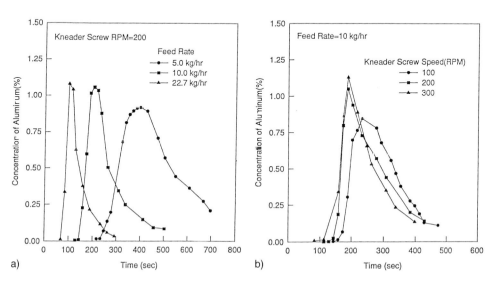

Fig. 4.49   Residence time distribution of a 46 mm modular Buss Kneader (and cross-head extruder) as a function of residence time distribution for (a) various feed rates and (b) various screw speeds

Kneader, the output of the Kneader oscillated. Attaching the crosshead extruder caused the output oscillation of the Kneader to be damped.

Lyu and White [48] and Shon et al. [49] have investigated residence time distributions in the Buss Kneader. The primary variable influencing residence time distributions in a Buss Kneader was found to be the throughput through the machine. As shown in Fig. 4.49, increasing throughput reduces residence time distribution. Residence time distributions are generally much broader than those of twin screw extruders [49].

4.2.4   Flow Simultations

4.2.4.1   General

The earliest flow simulations of fluid motions in a Buss Kneader took place in the late 1980s with the work of Booy and Kafka [51] and subsequently of Brzoskowski et al. [52]. These modeled detailed flow patterns and pressure fields around the pins (see also [52–54]). A broader view of the machine's behavior is taken in subsequent papers by Elemans and Meijer [45] in 1990 and by Lyu and White [46, 47, 55–57] in 1995 to 1997, which seek to simulate total machine response.

4.2.4.2   Flow Due to an Oscillating Screw

Flow in the screw channel is generally modeled by Elemans and Meijer [45] and by Lyu and White [46, 47] in terms of hydrothermal lubrication theory using a flattened screw channel that contains the coordinate axes. The latter authors were the first to analyze the influence of the axial motion of the screw relative to the barrel [46], in terms of model considering the coordinate axis fixed in the screw. The barrel tends to move relative to the screw. The velocity field is taken to have the form

$$\mathbf{v} = v_1(x_2, t)\mathbf{e}_1 + 0\mathbf{e}_2 + v_3(x_2, t)\mathbf{e}_3 \tag{4.18}$$

where 1 is the direction along the screw channel, 2 is the direction normal to the surface of the screw channel, and 3 is the transverse direction normal to the screw flights. The coordinate system is considered fixed into the screw channel. The equations of motion have the form

$$0 = -\frac{\partial p}{\partial x_1} + \frac{\partial \sigma_{12}}{\partial x_2} \tag{4.19a}$$

$$0 = -\frac{\partial p}{\partial x_3} + \frac{\partial \sigma_{32}}{\partial x_2} \tag{4.19b}$$

The boundary conditions are, following Lyu and White [46], of form

$$v_1(0) = v_3(0) = 0 \tag{4.20a}$$

$$v_1(H) = U_1(t) = \pi DN \cos\phi + \frac{dS(t)}{dt}\sin\phi \tag{4.20b}$$

$$v_3(H) = U_3(t) = -\pi DN \sin\varphi + \frac{dS(t)}{dt}\cos\phi \tag{4.20c}$$

where N represents the screw rotation rate, $S(t)$ the stroke of the oscillating screw, ϕ_0 is the screw helix angle, D the screw diameter, and H the channel depth.
A solution of Eqs. 4.19a, b for the velocity field in a screw channel is readily obtained for a Newtonian fluid where

$$\sigma_{12} = \eta\frac{\partial v_1}{\partial x_2}$$

$$\sigma_{32} = \eta\frac{\partial v_3}{\partial x_2} \tag{4.21}$$

where η_η is a constant viscosity. This has the form

$$v_1(x_2,t) = U_1(t)\frac{x_2}{H} - \frac{H^2}{2\eta}\left(\frac{\partial p}{\partial x_1}\right)\left[\frac{x_2}{H} - \left(\frac{x_2}{H}\right)^2\right] \tag{4.22a}$$

$$v_3(x_2,t) = U_3(t)\frac{x_2}{H} - \frac{H^2}{2\eta}\left(\frac{\partial p}{\partial x_3}\right)\left[\frac{x_2}{H} - \left(\frac{x_2}{H}\right)^2\right] \tag{4.22b}$$

The throughput Q is

$$Q = \int_0^H v_1 dx_2 = \frac{HWU_1(t)}{2} - \frac{H^3W}{12\eta}\frac{\partial p}{\partial x_1} \tag{4.23}$$

where $U_1(t)$ is given by Eq. 4.20b. Let a die be placed at the end of the screw with pressure drop characteristics

$$Q = \frac{k}{\eta}\Delta p \tag{4.24}$$

It can be seen that the output from the die oscillates with the screw movements. If we eliminate $\Delta_\Delta p$ between Eqs. 4.23 and 4.24 we obtain

$$Q = \frac{1}{2}\frac{HWU_1(t)}{1 + \dfrac{H^3W}{12\kappa L}} \tag{4.25}$$

The above argument shows the necessity of placing gear pump or starved crosshead extruder at the exit of the Kneader.

4.2.4.3   Flux Patterns and Pumping Characteristics of Kneader Elements

Elemans and Meijer [45] were the first to model flow in the different elements of the Buss Kneader. They neglected both the presence of the pins and the movement of the screw relation to the barrel. Lyu and White [46, 47] concluded that the first assumption was reasonable considering the results of earlier simulations of flow in pin barrel extruders [53, 54]. Lyu and White, however, include the effect of the oscillating screw. We may use Eq. 4.19 with the boundary conditions of Eq. 4.20, to compute flux fields in individual Kneader screw elements. This may be done for Newtonian fluids or non-Newtonian fluid models. Lyu and White [47] present an analysis for a power law fluid where the shear viscosity η of Eq. 4.19 is now taken as

$$\eta = K\left[\left(\frac{\partial v_1}{\partial x_2}\right)^2 + \left(\frac{\partial v_3}{\partial x_2}\right)^2\right]^{(n-1)/2} \tag{4.26}$$

They proceed by evaluating the fluxes q_1 and q_3 defined by

$$q_1 = \int_0^H v_1 dx_2 = \int_0^H \int \left(x_2 \frac{\partial p}{\partial x_1} + C_1\right)\frac{1}{\eta} dx_2 dx_2$$

$$q_3 = \int_0^H v_1 dx_2 = \int_0^H \int \left(x_2 \frac{\partial p}{\partial x_3} + C_3\right)\frac{1}{\eta} dx_2 dx_2 \tag{4.27}$$

where the second set of integrals arise from integrating Eqs. 4.21 and 4.19. The fluxes are balanced over a finite difference mesh as

$$q_1(i-1,j)\Delta x_3 + q_3(i,j+1)\Delta x_1$$
$$= q_1(i+1,j)\Delta x_3 + q_3(i,j-1)\Delta x_3 \tag{4.28}$$

Computed flux maps for EZ and KE elements are shown in Fig. 4.50 for a simplified flow model that excludes pins. These maps show backward leakages through the screw slices. The plots are for forward-moving screws. In Fig. 4.51 we plot the computed dimensionless backward leakage $Q_{leak}/U_1 WH$ as a function of dimensionless pressure gradient $\Delta PH^{n+1}/KU_1^n L$. The plot indicates the extensive backward leakage of the KE mixing element as compared to the EZ element. Backward leakage is greater for more highly non-Newtonian fluids with lower values of the power law index.

Screw characteristic curves relating throughput, pressure development, and screw speed have been determined for the EZ and KE elements for Newtonian and power law fluids. (Fig. 4.51). Different screw characteristic curves exist for forward- and backward-moving screws. The former exhibit better pumping behavior. The pumping of the EZ elements are

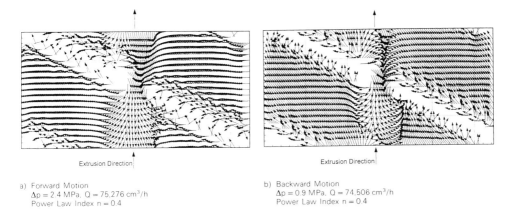

a) Forward Motion
Δp = 2.4 MPa, Q = 75,276 cm³/h
Power Law Index n = 0.4

b) Backward Motion
Δp = 0.9 MPa, Q = 74,506 cm³/h
Power Law Index n = 0.4

Fig. 4.50   Computed flux maps for EZ and KE modular Kneader elements showing forward stroke by Lyu and White

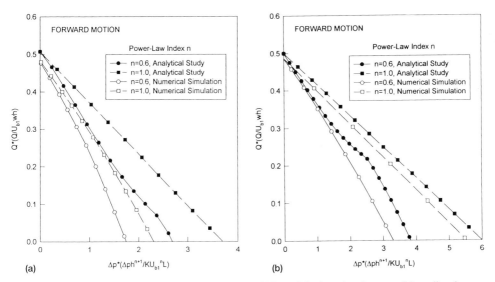

Fig. 4.51   (a) Predicted screw pumping characteristics of EZ modular kneader elements; (b) predicted screw pumping characteristics of KE modular kneader elements. The numerical calculation is superior to the analytical approach, which is one dimensional and neglects transverse shearring by Lyu and White

much better than the KE elements. The pumping character is reduced by decreasing power law exponents, i.e., by increasing non-Newtonian flow behavior. The influence of viscoelasticity on flow in a Buss Kneader has been considered by Lyn and White [55].

4.2.4.4   Theory of Composite Modular Machines

Theories of composite modular Buss Kneaders have been developed [45–47, 56, 57], based on presuming the same flow rate through each modular element and continuous pressure and temperature fields. If the throughput Q is known, screw characteristic curves allow one to calculate a pressure profile backward along the screw axis if this temperature profile is known. The methods used are similar to those developed earlier for modular, co-rotating twin screw extruders [58–60]. The procedures described by Lyu and White [46, 47, 55–57] involve beginning at the exit of a starved cross-head extruder attached to the Kneader and using screw characteristic curves to march backward through the die, cross-head extruder, and then the Kneader elements. One then marches forward using heat balances. Pressure profiles, regions of starvation, and temperature profiles are determined. Typical calculations are shown in Fig. 4.52.

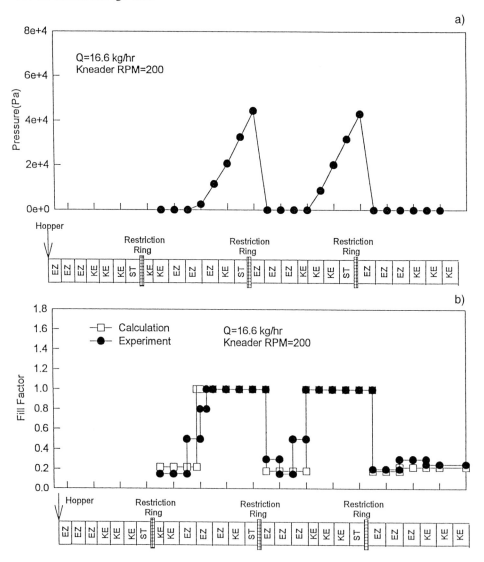

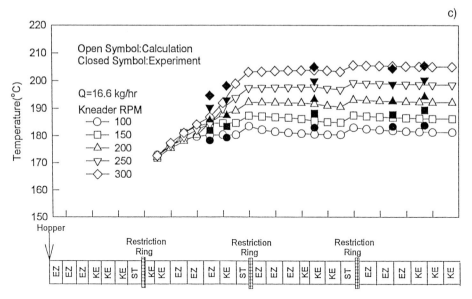

Fig. 4.52 (a) Calculated pressure profile for a modular Kneader during operation; (b) calculated fill factor profile for a modular Kneader during operation; (c) calculated temperature profile for a modular Kneader during operation based upon Lyn and White

4.2.5 Applications

The Buss Kneader is manufactured in sizes from 46 mm to 200 m diameter. Throughputs range from 10 kg/h to 3000 kg/h. There are over 3000 Kneaders in operation worldwide. The Kneader is considered to excel in areas involving thermally sensitive products including compounding polyvinyl chloride and thermoset powder coatings. Other applications include compounding engineering thermoplastics, carbon electrode paste, coatings, solid rocket propellents, polycondensation, and polymerization reactions.

References

1. List, H., Swiss Patent (filed 20 August 1945) 247, 704 (1947)
2. List, H., U.S. Patent (filed 19 August 1946) 2, 505, 125 (1950)
3. List, H., German Patent (filed 5 January 1949) 944, 727 (1956)
4. List, H., *Kunststoffe* (1950) 40, 185
5. Wurster, C., German Patent (filed 30 April 1901) 137, 813 (1901)
6. Anderson, F.B., U.S. Patent (filed 14 November 1930) 1, 848, 236 (1932)
7. Quillery, H.P.M., French Patent (filed 8 February 1939) 855, 885 (1940); British Patent (filed 8 February 1940) 635, 962 (1950)
8. Beck, H., German Patent (filed 16 December 1943) 858, 310 (1952)
9. Wunsche, A., German Patent (filed 12 September 1901) 131, 392 (1902)
10. Easton, R.W., British Patent (filed 25 September 1916) 109, 663 (1917)
11. Easton, R.W., U.S. Patent (filed 2 June 1920) 1, 468, 379 (1923)
12. Wiegand, S.L., U.S. Patent (filed 28 April 1874) 155, 662 (1874)
13. Holdaway, W.S., U.S. Patent (filed 2 June 1915) 1, 218, 602 (1917)

14. Montelius, C.O.J., U.S. Patent (filed 20 March 1925) 1, 698, 802 (1925)
15. Leistritz, P., Burghauser, F., German Patent (filed 24 April 1926) 453, 727 (1927)
16. Colombo, R., Italian Patent (filed 6 February 1939) 370, 578 (1939)
17. Anonymous (S.A. Liguna), Swiss Patent (filed 3 June 1941) 220, 550 (1942)
18. Colombo, R., German Patent (filed 20 July 1944) 883, 338 (1956)
19. Anonymous German Patent (filed 24 July 1943) 895, 058 (1953)
20. Meskat, W., German Patent (filed 17 October 1943) 852, 203 (1952)
21. Meskat, W., Erdmenger, R., German Patent (filed 7 July 1944) 862, 668 (1953)
22. Meskat, W., Erdmenger, R., German Patent (filed 28 July 1944) 872, 732 (1953)
23. Herrmann, H., *Schneckenmaschinen in der Verfahrenstechnik* (1972) Springer, Berlin
24. Anonymous German Patent (filed 31 January 1941) 750, 509 (1945)
25. Gubler, E., U.S. Patent (filed 6 February 1961) 3, 189, 324 (1965)
26. Sutter, F., U.S. Patent (filed 19 September 1963) 3, 219, 320 (1965)
27. Gresch, W., U.S Patent (filed 26 January 1966) 3, 367, 635 (1968)
28. Ruettener, E., Sutter, F., U.S. Patent (filed 18 November 1968) 3, 601, 370 (1971)
29. Geier, H.F., Irving, H.F., U.S. Patent (filed 9 March 1959) 3, 023, 455 (1962)
30. Schuur, G., U.S. Patent (filed 5 September 1963) 3, 224, 739 (1965)
31. List, H., U.S. Patent (filed 10 October. 1963) 3, 317, 959 (1967)
32. List, H., Ronner, F., U.S. Patent (filed 21 January 1965) 3, 347, 528 (1967)
33. Schneider, E., *Polit. Plastics* (1957) p. 481
34. Timm, Th., Stolzenberg, D., Fettback, H., *Kautschuk and Gummi* (1965) 18, p. 206
35. Todd, D.B., Hunt, J.W., *SPE ANTEC Tech. Papers* (1973) 19, p. 577
36. Stade, K., *Polym. Eng. Sci.* (1977) 17, p. 50
37. Stade, K., *Polym. Eng. Sci.* (1978) 18, p. 107
38. Jakopin, S., Franz, P., *Adv. Polym. Technol.* (1983) 3, p. 365
39. Kalyon, D.M., Bouazza, M., *SPE ANTEC Tech. Papers* (1983) 29, p. 778
40. Kalyon, D.M., Hallouch, M., *SPE ANTEC Tech. Papers* (1985) 31, p. 1206
41. Schnottale, P., *Kautsch Gummi Kunstst* (1985) 38, p. 116
42. Todd, D.B., *SPE ANTEC Tech. Papers* (1987) 33, p. 128
43. Franzen, B., Klason, C., Kubat, J., Kitano, T., *Composites* (1989) 20, p. 65
44. Thommen, H., *Plastverarbeiter* (1993) 44, (1), p. 12
45. Elemans, P.H.M., Meijer, H.E.H., *Polym. Eng. Sci.* (1990) 30, p. 893
46. Lyu, M.Y., White, J.L., *Int. Polym. Process.* (1995) 10, p. 305
47. Lyu, M.Y., White, J.L., *Int. Polym. Process.* (1996) 11, p. 208
48. Lyu, M.Y., White, J.L., *Polym. Eng. Sci.* (1998) 38, p. 1366
49. Shon, K., Chang, D., White, J.L., *Int. Polym. Process* (1999) 14, p. 44
50. Shon, K., White, J.L., *Polym. Eng. Sci.* (1999) 39, p. 1757
51. Booy, M.L., Kafka, F., *SPE ANTEC Tech. Papers* (1987) 33, p. 140
52. Brzoskowski, R., Kumazawa, T., White, J.L., *Int. Polym. Process.* (1991) 6, p. 136
53. Brzoskowski, R., White, J.L., Szydlowski, W., Nakajima, N., Min, K., *Int. Polym. Process* (1988) 3, p. 134
54. Shin, K.C., White, J.L., *Rubber Chem Technol* (1997) 70, p. 264
55. Lyu M.Y., White, J.L., *Polym. Eng. Sci.* (1997) 37, p. 623
56. Lyu, M.Y., White, J.L., *SPE ANTEC Tech. Papers* (1996) 42, p. 160
57. Lyu, M.Y., White, J.L., Int. Polym. Process. (1997) 12, p. 104
58. White, J.L., Szydlowski, W., *Adv. Polym. Technol.* (1987) 7. p. 419
59. Wang, Y., White, J.L., Szydlowski, W., *Int. Polym. Process.* (1989) 4, p. 262
60. Chen, Z., White, J.L., *Polym. Eng. Sci.* (1994) 9, p. 310

5 Single Screw Extruder Analysis and Design

Helmut Potente

5.1 Melt Conveying Section Analysis

5.1.1 Isothermal Analysis

5.1.1.1 Melt Conveying

This chapter deals with the mathematical and physical models for the melt conveying [1–22]. It is based on Chapter 2. But, first, some of the geometrical relations of the screw are discussed. The two main variables are the diameter D and the zone lengths L. The latter are generally characterized by the L/D ratio. Fig. 5.1 shows further important geometrical data. These are interrelated as follows

$$\tan \varphi = \frac{t}{\pi D} \tag{5.1}$$

$$D_{core} = D - 2H \tag{5.2}$$

$$W = t \cos \varphi - e \tag{5.3}$$

$$Z = \frac{L}{\sin \varphi} \tag{5.4}$$

where Z is the unwound screw channel length. For nonrectangular screw channel cross sections (Fig. 5.2), the Eq.s listed in Table 5.1 shall be used for the channel width.

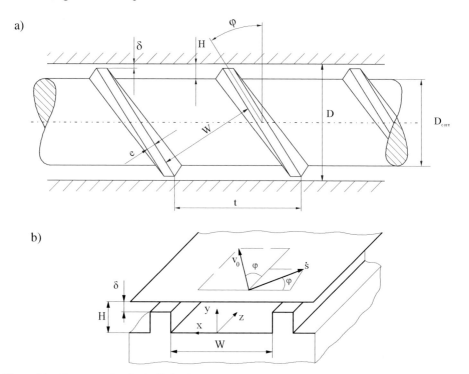

Figure 5.1   Geometry of a single-flighted screw and the unwound screw channel

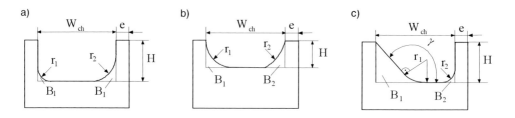

Figure 5.2   Channel geometries

Table 5.1   Eq.s for the Channel Width Correction

General Eq. for the equivalent flight width:	
	$W = W_{Ch} - \dfrac{B_1 + B_2}{H}$
Case a	$r_1 < H, \quad r_2 < H$
	$B_{1,2} = \dfrac{r^2}{4}(4 - \pi)$
Case b	$r_1 > H; \quad r_2 > H$
	$B_{1,2} = \dfrac{1}{2}(r + H)r \sin\left[\arccos\left(1 - \dfrac{H}{r}\right)\right] - \dfrac{\pi r^2}{360°}\arccos\left(1 - \dfrac{H}{r}\right)$
Case c	$r_1 < H; \quad r_2 < H; \quad \gamma > 90°$
	$B_{1,2} = r^2\left(\dfrac{1 + \cos\gamma}{\sin\gamma} - \pi\dfrac{180° - \gamma}{360°}\right) - \dfrac{1}{2}\dfrac{H^2}{\tan\gamma}$

For the further mathematical considerations we presume a fixed screw and a barrel rotating about the screw. An unwound screw channel is also assumed. We then have a rectangular channel across which a plate with the velocity v_0 (rotation velocity of the barrel) is moved (Fig. 5.1):

$$v_0 = \pi N D \tag{5.5}$$

This is called the channel model.

The following applies to the components in z- and x-coordinate direction:

Extruder: $\quad v_{0z} = v_0 \cos\varphi, \quad v_{0x} = v_0 \sin\varphi \tag{5.6}$

Injection molding machines: $\quad v_{0z} = v_0 \cos\varphi + \dot{s}\sin\varphi \tag{5.7}$

$$v_{0x} = v_0 \sin\varphi \quad \dot{s}\cos\varphi$$

where s is the axial retraction speed of the screw.

For a further mathematical and physical description of the melt flow, the Eq. of continuity and the Eq. of motion are required. With an incompressible polymer flow of density ρ, the continuity Eq. is as follows in an orthogonal Cartesian coordinates system:

$$\nabla \cdot \vec{v} = 0 \tag{5.8}$$

where v is the velocity vector. In this case, the motion Eq. can be written as

$$\nabla \cdot \underline{\sigma} + \rho\vec{g} = \rho\left(\frac{\partial\vec{v}}{\partial t} + \vec{v}\cdot\nabla\vec{v}\right) \tag{5.9}$$

where

$\underline{\sigma}$ = Cauchy stress tensor

ϱg = gravitational forces per unit volume

In nonisothermal processes, the energy Eq. has to be evaluated as the third conservation theorem.

Apart from the conservation theorems, it is necessary to specify a constitutive Eq. that describes the material behavior for a closed differential Eq. system. In the case of an incompressible fluid, the stress tensor σ can be divided up into a pressure-dependent and a shear-dependent term:

$$\underline{\sigma} = -p \cdot I + 2\eta \left(\dot{\gamma}, T \right) d \tag{5.10}$$

where

$$d = \frac{\left(\nabla \vec{v} + \left(\nabla \vec{v} \right)^{T} \right)}{2} \tag{5.11}$$

whereby the shear rate $\dot{\gamma}$ can be described by

$$\dot{\gamma} = \sqrt{2D{:}D} \tag{5.12}$$

In polymer processing, use is generally made of the power law for the temperature and shear rate-dependent viscosity η:

$$\eta\left(\dot{\gamma}, T \right) = K_{0T} e^{-\beta(T-T_0)} \dot{\gamma}^{n-1} \tag{5.13}$$

Neglecting the right side of the Eq. 5.9 and the gravitational influence, the following Eq. is yielded for a dimensionless flow:

$$0 = -\frac{\partial p}{\partial z} + K \frac{\partial}{\partial y} \left(\frac{\partial v_z}{\partial y} \right)^{n} \tag{5.14}$$

and, e.g., for a two-dimensional flow:

$$0 = -\frac{\partial p}{\partial z} + K \frac{\partial}{\partial y} \left\{ \left[\left(\frac{\partial v_x}{\partial y} \right)^{2} + \left(\frac{\partial v_z}{\partial y} \right)^{2} \right]^{(n-1)/2} \frac{\partial v_z}{\partial y} \right\} \tag{5.15}$$

$$0 = -\frac{\partial p}{\partial x} + K \frac{\partial}{\partial y} \left\{ \left[\left(\frac{\partial v_x}{\partial y} \right)^{2} + \left(\frac{\partial v_z}{\partial y} \right)^{2} \right]^{(n-1)/2} \frac{\partial v_x}{\partial y} \right\} \tag{5.16}$$

The boundary conditions are

$$y = 0; \; v_x = 0; \; v_z = 0 \tag{5.17}$$

$$y = H; \; v_x = v_{0x}; \; v_z = v_{0z} \tag{5.18}$$

For the one-dimensional case there is an analytical solution [1], but for the two- and three-dimensional cases there are only numerical solutions. Fig.s 5.3 and 5.4 show the solutions of these Eq.s for the dimensionless throughput (output), where

$$\pi_{\dot{V}} = \frac{\dot{V}}{\dfrac{iWH v_{0z}}{2}} \qquad \pi_p = \frac{H^{1+n}}{6K v_{0z}{}^{n}} \frac{\partial p}{\partial z} \tag{5.19}$$

where i = number of flights.

The extreme dependence on the power law exponent n is striking. Furthermore, the two-dimensional solution differs from the one-dimensional one, which can be explained by the fact that the differential Eqs. 5.15 and 5.16 are related via the viscosity.

For the further handling the Eq.s are linearized:

$$\pi_{\dot{V}} = \phi_1 - \phi_2 \pi_p \tag{5.20}$$

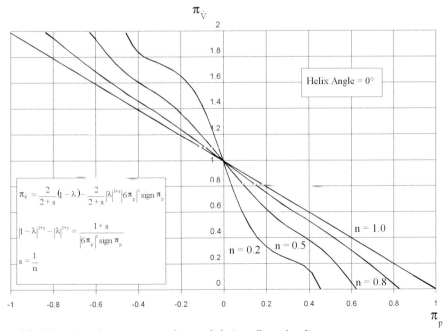

Figure 5.3   Mass throughput – pressure characteristic (one dimensional)

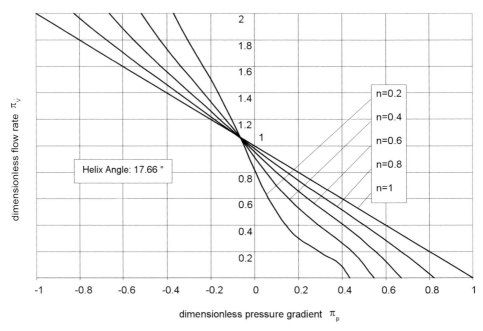

Figure 5.4   Dimensionless flow-rate versus dimensionless pressure gradient for isothermal flow (helix angle: 17.66°)

In the cases of the one-dimensional flow [2],

$$\phi_1 = 1 \qquad \phi_2 = n^{-0.94} \tag{5.21}$$

for the region

$$0.5 \leq \pi_{\dot{V}} \leq 1.5 \tag{5.22}$$

The following applies to the two-dimensional flow:

$$\phi_1 = \frac{n^A}{\cos^{nB} \varphi} \qquad \phi_2 = C \frac{\cos^{Dn} \varphi}{n^{E \sin \varphi + F \cos \varphi}} \tag{5.23}$$

for the region

$$\left. \begin{aligned} 0.8 &\leq t/D \leq 2 \\ 0.1 &\leq \pi_{\dot{V}} \leq 2 \\ 0.2 &\leq n \leq 1 \end{aligned} \right\} \tag{5.24}$$

In Table 5.2 the coefficients $A–F$ are listed [3]. Eq. 5.20 is applicable in the case of $W \gg H$.

Table 5.2   Exponents for the Parameters ϕ_1 and ϕ_2 of the Throughput Equations

	$0.8 \leq t/D \leq 1.2$			
	$0.2 \leq n \leq 0.5$		$0.5 \leq n \leq 1$	
	$0.1 \leq \pi_{\dot{V}} \leq 0.55$	$0.55 \leq \pi_{\dot{V}} \leq 2$	$0.1 \leq \pi_{\dot{V}} \leq 0.55$	$0.551 \leq \pi_{\dot{V}} \leq 2$
A	0.38497	0.11662	0.012498	0.07651
B	7.9159922	0.970032	0.180016	-0.12019
C	1.499341	1.699143	1.893166	1.648509
D	-2.99071	11.69126	12.936	10.49417
E	10.59296	16.15977	2.229467	3.577765
F	-3.50279	-4.65183	-0.47446	-0.66961
	$1.2 \leq t/D \leq 2$			
	$0.2 \leq n \leq 0.5$		$0.5 \leq n \leq 1$	
	$0.1 \leq \pi_{\dot{V}} \leq 0.55$	$0.55 \leq \pi_{\dot{V}} \leq 2$	$0.1 \leq \pi_{\dot{V}} \leq 0.55$	$0.551 \leq \pi_{\dot{V}} \leq 2$
A	0.27002	0.19065	0.09773	0.14805
B	0.703977	0.343839	-0.009767	-0.04172
C	1.092	1.16396	1.0135	0.99919
D	-1.65623	0.28374	0.028356	-0.000263
E	0.444159	0.006557	0.88602	0.874178
F	0.050242	0.779846	0.323872	0.70294

If, e.g., the throughput of a melt extruder (Fig. 5.5) shall be determined under isothermal conditions, Eq. 5.20 must be solved for the pressure gradient and integrated over z. The pressure difference across a zone is yielded. The total pressure difference across the extruder length will be

$$\Delta p = \sum_{i=1}^{3} \Delta p_i \qquad (5.25)$$

The throughput $\dot{V}$ is contained in the Eq.s for Δp_i. Due to the continuity Eq. it is identical for all zones. If Eq. 5.25 is solved for the throughput, we get

$$\pi_{\dot{V}} = \phi_1 X_1 - \frac{\phi_2}{X_2} \pi_{pz} \qquad (5.26)$$

In Tab. 5.3 the coefficients of this Eq. are listed. In case of a Newtonian melt ($n = 1$),

$$\phi_1 = \phi_2 = 1 \qquad (5.27)$$

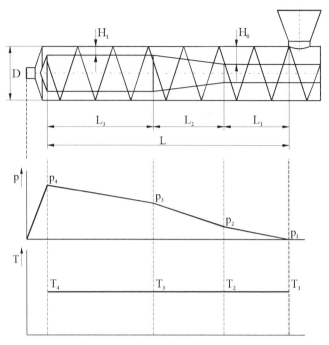

Figure 5.5   Geometrical dimensions, pressures and temperatures of a three-section extruder

If the condition $W \gg H$ is not applicable, correction factors must be introduced in Eq. 5.26, which will consider the flight influence:

$$\pi_{\dot{V}} = f_1 \phi_1 X_1 - f_2 \frac{\phi_2}{X_2} \pi_{pz} \qquad (5.28)$$

with

$$f_1 = 1 - 0.76 \exp\left(-0.4n\right)\frac{H}{W} \quad f_2 = \exp\left(-\frac{3}{4n}\frac{H}{W}\right) \qquad (5.29)$$

These are rough approximation Eq.s. A detailed consideration of these correction factors can be found in [4].

The leakage flow across the screw flights has been neglected in the calculation of the throughput. A consideration of the leakage flows is necessary, however, especially in case of a greater clearance between the screw flight and the barrel to allow a quite exact computation of the throughput. For this reason, a computation model for the melt throughput will be dealt with, where the leakage flow is taken into account and which is based on the current pressure/throughput Eq. [5]. The model shown in Fig. 5.6 is used. In the upper part of the Fig., the unwound screw channel is shown. The line CD is the perpendicular line to the screw axis and has the length πD. The volume flow balance for the balance area BCD is

$$\dot{V} = \dot{V}_z - \dot{V}_x \qquad (5.30)$$

Table 5.3   Isothermal Throughput Equations for Melt Extruders (Three-Section Screw)

$$\boxed{\pi_{\dot{V}} = \Phi_1 X_1 - \frac{\Phi_2}{X_2}\pi_p}$$

$$\pi_{\dot{V}} = \frac{\dot{V}}{i\,1/2\,WH_1 v_{0z}} \qquad \pi_p = \frac{H_1^{1+n}(p_4 - p_1)}{6KZ v_{0z}^{\,n}}$$

$$X_1 = \frac{\dfrac{L_1}{L}\left(\dfrac{H_1}{H_0}\right)^{1+n} + \dfrac{1}{n}\left(\dfrac{L_2}{L}\right)\left(\dfrac{H_1}{H_0}\right)^n \dfrac{(H_0/H_1)^n - 1}{(H_0/H_1)-1} + \dfrac{L_3}{L}}{\dfrac{L_1}{L}\left(\dfrac{H_1}{H_0}\right)^{2+n} + \dfrac{1}{1+n}\left(\dfrac{L_2}{L}\right)\left(\dfrac{H_1}{H_0}\right)^{1+n} \dfrac{(H_0/H_1)^{1+n} - 1}{(H_0/H_1)-1} + \dfrac{L_3}{L}}$$

$$X_2 = \frac{L_1}{L}\left(\frac{H_1}{H_0}\right)^{2+n} + \frac{1}{1+n}\left(\frac{L_2}{L}\right)\left(\frac{H_1}{H_0}\right)^{1+n} \frac{(H_0/H_1)^{1+n} - 1}{(H_0/H_1)-1} + \frac{L_3}{L}$$

$$Z = \frac{L}{\sin\varphi} = \frac{1}{\sin\varphi}(L_1 + L_2 + L_3)$$

$$W = \frac{\pi}{i}D\sin\varphi - e$$

$$v_{0z} = v_0 \cos\varphi = \pi ND\cos\varphi$$

Thus, the volume flow exceeding the balance boundary *CD* represents the total volume flow $\dot{V}$, $\dot{V}_z$ corresponds to the volumetric throughput of a screw zone in channel direction and $\dot{V}_x$ corresponds to the leakage flow. In the lower part of the Fig., the pressure distribution across the unwound screw channel is plotted. The pressure at point *C* is identical with the pressure at *D*, because *C* and *D* mark the same point. Presupposing that there is no pressure gradient in the *x* direction in the conveying channel, the pressure at point *A* is identical with the pressure at *C* and *D*, too. The pressure difference Δp_{AB} across the flight from *A* to *B* will then correspond with the pressure difference between *B* and *C* in the screw channel, i.e.

$$\Delta p_{AB} = \Delta p_{zc} = \Delta p_{zu} \tag{5.31}$$

Hence, we may say for the pressure gradient across the flight

$$\frac{\Delta p_{AB}}{e} = \frac{\Delta p_{AB}}{\Delta Z_u}\frac{\Delta Z_u}{e} = -\frac{\Delta p_{zu}}{\Delta Z_u}\frac{\pi D\cos\varphi}{e} \tag{5.32}$$

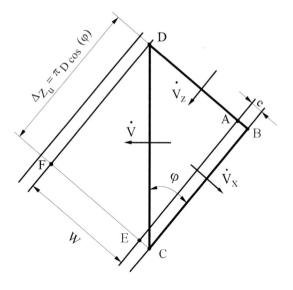

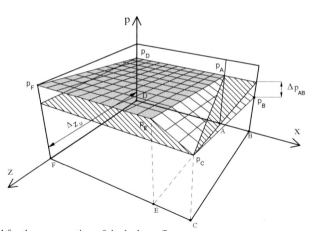

Figure 5.6   Model for the computation of the leakage flow

This Eq. results for the volume flow in the z direction according to Eq. 5.28 with constant channel depth ($X_1 = X_2 = 1$):

$$\dot{V}_z = \frac{1}{2} i W H v_{0z} \phi_1 f_1 - \frac{i H^{2+n} W v_{oz}^{1-n} \varphi_2 f_2}{12 K(T)} \frac{\Delta p_{zu}}{\Delta Z_u} \qquad (5.33)$$

and for the leakage flow in x direction:

$$\dot{V}_x = \frac{1}{2} \Delta Z_u \delta v_{0x} \phi_{1cl} + \frac{\delta^{2+n_{cl}} \Delta Z_u \phi_{2cl} \pi D \cos \varphi v_{0x}^{1-n_{cl}}}{12 K_{cl}(T) ei} \frac{\Delta p_{zu}}{\Delta Z_u} \qquad (5.34)$$

The index cl with n and K indicates that the coefficients of the power law in the flight clearance δ can differ from those in the channel.

With Eq. 5.30 it follows for the dimensionless throughput in a channel with constant flight depth that

$$\pi_V = \left(\phi_1 f_1 - \frac{\pi D \, \delta \phi_{1cl} \tan \varphi}{iWH} \right) - \left[\phi_2 f_2 + \frac{\delta^{2+n_{cl}}}{H^{2+n}} \frac{K(T)}{K_{cl}(T)} \frac{\phi_{2cl} \pi^2 D^2 \cos^{1+n} \varphi (\pi DN)^{n-n_{cl}}}{i^2 We \sin^{n_{cl}-1} \varphi} \right] \pi_{pz} \quad (5.35)$$

Screw zones with a variable flight depth are subdivided into elements of the length L_u with a constant flight depth H_i (Fig. 5.7). The volume flows and pressure differences, respectively, are calculated by means of the Eqs. 5.30–5.35 and the resulting pressure differences are summed up to the total pressure difference. An analoguous proceeding is necessary with multisection screws.

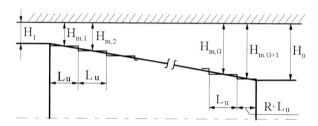

Figure 5.7   Subdivision of a compression zone in elements with constant channel depth

5.1.1.2   Power Consumption

To convey the melt, a certain work is necessary. It is equal to the product of the force F in stroke direction and the stroke s. Relating the work to the time, the power P is yielded. In general

$$P = Fv \quad (5.36)$$

For the further mathematical treatment we have to keep in mind that the screw is fixed and the barrel is rotating about the screw; i.e., only the shear stresses and velocities at the barrel surface $(y = H)$ are interesting.

Furthermore, it has to be considered that the necessary power is composed of a z and an x component:

$$P = Fv = F_{0z}v_{0z} + F_{0x}v_{0x} \quad (5.37)$$

i.e.,

$$P = \int_0^Z \int_0^W \left(\tau_{0x}v_{0x} + \tau_{0z}v_{0z} \right) dx\,dz \quad (5.38)$$

With the relations for Newtonian melts that result from the integration of the differential Eqs. 5.15 and 5.16 for $n = 1$,

$$\tau_{0z} = \eta \left(\frac{\partial v_z}{\partial y} \right)_{y=H} \tag{5.39}$$

$$\tau_{0x} = \eta \left(\frac{\partial v_x}{\partial y} \right)_{y=H} \tag{5.40}$$

$$v_z = v_{0z} \frac{y}{H} - \frac{1}{2\eta} y (H - y) \frac{\Delta p}{Z} \tag{5.41}$$

$$v_x = v_{0x} \frac{y}{H} \left(2 - \frac{3y}{H} \right) \tag{5.42}$$

the explicit expression of the power needed for the melt conveying zone is yielded from Eq. 5.38 if the zone has a constant flight depth H:

$$P_1 = \left(\frac{4\eta v_{0x}^2}{H} + \frac{\eta v_{0z}^2}{H} + \frac{v_{0z} H}{2} \frac{\Delta p}{Z} \right) ZW \tag{5.43}$$

or, if $\Delta p/Z$ is eliminated via Eq. 5.20,

$$P_1 = 4\eta Z \left(\frac{W}{H} v_0^2 - \frac{3 v_{0z} \dot{V}}{2H^2} \right) \tag{5.44}$$

with

$$v_0^2 = v_{0x}^2 + v_{0z}^2 \tag{5.45}$$

In case of a linear decrease of the flight depth H in z direction, the relation

$$H(z) = H_0 + \frac{H_1 - H_0}{Z} z \tag{5.46}$$

must be considered, too. In this case we have

$$P_2 = 4\eta Z \left(\frac{W v_0^2}{H_1 - H_0} \ln \frac{H_1}{H_0} - \frac{3 v_{0z} \dot{V}}{2H_0 H_1} \right) \tag{5.47}$$

Thus, all essential factors for the computation of the melt extruder have been determined. They apply to extruders with a negligible influence of the radial clearance ($\delta = 0$). As a rule, however, this cannot be neglected. In a first approach, we thus assume a mere drag flow and obtain

$$P_3 = \frac{\eta e Z v_0^{\,2}}{\delta} \tag{5.48}$$

For the total power input,

$$P = \sum_{j=1}^{l} i\left(P_{1j} + P_{3j}\right) + \sum_{k=1}^{n} i\left(P_{2k} + P_{3k}\right) \tag{5.49}$$

with
l = Number of zones with constant flight depth
n = Number of zones with variable flight depth
i = Number of channels

and thus for the torque

$$M = \frac{P}{2\pi N} \tag{5.50}$$

In power law melts, the shear loads in the x and y directions are correlated via the viscosity so that the shear loads at the wall are

$$\tau_{0x} = K \left[\left(\frac{\partial v_x}{\partial y}\right)^2 + \left(\frac{\partial v_z}{\partial y}\right)^2\right]^{(n-1)/2} \left.\frac{\partial v_x}{\partial y}\right|_{y=H} \tag{5.51}$$

$$\tau_{0z} = K \left[\left(\frac{\partial v_x}{\partial y}\right)^2 + \left(\frac{\partial v_z}{\partial y}\right)^2\right]^{(n-1)/2} \left.\frac{\partial v_z}{\partial y}\right|_{y=H} \tag{5.52}$$

However, the velocity profiles must be known to calculate the shear loads at the wall. There are no analytical solutions. For this reason the dimension analysis has been related with the finite element method and the regression analysis by Potente and Obermann [6] to calculate the power input. The dimension analysis yields

$$\pi_{Power} = f\left(\pi_{\dot{v}}, \frac{W}{H}, \frac{v_{0x}}{v_{0z}}, n\right) \tag{5.53}$$

with

$$\pi_{Power} = \frac{\Delta P_{channel} H^n}{KW\Delta\, zv_{0z}^{1+n}}\tag{5.54}$$

The two-and-a-half-dimensional FE computations in relation with the regression analysis yield the following Eq.s:

$$\pi_{Power} = c_1 + c_2\pi_{\dot{V}} + c_3\pi_{\dot{V}}^{2}\tag{5.55}$$

with

$$c_1 = 4\left[1+\left(\frac{v_{0x}}{v_{0z}}\right)^2\right]\exp\left[\left(0.408\frac{v_{0x}}{v_{0z}}+1\right)(n-1)\right]\tag{5.56}$$

$$c_2 = -3\left[1+0.61\frac{H}{W}+(1-n)\left(0.1665\frac{v_{0x}}{v_{0z}}-0.34\frac{H}{W}-0.534n\frac{v_{0x}}{v_{0z}}-0.93n-0.85\right)\right]\tag{5.57}$$

$$c_3 = (1-n)\left[0.18\ln\left(\pi\frac{v_{0x}}{v_{0z}}\right)-1.3n\right]\tag{5.58}$$

for the region

$$0<\pi_{\dot{V}}<2\quad 0.15<\frac{v_{0x}}{v_{0z}}<1\quad 3<\frac{W}{H}<\infty\tag{5.59}$$

The coefficient of determination is at 0.999. These Eq.s apply to the screw channel. For the radial clearance it is

$$\Delta P_{flight} = \frac{Ke\Delta zv_{0z}^{1+n}}{\delta^n}\left[1+\left(\frac{v_{0x}}{v_{0z}}\right)^2\right]^{(1+n)/2}\tag{5.60}$$

Multiple-flight multisection screws have to be treated analoguously with these equations, as has been done with Newtonian melts.

Fig. 5.8 shows the correlation between the dimensionless drive power and the dimensionless volume flow with the power law exponent n as parameter. With an increasing volume flow, the drive power decreases and may even reach negative values in case of a high throughput. This phenomenon, which at first glance seems paradoxical, can be explained by the different pressure conditions in the screw channel. If the pressure flow and the drag flow are oriented in the same direction, negative shear loads at the barrel wall may occur; which may lead to a negative drive power.

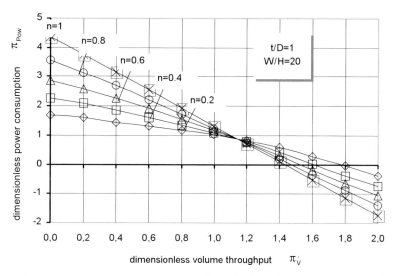

Figure 5.8   Dimensionless power consumption versus dimensionless volume throughput for melt conveying

5.1.2   Nonisothermal Analysis

5.1.2.1   Melt Conveying

The following approach was presented for the first time in [7] for the adiabatical Newtonian case and in [8, 9] for the nonisothermal power law case. It starts from the power Eq. where the balance area is inside the barrel:

$$P = \dot{m}c_v \, \Delta T + \dot{V} \Delta p + \dot{Q} \tag{5.61}$$

With approximately

$$\frac{\dot{Q}}{P} = \frac{1}{Br} = const \tag{5.62}$$

the power Eq. is

$$dP\left(1 - \frac{1}{Br}\right) = \rho c \, \dot{V} \, dT + \dot{V} \, dp \tag{5.63}$$

Based on the Eqs. 5.53 to 5.60 we have

$$dP = \psi_1 K \, dz \tag{5.64}$$

where ψ_1 is a function of $\pi \dot{v}$, n, v_{0x}, W, H, δ, e and v_{0z}. The throughput Eq. 5.20 can be written in analogy as

$$dP = \psi_2 K \, dz \tag{5.65}$$

The temperature dependence of K is

$$K = K_0 \exp\left[-\beta\left(T - T_0\right)\right] \tag{5.66}$$

When inserting Eq. 5.64 up to Eq. 5.66 in Eq. 5.63, the following is yielded after the integration, because ψ_1 and ψ_2 are independent of z:

$$\left[\frac{\left(1 - 1/Br\right)\psi_1}{\rho c \dot{V}} - \frac{\psi_2}{\rho c}\right]Z = \frac{\exp\left[\beta\left(T - T_0\right)\right] - 1}{\beta K_0} \tag{5.67}$$

T_0 is the initial temperature of the melt. For the temperature increase in z direction we thus have from Eq. 5.67

$$\Delta T = T_b - T_0 = \frac{1}{\beta}\ln\left\{1 + \left[\frac{\left(1 - 1/Br\right)\psi_1}{\rho c \dot{V}} - \frac{\psi_2}{\rho c}\right]\beta\, K_0 Z\right\} \tag{5.68}$$

After the transformation and integration we get from Eq. 5.20

$$\Delta p = p_Z - p_0 = \frac{1}{\phi_2}\left(\phi_1 - \pi_{\dot{V}}\right)\frac{6 v_{0z}}{H^{1+n}}^{n}\int_0^Z K dz \tag{5.69}$$

The following relation can be derived from Eqs. 5.63 to 5.65:

$$\int_0^Z K dz = \frac{\Delta T}{\left[\dfrac{\left(1 - 1/Br\right)\psi_1}{\rho c \dot{V}} - \dfrac{\psi_2}{\rho c}\right]} \tag{5.70}$$

and hence with Eq. 5.67

$$\int_0^Z K dz = \frac{\beta K_0 Z \Delta T}{\exp\left(\beta \Delta T\right) - 1} \tag{5.71}$$

By substituting Eq. 5.71 with Eq. 5.69 we get

$$\pi_{\dot{V}} = \phi_1 - \phi_2 \frac{\exp\left(\beta \Delta T\right) - 1}{\beta \Delta T}\pi_p \tag{5.72}$$

Which applies if $\dot{Q}/P = 1/Br = $ const. In the isothermal case, the Equation enters into Eq. 5.20.

The temperature must be calculated by iteration with Eq. 5.68 and Eq. 5.72. The nonisothermal three-section screw is principally treated in the same way as the isothermal screw; i.e., the pressure differences of the single zones are summed up at which there are additional temperature terms in the equations. At first, a linear increasing temperature profile is assumed over the screw length Fig. 5.9. The equations in Table 5.4 are yielded. The coefficient K_0 refers to the initial temperature. In Table 5.12 K is referred to the melt temperature in front of the screw. This changes the explicit expressions for the Ω functions. The latter can be an advantage for the integral treatment of plasticating extruders. Since the temperature calculation presented in this paragraph is only a special case, the next section will especially deal with the temperature calculation.

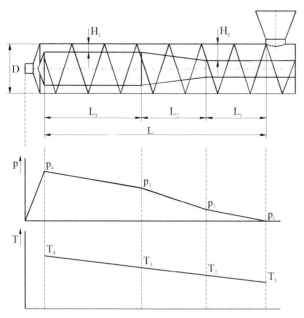

Figure 5.9   Geometrical dimensions, pressures and temperatures of a three-secion screw

5.1.2.2   Temperature

Since it is a three-dimensional flow profile, the melt temperature can be calculated adequately only by means of the 3D FE method. However, this method is still very time-consuming. 2D FE computations or 2½D FE computations do not provide satisfactory solutions. In many cases, a one-dimensional solution suffices for approximate solutions, where an average dissipation term $\overline{\tau\dot{\gamma}}$ is determined across the screw channel cross section. This is the way that will be described here.

The temperature is calculated on the basis of the channel model. The following assumptions are made for the computation of the temperature profile:

- The screw channel is regarded as a flat channel, i.e., $W \gg H$. The influence of the flights can thus be neglected.
- The melt does not slip at the wall.
- The flow is laminar and incompressible ($c = c_v = c_p$).
- The flow behavior of the melt shall follow the power law $\tau = K\dot{\gamma}^n$.
- All material data except for the viscosity are regarded as independent of the temperature. If the temperature range is not too great, this assumption is admissible for polymer melts as a useful approximation. This is especially applicable for the heat and temperature conductivity.

Under these circumstances, the differential equation for the temperature is

$$\rho c \bar{v}_z \frac{\partial T}{\partial z} = \lambda \frac{\partial^2 T}{\partial y^2} + \left(\overline{\tau \dot{\gamma}}\right)_0 e^{-\beta(T-T_0)} \tag{5.73}$$

where $\left(\overline{\tau \dot{\gamma}}\right)_0$ is the average dissipated energy per volume unit at the temperature T_o across the channel crosssection. The exponential function describes the temperature dependence of the viscosity. By introducing dimensionless characteristic values, the DE can be written as follows:

$$Gz \frac{\partial \Theta}{\partial \zeta} = \frac{\partial^2 \Theta}{\partial \xi^2} + Br e^{-\beta(T-T_0)} \tag{5.74}$$

with

$$\Theta = \frac{T - T_b}{T_b} \qquad \xi = \frac{y}{H} \qquad \zeta = \frac{z}{Z}$$

$$Gz = \frac{\rho c \bar{v}_z H^2}{\lambda Z} = \frac{\dot{m}H}{\rho a W Z} \tag{5.75}$$

$$Br = \frac{\left(\overline{\tau \dot{\gamma}}\right)_0 H^2}{\lambda T_b} \approx \frac{K_0 \left(v_{0x}^2 + v_{0z}^2\right)^{(1+n)/2} H^{1-n}}{\lambda T_b}$$

Table 5.4   Nonisothermal Throughput Eq. for Melt Extruders (Three-Section Screw)

$$\pi_{\dot{V}} = \Phi_1 X_1 - \frac{\Phi_2}{X_2} \pi_p$$

$$\pi_{\dot{V}} = \frac{2\dot{V}}{WH_1 v_{0z}} \qquad \pi_p = \frac{\Delta p H_1^{1+n}}{6K(T_1)v_{0z}^n Z} \qquad \Delta p = p_4 - p_1$$

$$X_1 = \frac{\dfrac{L_1}{L}\Omega_1\left(\dfrac{H_1}{H_0}\right)^{1+n} + \dfrac{1}{n}\Omega_2\dfrac{L_2}{L}\left(\dfrac{H_1}{H_0}\right)^n \dfrac{(H_0/H_1)^n - 1}{(H_0/H_1) - 1} + \dfrac{L_3}{L}\Omega_3}{\dfrac{L_1}{L}\Omega_1\left(\dfrac{H_1}{H_0}\right)^{2+n} + \dfrac{1}{1+n}\Omega_2\dfrac{L_2}{L}\left(\dfrac{H_1}{H_0}\right)^{1+n} \dfrac{(H_0/H_1)^{1+n} - 1}{(H_0/H_1) - 1} + \dfrac{L_3}{L}\Omega_3}$$

$$X_2 = \frac{L_1}{L}\Omega_1\left(\frac{H_1}{H_0}\right)^{2+n} + \frac{1}{n+1}\Omega_2\frac{L_2}{L}\left(\frac{H_1}{H_0}\right)^{1+n} \frac{(H_0/H_1)^{1+n} - 1}{(H_0/H_1) - 1} + \frac{L_3}{L}\Omega_3$$

$$\Omega_1 = \frac{\beta(T_2 - T_1)}{e^{\beta(T_2 - T_1)} - 1} \qquad \Omega_2 = \frac{\beta(T_3 - T_2)}{e^{\beta(T_3 - T_1)} - e^{\beta(T_2 - T_1)}} \qquad \Omega_3 = \frac{\beta(T_4 - T_3)}{e^{\beta(T_4 - T_1)} - e^{\beta(T_3 - T_1)}}$$

The barrel wall temperature T_b, the unwound channel length Z, and the channel height H and channel width W of the considered screw section enter into the characteristic value. Except for the viscosity, the material data are regarded as invariant. Furthermore, $\bar{v}_z$ is the average flow rate in z direction, v_{0x} and v_{0z} are the peripheral speeds at the barrel, $\dot{m}$ is the mass throughput and K_0 is the coefficient of the power law at the temperature T_0. For the calculation, the average temperature is chosen for T_0 at the starting cross section.

Since a closed solution of the DE is no longer possible due to the exponential function term, it is linearized in the interesting domain:

$$e^{-\beta(T-T_0)} = e^{-\beta T_b \Theta} e^{-\beta(T_b-T_0)} = C_1 - C_2 \beta(T-T_b) \tag{5.76}$$

The exponential function is approximated with the secant between the temperatures T_0 and T_b. The corresponding constant values are

$$C_1 = e^{-\beta(T_b-T_0)} \qquad C_2 = \frac{1 - e^{-\beta(T_b-T_0)}}{\beta(T_b-T_0)} \tag{5.77}$$

The DE is described as

$$\frac{\partial^2 \Theta}{\partial \xi^2} - Gz \frac{\partial \Theta}{\partial \xi} - C_2 \beta T_B Br \Theta = -C_1 Br \tag{5.78}$$

Due to the corresponding boundary and initial conditions, the Eq. 5.78 is adapted to the desired process control. The conditions are as follows:

- The barrel wall temperature is invariant across the considered channel length (isothermal barrel), i.e.;

$$T_b = const. \qquad \Theta = 0 \qquad \text{for } \xi = 1 \text{ and } 0 \le \zeta \le 1 \tag{5.79}$$

- There is no heat transfer to the screw (adiabatical screw), i.e.;

$$\frac{\partial \Theta}{\partial \xi} = 0 \qquad \text{for } \xi = 0 \tag{5.80}$$

- Initial condition: The temperature distribution at the start is preset by an appropriate initial value function $\Theta_0(\xi)$ (see Fig. 5.10):

$$\zeta = 0 \qquad T_0 = T_0(\xi) \qquad \Theta_0 = \Theta_0(\xi) \tag{5.81}$$

Since it is a linear DE, a solution approach according to the principle of superposition can be chosen:

$$\Theta(\xi, \zeta) = \Theta_1(\xi, \zeta) + \Theta_2(\xi, \zeta) \tag{5.82}$$

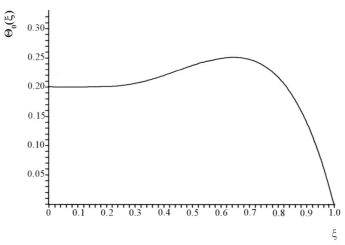

Figure 5.10   Initial dimensionless temperature profile of 4th order

DE for Θ_1:

$$Gz\frac{\partial\Theta_1}{\partial\zeta} = \frac{\partial^2\Theta_1}{\partial\xi^2} - C_2\beta T_b Br\Theta_1 + C_1 Br \qquad (5.83)$$

The boundary condition is equal to the boundary condition of Θ.

The initial condition says: The temperature T_0 of the melt at the beginning of the considered section is equal to the barrel wall temperature

$$\zeta = 0 \qquad T_0 = T_b \qquad \Theta = 0 \qquad (5.84)$$

Solution for Θ_1:

$$\Theta_1 = \frac{C_1}{C_2\beta T_b}\left[1 - \frac{\cosh\left(\xi\sqrt{C_2\beta T_b Br}\right)}{\cosh\sqrt{C_2\beta T_b Br}}\right] - \frac{16}{\pi^3}C_1 Br$$

$$\sum_{m=0}^{\infty}\left\{\frac{(-1)^m}{(1+2m)}\frac{\cos\left[(1+2m)\frac{\pi}{2}\xi\right]}{\left(\frac{2}{\pi}\right)^2 C_2\beta T_b Br + (1+2m)^2}\right.$$

$$\left.\exp\left\{-\left[(1+2m)^2 + \left(\frac{2}{\pi}\right)^2 C_2\beta T_b Br\right]\left(\frac{\pi}{2}\right)^2\frac{\zeta}{Gz}\right\}\right\} \qquad (5.85)$$

DE for Θ_2:

$$Gz\frac{\partial\Theta_2}{\partial\zeta} = \frac{\partial^2\Theta_2}{\partial\xi^2} - C_2\beta T_b Br\Theta_2 \qquad (5.86)$$

The boundary condition is equal to the boundary condition of Θ.

The initial condition says: At the point $\zeta = 0$ the starting profile Θ $(\xi, 0) = \Theta_0(\xi)$ be preset. A function representing the expected temperature distribution shall be chosen for the starting profile. This function will be adapted to the corresponding boundary conditions it has to meet. Fig. 5.10 shows a polynomial approach of 4th order as starting profile. The thus formulated problem has been solved by [10] by means of the Fourier method.

Solution for Θ_2:

$$\Theta_2 = \sum_{m=0}^{\infty} \left\{ \exp\left\{-\left[C_2 \beta T_b Br + \left(\frac{\pi}{2} + m\pi\right)^2\right]\frac{\zeta}{Gz}\right\} \cos\left[\left(\frac{\pi}{2} + m\pi\right)\xi\right]\alpha_m \right\} \tag{5.87}$$

with

$$\alpha_m = 2\int_0^1 \left\{ \Theta_0(\xi) \cos\left[\left(\frac{\pi}{2} + m\pi\right)\xi\right] \right\} d\xi \tag{5.88}$$

Of course, the temperature profile versus the channel height can show a maximum or a minimum, respectively, between the screw and the barrel wall, as can, e.g., be imagined in case of high dissipation in the screw channel. To describe such profiles, a polymonimal of 4th order has been chosen for the starting profile.

The coefficients of the general approach, on the other hand, are determined by the boundary conditions and by means of further points on the profile. These points are chosen at a regular distance at $\xi = 0$, $1/3$ and $2/3$. Hence, the function values $\Theta(\xi)$ must be calculated at these points; Observing the boundary conditions, the following is yielded for the initial temperature profile:

$$\Theta_0(\xi) = \left[-24.75\Theta(0) + 40.5\Theta(1/3) - 20.25\Theta(2/3) + 4.5\Theta(1)\right]\xi^4$$
$$+ \left[45\Theta(0) - 67.5\Theta(1/3) + 27\Theta(2/3) - 4.5\Theta(1)\right]\xi^3 \tag{5.89}$$
$$+ \left[-21.25\Theta(0) + 27\Theta(1/3) - 6.75\Theta(2/3) + 1\Theta(1)\right]\xi^2 + \Theta(0)$$

By means of this function of the initial temperature profile the terms α_m are computed, which are substituted in the solution for $\Theta_2(\xi, \zeta)$:

$$\alpha_m = \left[\frac{4\Theta_0(1)}{(\pi + 2m\pi)} + \frac{11120\Theta_0(0) - 21600\Theta_0(1/3) + 15120\Theta_0(2/3) - 4640\Theta_0(1)}{(\pi + 2m\pi)^3}\right.$$
$$\left. - \frac{38016\Theta_0(0) - 62208\Theta_0(1/3) + 31104\Theta_0(2/3) - 6912\Theta_0(1)}{(\pi + 2m\pi)^5}\right] \sin\left(\frac{\pi}{2} + m\pi\right) \tag{5.90}$$
$$+ \frac{8640\Theta_0(0) - 12960\Theta_0(1/3) + 5184\Theta_0(2/3) - 144\Theta_0(1)}{(\pi + 2m\pi)^4}$$

By means of the Eqs. 5.82, 5.85, 5.87 and 5.90, the temperature Θ can be completely calculated, depending on ξ and ζ. The temperature profile at the end of the screw section will be evaluated for $\zeta = 1$ at different points ξ. The integration over ξ yields the average temperature $\overline{\Theta}$, depending on the three temperatures $\Theta(0)$, $\Theta(^1/_3)$, and $\Theta(^2/_3)$ that have to be computed:

$$\overline{\Theta} = 0.21667\Theta(0) + 0.225\Theta(1/3) + 4.5\Theta(2/3) + 0.10833\Theta(1) \tag{5.91}$$

In injection molding there is a downtime phase for the cooling process besides the plasticating phase. The following assumptions are made for the description of the temperature during this phase:

- The screw channel that is filled with melt is regarded as an infinite plate with a finite thickness.
- A one-dimensional view is chosen; i.e., the temperature field is dependent only on the downtime t_{st} and on the y coordinate.
- The melt in the screw channel is characterized by homogeneous and isotropic material features.
- At the beginning of the downtime, $t_{st} = 0$, the temperature over the screw channel height $T_{st,0}$ is equal to the average temperature after the plasticating phase.

The descriptive DE says

$$\frac{\partial T}{\partial t} = a \frac{\partial^2 T}{dy^2} \tag{5.92}$$

To solve the DE, an adiabatical screw and an isothermal barrel wall are presumed. The following solution is yielded:

$$\theta_{st} = \frac{4}{\pi} \sum_{n=1}^{\infty} \frac{(-1)^{n-1}}{2n-1} \cos\left(\frac{2n-1}{2}\pi\xi\right) \exp\left[-\left(\frac{2n-1}{2}\pi\right)^2 Fo\right] \tag{5.93}$$

with

$$\theta_{st} = \frac{T_{st} - T_b}{T_{st,0} - T_b} \qquad \xi = \frac{y}{H} \qquad Fo = \frac{at_{st}}{H^2} \tag{5.94}$$

After the integration over the dimensionless channel height ξ, the average dimensionless melt temperature

$$\overline{\theta}_{st} = \frac{8}{\pi^2} \sum_{n=1}^{\infty} \frac{1}{(2n-1)^2} \exp\left[-\left(\frac{2n-1}{2}\pi\right)^2 Fo\right] \tag{5.95}$$

is yielded with

$$\overline{\theta}_{st} = \frac{\overline{T}_{st} - T_b}{T_{st,0} - T_b} \tag{5.96}$$

5.1.2.3   Power Consumption

The power is computed with the Eqs. 5.54 to 5.60. For this, the screw has to be subdivided into cylindrical elements, and a temperature-dependent K value has to be introduced in the characteristic value (Eq. 5.54), the temperature being computed with the Eq.s of the preceding section (Section 5.1.2.2). The total power required is yielded by summing up the power values of the single elements.

5.1.2.4   Melts That Slip at the Wall

5.1.2.4.1   One-Dimensional Treatment

So far, it has been assumed that the melt does not slip at the barrel or at the screw wall. There are, however, materials that do not meet this condition. Typical examples are PVC, highly molecular PE, elastomers, polymer suspensions, ceramic materials, and food products. These materials slip at the wall [12–22] as is illustrated, e.g., in Fig. 5.11. This affects the throughput, the pressure profile, the melt temperature, the power consumption, and the homogenization.

The slip at the wall can be described by the following approach [20]:

$$v_{sl}^* = a \sinh\left[b\left(\frac{\tau_{sl}^*}{\tau_c^*} - 1 \right) \right]$$

(5. 97)

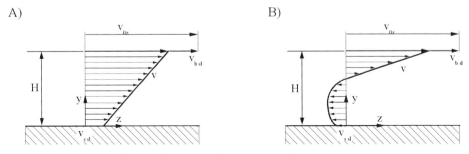

Figure 5.11   Velocity distribution of the longitudinal flow with slip at the wall A) unchoked extruder, B) choked extruder

Where τ_c^* is the critical dimensionless shear stress where the slipping starts; τ_{sl} is the slip shear stress, v_{sl}^* the dimensionless velocity difference between the wall velocity and the melt velocity at the wall, and a and b are constants. For reasons of clearness we use the first approximation here:

$$\tau_{sl}^* = \tau_c^*\left(1 + \frac{v_{sl}^*}{ab} \right) = \tau_c^* + kv_{sl}^*$$

(5.98)

Furthermore, we proceed from the simplest form of the DE (Eq. 5.9), the one-dimensional form:

$$\frac{\partial \tau^*}{\partial \xi} = 6\pi_p \tag{5.99}$$

with

$$\xi = \frac{y}{H} \qquad v_{sl}^* = \frac{v_{sl}}{v_0} \qquad \tau^* = \frac{\tau}{K(v_0/H)^n} \qquad \pi_p = \frac{H^{1+n}}{6Kv_0^n}\frac{\partial p}{\partial z} \tag{5.100}$$

The integration yields

$$\tau^*(\xi) = \tau_s^* + 6\pi_p \xi \tag{5.101}$$

For $\xi = 1$,

$$\pi_p = \frac{\tau_b^* - \tau_s^*}{6} \tag{5.102}$$

In the Newtonian case ($n = 1$, $K = \eta$) the shear stresses and shear rates are

$$\tau^* = \frac{\tau}{\eta(v_0/H)} = \frac{dv^*}{d\xi} \qquad \tau_s^* = 1 - 3\pi_p \qquad \tau_b^* = 1 + 3\pi_p \tag{5.103}$$

Eq. (5.101) thus is

$$\frac{dv^*}{d\xi} = \tau_s^* + 6\pi_p \xi \tag{5.104}$$

For further handling of the problem, we have to distinguish two cases: $\tau_c^* \geq 1$ and $\tau_c^* < 1$. With $\tau_c^* \geq 1$, no slip at the wall can be stated in the beginning. With an increasing $|\pi_p|$ the slip at the wall starts on one side and continues on both sides with further increasing $|\pi_p|$. With $\tau_c^* < 1$, two-sided slip at the wall can be stated in the beginning. With an increasing $|\pi_p|$ it turns into a one-sided slip at the wall and turns again into a two-sided slip when $|\pi_p|$ is further increased.

First, we will consider the case $\tau_c^* \geq 1$. With a positive π_p the slip starts at the wall set in motion, thus the following boundary conditions apply:

$$\xi = 1 \qquad v^* = 1 - v_{b,sl}^* \qquad \tau_b^* = \tau_c^* + kv_{b,sl}^* \tag{5.105}$$

with $v_{b,sl}^*$ = dimensionless rate of slip at the barrel.

The solution of the DE thus is

$$v^* = \tau_s^*(\xi - 1) + 3\pi_p(\xi^2 - 1) + (1 - v_{b,sl}^*) \tag{5.106}$$

With the boundary conditions,

$$\xi = 0 \qquad v^* = 0 \qquad \tau_s^* = \tau_b^* - 6\pi_p \tag{5.107}$$

Eq. 5.106 yields for the rate of slip at the barrel wall:

$$v_{b,sl}^* = \frac{1 + 3\pi_p - \tau_c^*}{1 + k} = \frac{3}{1 + k}\left(\pi_p - \pi_{p1}\right) \tag{5.108}$$

and by integration for the volume flow:

$$\pi_{\dot{V}} = 1 - \pi_p - \frac{3}{1+k}\left(\pi_p - \pi_{p1}\right) \tag{5.109}$$

These Equations are applicable in the following domain:

$$\left|\pi_p\right| \geq \left|\pi_{p1}\right| \quad \text{and} \quad \left|\pi_p\right| \leq \left|\pi_{p2}\right| \tag{5.110}$$

$$\pi_{p1} = \pm\frac{1}{3}\left(\tau_c^* - 1\right) \qquad \pi_{p2} = \pm\frac{1}{3}\left(\tau_c^* + \frac{k}{2+k}\right)$$

When increasing π_p, there is also slip at the screw wall. The following boundary conditions are applicable:

$$\xi = 0 \qquad v^* = -v_{s,sl}^* \qquad \tau_s^* = -\left(\tau_c^* + kv_{s,sl}^*\right) \tag{5.111}$$

Thus, Eq. 5.106 yields

$$v_{s,sl}^* = -\left[\frac{k - 3k\pi_p + k\tau_c^* - 6\pi_p + 2\tau_c^*}{k(2+k)}\right] \tag{5.112}$$

$$v_{b,sl}^* = \frac{6\pi_p}{k} - \frac{2\tau_c^*}{k} - v_{s,sl}^* \tag{5.113}$$

and for the dimensionless volume flow:

$$\pi_{\dot{V}} = 1 - \pi_p - \frac{6}{k}\left(\pi_p - \pi_{p2}\right) - \frac{3}{1+k}\left(\pi_{p2} - \pi_{p1}\right) \tag{5.114}$$

for $\left|\pi_p\right| \geq \left|\pi_{p2}\right|$

In case of $\tau_c^* < 1$ the following boundary conditions

$$\xi = 0 \qquad v^* = -v_{s,sl}^* \qquad \tau_s^* = \tau_c^* + kv_{s,sl}^* \tag{5.115}$$

entail the following solutions:

$$v_{s,sl}^{*} = \frac{k - k\tau_c^{*} - 3k\pi_p - 6\pi_p}{k(2+k)} \tag{5.116}$$

$$v_{b,sl}^{*} = \frac{6}{k}\pi_p + v_s^{*} \tag{5.117}$$

$$\pi_{\dot{V}} = 1 - \pi_p - \frac{6}{k}\pi_p \tag{5.118}$$

They are applicable for

$$\left|\pi_{p1}\right| \leq \left|\pm\frac{k}{3(2+k)}\left(1 - \tau_c^{*}\right)\right| \tag{5.119}$$

With a further pressure increase, there is again no slip at one side of the wall and Eq. 5.108 for the rate of slip at the wall and thus for the dimensionless throughput applies with the boundary condition Eq. 5.107:

$$\pi_{\dot{V}} = 1 - \pi_p - \frac{3}{1+k}\left(\pi_p - \pi_{p1}\right) - \frac{6}{k}\pi_{p1} \tag{5.120}$$

within the domain

$$\left|\pi_p\right| \geq \left|\pi_{p1}\right| \quad \text{and} \quad \left|\pi_p\right| \leq \left|\pi_{p2}\right| \tag{5.121}$$
$$\pi_{p2} = \pm\frac{1}{3}\left(\tau_c^{*} + \frac{k}{2+k}\right)$$

Finally, there will again be slip at two sides of the wall and the following Eq. is applicable:

$$\pi_{\dot{V}} = 1 - \pi_p - \frac{6}{k}\left(\pi_p - \pi_{p2}\right) - \frac{3}{1+k}\left(\pi_{p2} - \pi_{p1}\right) - \frac{6}{k}\pi_{p1} \tag{5.122}$$

for $\left|\pi_p\right| \geq \left|\pi_{p2}\right|$. Fig. 5.12 shows a scheme illustrating these correlations once again.

There are the following relations for power law melts:

$$\tau_b^{*} - \tau_s^{*} = \left|6\pi_p\right|sign\pi_p \tag{5.123}$$

$$\tau_s^{*} = -\lambda\left|6\pi_p\right|sign\pi_p \tag{5.124}$$

$$\tau^{*}(\xi) = \tau_s^{*} + \left(\tau_b^{*} - \tau_s^{*}\right)\xi \tag{5.125}$$

and thus

$$\frac{dv^*}{d\xi} = \left[\tau_s^* + \left(\tau_b^* - \tau_s^*\right)\xi\right]^{1/n} \tag{5.126}$$

λ being the place where the shear rate becomes zero. The further proceeding is identical with that for Newtonian melts.

We now consider the influence of slip at the wall on the melt temperature. For this, we proceed from the temperature gradient in channel direction [22]:

$$\frac{dT}{dz} = \frac{dT}{dz}\bigg|_{conduction} + \frac{dT}{dz}\bigg|_{shearing} \tag{5.127}$$

The following applies to the shearing:

$$\frac{dT}{dz}\bigg|_{shearing} \sim \overline{\tau\dot\gamma} \sim \overline{\tau}^2 \tag{5.128}$$

with

$$\overline{\tau}^2 = \int_0^1 \tau(\xi)^2 \, d\xi \tag{5.129}$$

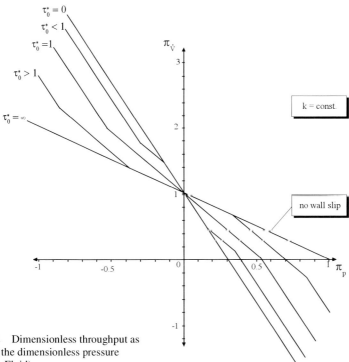

Figure 5.12   Dimensionless throughput as function of the dimensionless pressure (Newtonian Fluid)

Eq. 5.101 yields

$$\overline{\tau} = \sqrt{\frac{\tau_s^2 + \tau_s \tau_b + \tau_b^2}{3}} \qquad\qquad (5.130)$$

Now, a correction factor is introduced:

$$f_{corr} = \sqrt{\frac{\left(\tau_s^2 + \tau_s \tau_b + \tau_b^2\right)_{Slip}}{\left(\tau_s^2 + \tau_s \tau_b + \tau_b^2\right)_{NoSlip}}} \qquad\qquad (5.131)$$

With this correction factor, the following equations is yielded for the temperature gradient under slipping conditions:

$$\left.\frac{dT}{dz}\right|_{Shear,Slip} = \left.\frac{dT}{dz}\right|_{Shear,No\,Slip} f^2 \qquad\qquad (5.132)$$

Via the equations written in Section 5.1.2.2, the melt temperature under slipping conditions is obtained.

5.1.2.4.2   Two-Dimensional Treatment

The two-dimensional case is

$$\tau_z(\xi) = \tau_{sz} + \left(\tau_{bz} - \tau_{sz}\right)\xi \qquad\qquad (5.133)$$

$$\tau_x(\xi) = \tau_{sx} + \left(\tau_{bx} - \tau_{sx}\right)\xi \qquad\qquad (5.134)$$

This yields

$$\tau^2(\xi) = \tau_s^2(\xi) + \tau_b^2(\xi) \qquad\qquad (5.135)$$

and thus for the average value

$$\overline{\tau} = \sqrt{\frac{1}{3}\left[\tau_s^2 + \tau_{sz}\tau_{bz} + \tau_{sx}\tau_{bx} + \tau_b^2\right]} \qquad\qquad (5.136)$$

For Newtonian melt we get

$$\tau_{bz}{}^* = 4 - 3\pi_{\dot{V}} \qquad \tau_{bx}{}^* = 4 \qquad\qquad (5.137)$$

$$\tau_{sz}{}^* = 3\pi_{\dot{V}} - 2 \qquad \tau_{sx}{}^* = -2 \qquad\qquad (5.138)$$

or

$$\tau_{sz}^{\ *} = \tau_{bz}^{\ *} - 6\left(1 - \pi_{\dot{V}}\right) = \tau_{bz}^{\ *} - 6\pi_p \qquad (5.139)$$

$$\tau_{sx}^{\ *} = \tau_{bx}^{\ *} - 6 \qquad (5.140)$$

respectively. Thus, we obtain

$$\tau_b^{\ *} = \sqrt{\left(1 + 3\pi_p\right)^2 + \left(4\frac{v_{0x}}{v_{0z}}\right)^2} \qquad (5.141)$$

$$\tau_s^{\ *} = \sqrt{\left(1 - 3\pi_p\right)^2 + \left(2\frac{v_{0x}}{v_{0z}}\right)^2} \qquad (5.142)$$

Slip at the barrel wall will occur with

$$\pi_{p1}^{\ *} = \frac{1}{3}\left[\sqrt{\tau_c^{*2} - \left(4\frac{v_{0x}}{v_{0z}}\right)^2} - 1\right] \qquad (5.143)$$

and slip at the screw wall with

$$\pi_{p2}^{\ *} = \frac{1}{3}\left[1 - \sqrt{\tau_c^{*2} - \left(2\frac{v_{0x}}{v_{0z}}\right)^2}\right] \qquad (5.144)$$

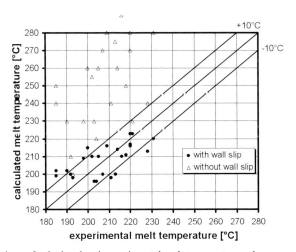

Figure 5.13   Comparison of calculated and experimental melt temperature values

These values have to be substituted in the throughput Eq. 5.109. Furthermore, we obtain the correction factor for the temperature by means of Eqs. 5.136 and 5.131.

A graphical illustration of the comparison of calculated and measured mean temperatures for a melt with slip can be found in Fig. 5.13. It is obvious that the consideration of the slip condition considerably improves the quality of calculation results.

5.2   Melting Analysis

5.2.1   Starting Point

Melting is one of the central tasks of a plasticating extruder. Therefore, screws have to be estimated according to the capacity of melting solid material [23–46]. Visual qualitative analyses of the melting process have led to the known melting models shown in Fig. 5.14 [23–26]. To a large extent, the Maddock model has prevailed for melts that do not slip at the wall. Therefore, all further considerations are based on this model.

Fig. 5.15 [37] shows a scheme of the complete melting section. Solid A is completely covered by melt and flows on the melt films C_1 and C_2 on the root surface and screw flank. The thickness of these melt films is permanently increasing in transport direction due to the heating by the screw. However, the melting of the solid mainly takes place at the inside barrel wall. The solid is heated through conduction and dissipation in a melt film with a thickness of only a few tenths of a millimeter, B. The relative velocity v_r between the barrel and the solid leads to an immediate transport of the melt. It accumulates in the melt vortex E at the leading edge of flight. Little of the solid is also melted at the side that faces the melt vortex. Furthermore, a small part is returned from E via the clearance F to the melt film B (leakage flow). In the screw channel direction the melt vortex becomes broader in case of a constant flight depth H. It presses the solid to the leading edge of flight. Toward the end of the melting section, the solid extrudate breaks into single pieces. Therefore, an exact definition of the end of the melting zone is not possible.

This first basic theoretical analysis of the melting process was made by Tadmor [24], who developed the classic Tadmor model. It neglects the melt films at the screw and assumes a constant melt film thickness at the barrel wall at right angles to the screw channel direction. The viscosity of the melt is thus temperature-dependent and is based on the assumption of a linear temperature profile across the melt film thickness.

A closed solution of the Tadmor model is possible only for Newtonian melts. With power law melts it can be solved only by iterations because the melt film thickness is once again implicitly comprised in the Eq.s for the melt film thickness and the melt rate.

A series of mathematical models has been developed on the basis of the Tadmor model. The most comprehensive one is the five-section model by Lindt [30]. In contrast to his original model [26] it is again based on the Maddock model. A critical assessment of all known models until 1985 can be found in [31]. The comprehensive nonlinear models can only be solved numerically.

In the years 1982/88, Chung and his staff developed an analytical model for the melting process [28, 35]. In this model, only the melting process at the barrel wall is considered. Assuming a temperature-dependent viscosity and a negligible Brinkman number, the

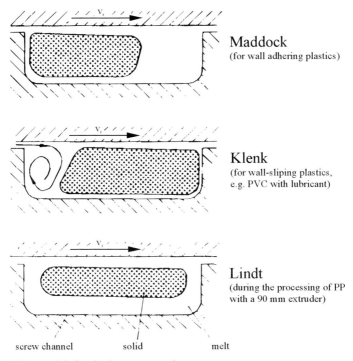

Maddock
(for wall adhering plastics)

Klenk
(for wall-sliping plastics,
e.g. PVC with lubricant)

Lindt
(during the processing of PP
with a 90 mm extruder)

screw channel   solid   melt

Figure 5.14   Melting models for single screw extruders

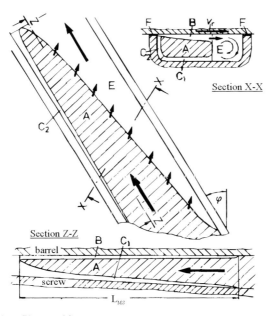

Section X-X

Section Z-Z

barrel

screw

L_{MS}

Figure 5.15   Solid bed profile at melting

proportionality $\delta \sim \sqrt{x}$ has been determined for the melt film thickness δ depending on the point x. The essential equations are referenced in Rauwendaal [32]. Since in case of power law melts iterations are not necessary in the Chung model, melting calculations can be performed very quickly.

Starting in the middle of the 1980s, Potente and his staff [33, 34, 40, 42–44] developed a melting model for power law melts on the basis of the Tadmor model. This model considers the premelting and melting as well as the acceleration of the solid bed in the compression zone. It allows the computation of the place of the formation of the melt vortex. Here we will confine ourselves to this model. We first deal with the melting and then with the premelting (Fig. 5.16).

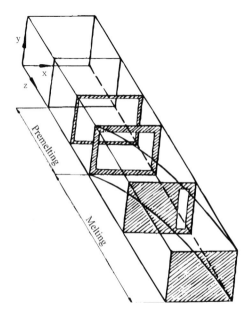

Figure 5.16   Premelting and melting section

5.2.2   Melting

5.2.2.1   Requirements for the Mathematical Treatment

1. The melt film at the screw (Fig. 5.14) has no considerable influence on the increase of the melt pool.
2. At the barrel wall there is a melt film that follows the function

$$\delta = \delta_0 \left(\frac{x}{W} \right)^c \qquad (5.145)$$

the exponent c being unknown in the beginning.

3. A fully developed temperature profile with regard to the barrel is available in the solid bed; i.e., the temperature profile does not change in screw channel direction (z direction, Fig. 5.17).

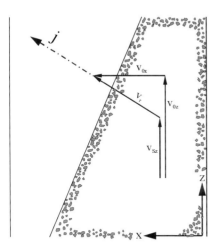

$$V_r = \left[(V_{0z} - V_{Sz})^2 + V_{0x}^2 \right]^{\frac{1}{2}}$$

Figure 5.17   Scheme for the explanation of the velocity components in the melt film at the barrel

Then, the energy Eq. is

$$-\frac{v_{Sy}}{a_S}\frac{\partial T}{\partial y} - \frac{\partial^2 T}{\partial y^2} = 0 \tag{5.146}$$

and its solution is

$$T - T_S = \left(T_{Fl} - T_S \right) e^{-\frac{v_{Sy} y}{a_S}} \tag{5.147}$$

where T_{Fl} is the flow temperature, T_S the solid bed temperature in the center of the solid bed, a_S the thermal diffusivity of the solid, and v_{Sy} the melting velocity in y direction. The heat flow $\dot{q}_S$ at $y = 0$ is

$$\dot{q}_{S(y=0)} - \frac{\lambda_S \left(T_{Fl} - T_S \right)}{a_S} v_{Sy} = \rho_S v_{Sy} \Delta h_S \tag{5.148}$$

where λ_S is the heat conductivity, ρ_S the solids density, and Δh_S the enthalpy increase of the solid.

4. A fully developed temperature profile is available at the barrel wall in the melt film; i.e., the temperature is only a function of the y coordinate. On this assumption, the melt film thickness Eq. 5.145 is no function of the z coordinate; i.e., the melt film thickness does not change in channel direction at the point x.

5. There is a mere drag flow at the barrel wall in the melt film.
6. A power law melt is presumed:

$$\tau = K\left(T_{Fl}\right)\dot{\gamma}^{n}e^{-\beta\left(T_{M}-T_{Fl}\right)}$$

(5.149)

7. The viscosity is temperature-dependent and follows a linear temperature profile in the melt film at the barrel wall:

$$T - T_{Fl} = \left(T_{b} - T_{Fl}\right)\frac{y}{\delta}$$

(5.150)

8. The barrel wall temperature T_{b} is always higher than the flow temperature T_{Fl}.
9. The melt does not slip at the wall.

The melting model in Fig. 5.18a is replaced by the melting model in Fig. 5.18b; i.e., an average melt film thickness is used for computation:

$$\bar{\delta} = \frac{1}{X}\int_{0}^{X}\delta(x)dx = \frac{\delta(X)}{1+c}$$

(5.151)

5.2.2.2   Mathematical Treatment Regardless of the Leakage Flow Across the Screw Flight

5.2.2.2.1   Velocity Profile in the Melt Film at the Barrel Wall (Fig. 5.18)

The Eq. of motion is

$$\frac{\partial \tau_{yj}}{\partial y} = 0$$

(5.152)

i.e.,

$$\tau_{yj} = c_{1}$$

(5.153)

With Eq. 5.149 yields the following for the velocity gradient in direction of the relative velocity:

$$\frac{\partial v_{j}}{\partial y} = \left[\frac{c_{1}}{K\left(T_{Fl}\right)}\right]^{1/n}e^{\left(\beta/n\right)\left(T-T_{Fl}\right)}$$

(5.154a)

With the temperature profile Eq. 5.150 assumed for the viscosity one gets

$$\frac{\partial v_{j}}{\partial \xi} = \bar{\delta}c_{2}e^{A\xi}$$

(5.154b)

with

$$A = \frac{\beta}{n}\left(T_{b} - T_{Fl}\right)$$

(5.155)

$$c_2 = \left[\frac{c_1}{K(T_{Fl})} \right]^{1/n} \tag{5.156}$$

$$\xi = \frac{y}{\overline{\delta}} \tag{5.157}$$

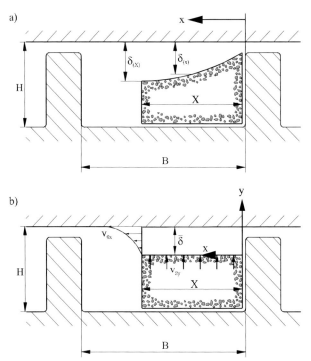

Figure 5.18   Melting model

The integration of Eq. 5.154b yields

$$v_j = \frac{\overline{\delta}\, c_2}{A} e^{A\xi} + c_3 \tag{5.158}$$

Where the constants c_2 and c_3 are determined by means of the boundary conditions $v_j(0) = 0$; $v_j(1) = v_r$. The result is

$$c_2 = \frac{A v_r}{\overline{\delta}\left(e^A - 1\right)} \tag{5.159}$$

$$c_3 = -\frac{\overline{\delta} c_2}{A} \tag{5.160}$$

By substitutions in Eq. 5.158 the Eq. for the velocity profile in direction of the relative velocity Fig. 5.14 is

$$\frac{v_j}{v_r} = \frac{e^{A\xi} - 1}{e^A - 1} \tag{5.161}$$

and analoguously in x direction

$$\frac{v_x}{v_{0x}} = \frac{e^{A\xi} - 1}{e^A - 1} \tag{5.162}$$

Furthermore, for the shear rate in direction of the relative velocity we have

$$\frac{\partial v_j}{\partial y} = \frac{v_r A e^{A\xi}}{\overline{\delta}\left(e^A - 1\right)} \tag{5.163}$$

5.2.2.2.2   Melting Velocity

To determine the melting velocity v_{Sy}, the volume flow that enters the melt pool from the melt film at the barrel wall must be determined first. We obtain

$$\dot{V}_x = \overline{\delta}\int_0^1 v_x d\xi = \frac{\overline{\delta}v_{0x}}{1 - e^A}\int_0^1 \left(1 - e^{A\xi}\right) d\xi \tag{5.164}$$

The integration yields

$$\dot{V}_x = \frac{\overline{\delta}v_{0x}}{2} k_1 \tag{5.165}$$

with

$$k_1 = 2\left(\frac{1}{1 - e^A} + \frac{1}{A}\right) \tag{5.166}$$

Due to the continuity Eq. we can say

$$\frac{1}{2}k_1 \rho_M v_{0x}\overline{\delta} = \rho_S X \overline{v}_{Sy} \tag{5.167}$$

The average melting velocity results in

$$\overline{v}_{Sy} = \frac{k_1}{2}\frac{\rho_M}{\rho_S}v_{0x}\frac{\overline{\delta}}{X} \tag{5.168}$$

5.2.2.2.3   Temperature Profile in the Melt Film at the Barrel Wall

Because of the conditions, the energy Eq. for the melt film at the barrel wall is

$$\lambda_M \frac{\partial^2 T}{\partial y^2} + \tau_{yj} \frac{\partial v_j}{\partial y} = 0 \tag{5.169}$$

With the boundary conditions $T(0) = T_{Fl};\; T(\overline{\delta}) = T_b$ the following Eq. is yielded for the temperature:

$$\frac{T - T_{Fl}}{T_b - T_{Fl}} = \xi + Br \left\{ \frac{1}{A^2} \left(\frac{A}{e^A - 1} \right)^{1+n} \left[1 - e^{A\xi} - \xi \left(1 - e^A \right) \right] \right\} \tag{5.170}$$

with

$$Br = \frac{K \left(T_{Fl} \right) v_r^{1+n} \overline{\delta}^{1-n}}{\lambda_M \left(T_b - T_{Fl} \right)} \tag{5.171}$$

5.2.2.2.4   Heat Flows at the Boundary Layer Between the Melt film and the Solid Bed

For the heat flows at the boundary layers between the melt film and the solid we obtain

$$\dot{q}_{M(y=0)} = \dot{q}_{S(y=0)} + \rho_S \overline{v}_{Sy}\, \Delta h_{Mel} \tag{5.172}$$

the last term being the melting rate. With

$$\dot{q}_{M(y=0)} = +\lambda_M \left. \frac{dT}{dy} \right|_{y=0} \tag{5.173}$$

Eqs. 5.170, 5.148, and 5.168, Eq. 5.172 is

$$2 + k_2\, Br = \frac{\rho_M k_1 v_{0x}\, \Delta h}{\lambda_M \left(T_b - T_{Fl} \right)} \frac{\overline{\delta}^2}{X} \tag{5.174}$$

with

$$k_2 = \frac{2}{A^2} \left(\frac{A}{e^A - 1} \right)^{1+n} \left(e^A - A - 1 \right) \tag{5.175}$$

When introducing the following normalizations or abbreviations, respectively,

$$\psi = \frac{\delta}{\delta_0} = \frac{\overline{\delta}}{\overline{\delta}_0} \qquad y = \frac{X}{W} \tag{5.176}$$

$$Br_{01} = k_2 Br_0 = \frac{k_2 v_r^{1+n} \overline{\delta}_0^{1-n} K(T_{Fl})}{\lambda_M (T_b - T_{Fl})} \tag{5.177}$$

$$\pi_0 = \frac{\rho_M k_1 v_{0x} \Delta h \overline{\delta}_0^2}{\lambda_M (T_b - T_{Fl}) W} \tag{5.178}$$

Eq. 5.174 is

$$2 + Br_{01} \psi^{1-n} = \pi_0 \frac{\psi^2}{y} \tag{5.179}$$

The differences indicated in Fig. 5.19 have to be substituted as enthalpy difference Δh.

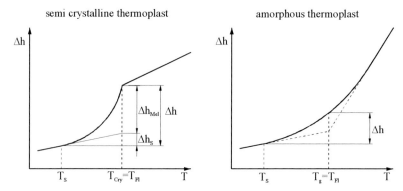

Figure 5.19   Enthalpy differences for semi-crystalline and amorphous materials

5.2.2.2.5   Determination of the Exponent c and the Melt Film Thickness δ_0

For

$$\psi = y = 1 \tag{5.180a}$$

Eq. 5.179 is

$$2 + Br_{01} = \pi_0 \tag{5.180b}$$

By dividing the Eq. 5.179 by Eq. 5.180b we obtain

$$y = \left[\frac{2 + Br_{01}}{2 + Br_{01} \psi^{1-n}} \right] \psi^2 \tag{5.181}$$

As becomes obvious in Fig. 5.20, where Eq. 5.181 is plotted in double logarithmic representation, Eq. 5.181 can be respresented with a quite exact approximation through the relation

$$\psi = y^c \tag{5.182}$$

with

$$\frac{1}{2} \leq c \leq \frac{1}{1+n} \tag{5.183}$$

The exponent c and δ_0 results from

$$c = \frac{\lg(\psi_1/\psi_2)}{\lg(y_1/y_2)} \qquad \delta_0 = \delta_1 y_1^{-c} \tag{5.184}$$

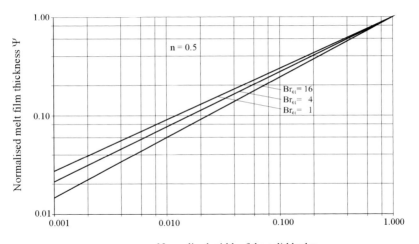

Figure 5.20   Normalized melt film thickness depending on the normalised solid bed width

5.2.2.2.6   Solid Bed Profile and Melting Length Regardless of the Melting Film at the Screw

The following differential Eq. can be yielded from the mass balance for the solid area (Fig. 5.21), neglecting the leakage flow $\dot{m}_{cl}$:

$$-\frac{d}{dz}(\rho_S v_{Sz} X H_S) = \frac{1}{2} \rho_M v_{0x} k_1 \delta \tag{5.185}$$

where the left side of the Eq. represents the change of the density, the solid bed velocity, and the width and height of the solid bed in channel direction; the right side shows the melted material entering the melt pool.

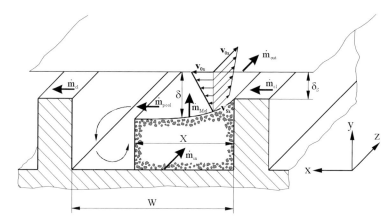

Figure 5.21   Melt flows at melting

With the following normalization

$$\psi = \frac{\delta}{\delta_0} \qquad u = \frac{H_S}{H_{S0}} = 1 - \beta\zeta \qquad y = \frac{X}{W} \tag{5.186}$$

$$\zeta = \frac{z}{D} \qquad w = \frac{\rho_S v_S}{\rho_{S0} v_{S0}} \tag{5.187}$$

Eq. 5.185 yields

$$-\frac{d}{d\zeta}(uwy) = \pi_1 \psi \tag{5.188}$$

with

$$\pi_1 = \frac{\rho_M v_{0x} k_1 \delta_0 D}{2\dot{m}} \tag{5.189}$$

with

$$\psi = y^c \tag{5.190}$$

the DE will be

$$-\frac{d}{d\zeta}(uwy) = \pi_1 y^c \tag{5.191}$$

The solutions are listed in Tables 5.5 and 5.6.

Table 5.5   Solutions of the Solid Bed DE, No Leakage Flow

Conditions :	$u = const.$ $a = 0$ $\Psi_{cl} = 0$

$$y = \left[1 - (1-c)\pi_1\zeta\right]^{1/(1-c)}$$

$$\zeta_1 = \frac{1}{\pi_1(1-c)}$$

Conditions :	$u = 1 - \beta\zeta$ $a = 0$ $\Psi_{cl} = 0$

$$y = \left[\frac{\pi_1}{\beta} + \left(1 - \frac{\pi_1}{\beta}\right)\frac{1}{(1-\beta\zeta)^{1-c}}\right]^{1/(1-c)}$$

$$\zeta_1 = \frac{1}{\beta}\left[1 - \left(\frac{\pi_1 - \beta}{\pi_1}\right)^{1/(1-c)}\right]$$

Table 5.6   Equations for the Computation of the Solid Bed Profile in the Melting Zone of a Three-Section Screw, No Leakage Flow

Section I Differentialeq.	$-\dfrac{\partial y}{\partial \zeta_0} = \pi_1 y^c$; $\pi_1 = \dfrac{\rho_B k_1 v_{0x}\delta_0 D}{2\dot{m}}$
Solution	$y = \left[1 - (1-c)\pi_1\zeta_0\right]^{1/(1-c)}$ $\Psi = \left[1 - (1-c)\pi_1\zeta_0\right]^{c/(1-c)}$
Section II Differentialeq.	$-\dfrac{d(uy)}{d\zeta_1} = \pi_1\Psi_1 y^c$ $\Psi_1 = \dfrac{\delta_1}{\delta_0}$
Solution	$y = \left[\dfrac{\pi_1\Psi_1}{\beta} + \left(y_1^{1-c} - \dfrac{\pi_1\Psi_1}{\beta}\right)\dfrac{1}{(1-\beta\zeta_1)^{1-c}}\right]^{1/1-c}$ $\Psi = \left[\dfrac{\pi_1\Psi_1}{\beta} + \left(y_1^{1-c} - \dfrac{\pi_1\Psi_1}{\beta}\right)\dfrac{1}{(1-\beta\zeta_1)^{1-c}}\right]^{c/1-c}$
Section III Differentialeq.	$-\dfrac{dy}{d\zeta_2} = \pi_1\dfrac{\Psi_2}{u_2}y^c$ $u_2 = \dfrac{H_1}{H_2}$ $\Psi_2 = \dfrac{\delta_2}{\delta_0}$
Solution	$y = \left[y_2^{1-c} - (1-c)\pi_1\dfrac{\Psi_2}{u_2}\zeta_2\right]^{1/(1-c)}$ $\Psi = \left[y_2^{1-c} - (1-c)\pi_1\dfrac{\Psi_2}{u_2}\zeta_2\right]^{c/(1-c)}$

5.2.2.3   Mathematical Treatment in Consideration of the Leakage Flow Across the Screw Flight

The influence of the leakage flow across the screw flight only shows in the mass balance. The solution is [*]

$$\frac{X}{W} = \frac{\dfrac{k_1}{(1+c)^2} \dfrac{\rho_M v_{0x} \Delta h}{\lambda_M (T_b - T_{Fl}) W}}{2 + \dfrac{k_2 K(T_{Fl}) v_r^{1+n}}{\lambda_M (T_b - T_{Fl})} \left(\dfrac{\delta}{1+c}\right)^{1-n}} \left(1 - \frac{\bar{\delta}_{cl}}{\bar{\delta}}\right) \delta^2 \tag{5.192}$$

and written as Eq. 5.181:

$$y = \left(\frac{2 + Br_{01}}{2 + Br_{01}\psi^{1-n}}\right)\left(\frac{\psi - \psi_{cl}}{1 - \psi_{cl}}\right)\psi \tag{5.193}$$

Where $\bar{\delta}_{cl}$ is the mean clearance or ψ_{cl} is the mean clearance normalized with $\bar{\delta}_0$. When plotting $(\psi - \psi_{cl})(1 - \psi_{cl})$ in a double logarithmic representation versus y, well approximated lines are obtained, which can be further approximated by

$$\frac{\psi - \psi_{cl}}{1 - \psi_{cl}} = y^{c'} \tag{5.194}$$

The following will then apply to the exponent c':

$$c' = \frac{\left(\dfrac{\psi_1 - \psi_{cl}}{1 - \psi_{cl}}\right) \Big/ \left(\dfrac{\psi_2 - \psi_{cl}}{1 - \psi_{cl}}\right)}{\lg(y_1/y_2)} \tag{5.195}$$

and to the initial melt film thickness:

$$\delta'_0 = (\delta_1 - \delta_{cl})y_1^{-c'} + \delta_{cl} \tag{5.196}$$

The mass balance for the solid area (Fig. 5.22) yields the following differential equation:

$$-\frac{d}{dz}(\rho_S v_{Sz} X H_S) = \frac{1}{2}\rho_M v_{0z} k_1(\delta - \delta_{cl}) \tag{5.197}$$

---

[*] Note: It has been assumed that the clearance profile at the screw flight corresponds with that of the melt film profile.

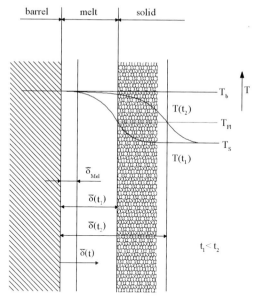

Figure 5.22   Melt film thickness and temperature profile for time t_1 and t_2

With the following normalizations

$$\psi = \frac{\delta}{\delta_0} = \frac{\overline{\delta}}{\overline{\delta}_0} \qquad \psi_S = \frac{\delta_{cl}}{\delta_0} = \frac{\overline{\delta}_{cl}}{\overline{\delta}_0} \qquad u = \frac{H_S}{H_{S0}} = 1 - \beta\zeta \qquad y = \frac{X}{W} \qquad (5.198)$$

$$\zeta = \frac{z}{D} \qquad w = \frac{\rho_S v_S}{\rho_{S0} v_{S0}} \qquad (5.199)$$

Eq. 5.197 yields

$$-\frac{d}{d\zeta}\left(uwy\right) = \pi_1\left(\psi - \psi_{cl}\right) \qquad (5.200)$$

with

$$\pi_1 = \frac{\rho_M v_{0x} k_1 \delta_0 D}{2\dot{m}} \qquad (5.201)$$

and

$$\left(\psi - \psi_{cl}\right) = \left(1 - \psi_{cl}\right)y^{c'} \qquad (5.202)$$

the DE will be

$$-\frac{d}{d\zeta}\left(uwy\right) = \pi_1\left(1 - \Psi_{cl}\right)y^{c'} \qquad (5.203)$$

For w we introduce

$$w = u^{-a} \tag{5.204}$$

Where a is a measure for the acceleration or deceleration of the solid bed in the compression zone.

We can say that

$$a < 0 \text{ solid bed compression (deceleration)} \tag{5.205}$$

$$a = 0 \qquad w = 1 \tag{5.206}$$

$$a = 1 \qquad w \sim \frac{H_{S0}}{H_S} \tag{5.207}$$

It can be shown that the differential Eq. (5.203) in its explicit form is always a Bernoulli Eq. and can thus be solved. The initial condition

$$\zeta = 0 \qquad \psi = \psi_1 \qquad y = y_1 \tag{5.208}$$

gives

$$y = \left\{ \left[\frac{\psi_1(1-c')\pi_1(1-\psi_{cl})}{\beta[(1-c')(1-a)+a]} \right] u^a + \left[\left(y_1^{1-c'} - \frac{\psi_1(1-c')\pi_1(1-\psi_{cl})}{\beta[(1-c')(1-a)+a]} \right) \frac{1}{u^{(1-c')(1-a)}} \right] \right\} \frac{1}{1-c'} \tag{5.209}$$

This is the most general form of the melting Eq. for a continuous process.

5.2.2.4   Mathematical Treatment of Melting During the Screw Downtime at Injection Molding

For the downtime of the screw, the melting can exclusively be explained by the mere heat conduction at the barrel wall to the solid bed. The melting at the root surface and at the screw flight is neglected. For the computation of the melting during the downtime, we refer to references [42, 45].

During the downtime phase of a cycle, the solid is continuously melted. The average thickness of the melt film $\bar{\delta}$, which separates the barrel wall from the solid, increases with time by the accumulation of the newly melted material. The heat that is responsible for the melting is transported from the hot barrel wall through an increasing melt layer to the surface of the solid bed. This heat transport problem corresponds with the classic Neumann problem of heat conduction with a moving boundary and a change of the physical state. The increasing melt layer thicknesses and the related coordinate displacement as well as the temperature profiles for the time t_1 and t_2 are represented in Fig. 5.22.

The average melt film thickness $\overline{\delta}$ can be determined as a function of time [42, 45]

$$\overline{\delta} = K \sqrt{t}$$

(5.210)

where K must be determined by iterations by

$$-\frac{1}{2}\rho_S \Delta h_{Mel} K = \frac{\left(T_{Fl} - T_b\right)\lambda(T)\exp\left(\left(K\frac{\rho_S}{\rho(T)}\right)^2 \middle/ 4a_M\right)}{\sqrt{\pi a_M}\, erf\left(K\frac{\rho_S}{\rho(T)} \middle/ 2\sqrt{a_M}\right)} - \frac{\left(T_S - T_{Fl}\right)\lambda_S \exp\left(\dfrac{-K^2}{4a_S}\right)}{\sqrt{\pi a_M}\left[1 - erf\left(\dfrac{K}{2\sqrt{a_M}}\right)\right]}$$

(5.211)

Equation (5.210) is applicable only if the initial melt thickness is $\overline{\delta} = 0$. Actually, the downtime phase starts with the average film thickness $\overline{\delta}_{Mel}$ that results from the melting profile of the plasticating phase. Despite this, for reasons of simplification we assume in analogy to [45] that the melting during the downtime takes place according to Eq. 5.210. After the downtime t_{st} we can say for the melt film thickness, considering $\overline{\delta}_{Mel}$ for the total melt film thickness,

$$\overline{\delta} = K\sqrt{t_{st} + \left(\frac{\overline{\delta}_{Mel}}{K}\right)^2}$$

(5.212)

Fig. 5.23 shows the solid bed width at the end of the rotation phase (a), at the end of the downtime phase (b), and at the beginning of the new rotation phase (c). The crucial assumption in the model is that the average melt film thicknesses $\overline{\delta}$ at the end of the rotation and at the beginning of the new rotation phase are identical. The reason is a rearrangement of the solid bed. For the new solid bed width we obtain

$$X_{st} = X_{plast}\frac{H - \overline{\delta}}{H - \overline{\overline{\delta}}_{Mel}}$$

(5.213)

Thus, the conditions of a continuous process at any point are also applicable at the beginning of a metering phase; i.e., it is possible to use, e.g., Eq. 5.209.

5.2.2.5   Computation of the Melted Material

When considering barrier screws, it is frequently insufficient to refer only to the profile of the solid bed width as a measure of the melting behavior. Among other things, the aim of a barrier zone design is to keep the solid bed as wide as possible (in the ideal case ($X/W_s = 1$). In this case, a decrease of the solid bed width can rarely be stated so that the value of such a statement is not satisfactory. It is far more revealing to consider the profile of the melted

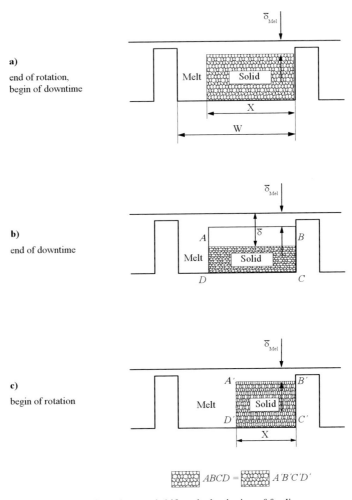

Figure 5.23   Melting in the downtime phase and shift at the beginning of feeding

material $\dot{m}_M$ along the screw channel, which shall be computed in the following. Proceeding from the differential Eq. for the melting (Eq. 5.203) while neglecting the leakage flow and in consideration of the change of the solid bed channel width we obtain

$$-\frac{d}{d\zeta}\left(uvwy\right)=\pi_1\psi \tag{5.214}$$

with

$$v=\frac{W\left(\zeta\right)}{W_{S0}} \tag{5.215}$$

The integration of Eq. 5.214 under the condition of Eq. 5.182

$$-\int_0^{\zeta} d(uvwy) = \pi_1 \int_0^{\zeta} y^c d\zeta \tag{5.216}$$

gives a solution for the left side:

$$1 - uvwy = \pi_1 \int_0^{\zeta} y^c \, d\zeta \tag{5.217}$$

The solution of the integral on the right side of Eq. 5.216 results from the melt flow that melts in a differential section $d\zeta$ in the solid bed:

$$\dot{m}_M(\zeta) = \frac{1}{2} \rho_M v_{0x} k_2 D \delta_0 \int_0^{\zeta} y^c \, d\zeta \tag{5.218}$$

With Eq. 5.218 and Eq. 5.217 we obtain for the melted material

$$\dot{m}_M(\zeta) = \dot{m}(1 - uvwy) \tag{5.219}$$

Assuming an invariant solid flow per unit area $(w = const)$ and a constant solid channel width $(v = const)$, Eq. 5.219 is simplified:

$$\dot{m}_M(\zeta) = \dot{m}(1 - uy) \tag{5.220}$$

5.2.2.6   Premelting

Only the extrusion process is considered. We proceed from the model shown in Fig. 5.24. At first, the porosity of the solid bed is neglected. In this case, the differential Eq. describing the heating process of the solid bed is

$$\frac{v_{Sz} \partial T}{a_S \partial z} - \frac{\overline{v}_{Sy} \partial T}{a_S \partial y'} - \frac{\partial^2 T}{\partial y'^2} = 0 \tag{5.221}$$

Under the initial and boundary conditions

$$z = 0 \qquad y' = 0 \qquad T = T_S \tag{5.222}$$

in y' direction of semi-inifinite space

$$y' = 0 \qquad x > 0 \qquad T = T_{Fl} \tag{5.223}$$

the following expression is yielded for the heat flow at the point $y' = 0$:

$$\dot{q}_{S(y'=0)} = +\frac{\overline{v}_{Sy}}{2}\rho_S \Delta h_S \left[1 + erf\left(\frac{\overline{v}_{Sy}}{2}\right)\sqrt{\frac{Z}{a_S \overline{v}_{Sz}}}\right] + \frac{\lambda_S(T_{Fl}-T_S)}{\sqrt{\frac{\pi a_S Z}{v_{Sz}}}} e^{\left[-\left(\frac{\overline{v}_{Sy}}{2}\sqrt{\frac{Z}{a_S \overline{v}_{Sy}}}\right)^2\right]} \quad (5.224)$$

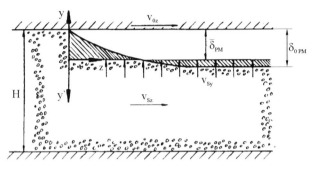

Figure 5.24   Premelting model

With the surface porosity (Fig. 5.25)

$$\varepsilon = 1 - \varepsilon_S \quad (5.225)$$

and

$$\overline{v}_{Sy} = \frac{1}{2}k_1 \frac{\rho_M}{\rho_S}\left(\frac{v_{0z}}{v_{Sz}}-1\right)\frac{v_{Sz}\overline{\delta}_{PM}}{\varepsilon_S Z} \quad (5.226)$$

the following expression is yielded for the heat flow $\dot{q}_S$:

$$\dot{q}_{S(y=0)} = \frac{\lambda_S(T_{Fl}-T_S)}{\sqrt{\pi a_S}}\sqrt{\frac{v_{Sz}}{\varepsilon_S Z}}$$

$$\exp\left\{-\left[\frac{k_1}{4}\frac{\rho_M}{\rho_S}\left(\frac{v_{0z}}{v_{Sz}}-1\right)\sqrt{\frac{v_{Sz}}{a_S \varepsilon_S Z}}\overline{\delta}_{PM}\right]^2\right\} + \frac{k_1}{4}\rho_M\left(\frac{v_{0z}}{v_{Sz}}-1\right) \quad (5.227)$$

$$\Delta h_S \frac{\overline{\delta}_{PM}v_{Sz}}{\varepsilon_S Z}\left\{1 + erf\left[\frac{k_1}{4}\frac{\rho_M}{\rho_S}\left(\frac{v_{0z}}{v_{Sz}}-1\right)\sqrt{\frac{v_{Sz}}{a_S \varepsilon_S Z}}\overline{\delta}_{PM}\right]\right\}$$

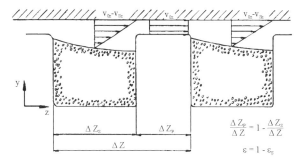

Figure 5.25   Premelting model in consideration of the porosity

Due to the differentiation of Eq. 5.170 we obtain for the melt film according to Section 5.2.2.2

$$\dot{q}_{M(y=0)} = \frac{\lambda_M \left(T_b - T_{Fl}\right)}{2\delta_{PM}}\left[2 + \frac{k_2 K\left(T_{Fl}\right)v_r^{1+n}\overline{\delta}_{PM}^{1-n}}{\lambda_M \left(T_b - T_{Fl}\right)}\right] \tag{5.228}$$

Thus, the following applies to the total energy balance at the boundary layer between the melt film and the solid:

$$\dot{q}_{M(y=0)} = \dot{q}_{S(y=0)} + \rho_S \overline{v}_{Sy} \Delta h_{Mel} \tag{5.229}$$

At the end of the premelting zone, the melt film thickness

$$\delta_{0,PM} = \left(1+c\right)\overline{\delta}_{PM} \tag{5.230}$$

can be stated (Fig. 5.24).

This film thickness changes into a melt film thickness depending on the x coordinate with the formation of the melt pool, so that (Fig. 5.26)

$$\delta_{0,PM} = \frac{\delta_0}{1+c} \tag{5.231}$$

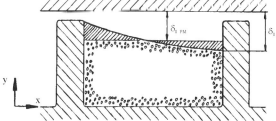

Figure 5.26   Scheme for the computation of $\delta_{0,PM}$

For the average melt film thickness we thus can say

$$\overline{\delta}_{PM} = \frac{\delta_0}{(1+c)^2} \tag{5.232}$$

This expression has to be substituted in Eqs. 5.226 to 5.228. Presupposing that the same exponent c is applicable to the premelting and the melting, the premelting length z can be computed by iterations from Eqs. 5.226 to 5.232.

5.2.2.7   Approaches for the Computation of the Melt Films at the Screw

So far, the melting process has been considered with regard to the melt film at the barrel wall. We will now consider the melt film at the screw. For this, we make the following assumptions:

1. The place of the melt film formation at the root surface is identical with the beginning of the premelting (Fig. 5.27).
2. The relative velocity is equal to the solid bed velocity v_{Sz} in z direction (Fig. 5.27).
3. The melting velocity in the boundary layer of solid/melt film is equal at the screw flight and at the root surface.
4. The mold exponent c for the barrel melt film thickness corresponds with that describing the melt film thickness at the root surface in longitudinal channel direction.

These assumptions yield the following relation for the melt film thickness δ_0, epending on Z_{Mel} (Fig. 5.27):

$$1 = \frac{\dfrac{k_1^*}{(1+c)^2} \dfrac{\rho_M v_{Sz} \Delta h}{\lambda_S (T_s - T_{Fl})} \dfrac{\delta_0^2}{Z_{Mel}}}{2 + \dfrac{k_2^*}{(1+c)^{1-n}} \dfrac{K(T_{Fl}) v_{Sz}^{1+n} \delta_0^{1-n}}{\lambda_M (T_s - T_{Fl})}} \tag{5.233}$$

This is identical with the Eq.s for the melt film at the barrel wall. For v_{0x} and v_r only v_{Sz} has to be substituted and the sign for A in the k factors has to be changed so that

$$k_1^* = 2\left(\frac{1}{1-e^{-A}}\right) - \frac{1}{A} \tag{5.234}$$

$$k_2^* = \frac{2}{A^2}\left(\frac{A}{1-e^{-A}}\right)^{1+n} \left(e^{-A} + A - 1\right) \tag{5.235}$$

The melt film at the screw reduces again the solid bed length.

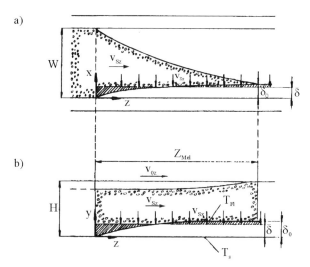

Figure 5.27   Model for the melt film at the screw a) screw flight b) root surface

5.2.3   Pressure-Throughput Behavior and Power Comsumption in the Melting Zone

Fig. 5.28 shows a scheme of the pressure profile in the melt films and in the melt pool by means of a volume element. It must be known for a correct pressure-throughput computation and thus for a coupling of the function zones (solid bed transport, melting, melt transport). However, three-dimensional finite element computations have shown that there are only slight differences between the pressure gradients across the screw channel length, in the melt films and in the melt pool [46]. Therefore, an average dimensionless pressure gradient in channel direction is presumed.

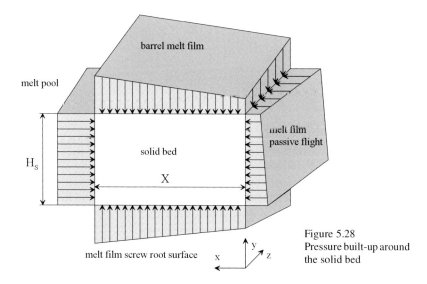

Figure 5.28
Pressure built-up around the solid bed

The total volume flow results from the sum of the volume flows in the melt films and in the melt pool as well as the solid bed melt flow normalized with the melt density. We obtain

$$\dot{V} = \sum_{i=1}^{4} \dot{V}_{iM} + \dot{V}_S \frac{\rho_S}{\rho_M} \qquad (5.236)$$

with

$$\dot{V}_S = XH_S v_{Sz} = X\left(H - \overline{\delta} - s\right)v_{Sz} \qquad (5.237)$$

where the volume flows in the corners of the screw channel (transition to the single sections) have been neglected due to the velocity conditions.

First, we consider a one-dimensional Newtonian flow with
Large X/W ratio ($X/W \geq 0.4$):

$$\pi_{\dot{V}} = \frac{2\dot{V}}{WHv_{0z}} = \frac{\overline{\delta}X}{WH} + \frac{v_{Sz}}{v_{0z}}\left(\frac{X}{W} + \frac{H_S}{H}\right) - 2\frac{X}{W}\frac{H_S}{H}\frac{v_{Sz}}{v_{0z}}\left(1 - \frac{\rho_S}{\rho_M}\right)$$
$$- \pi_p' \frac{1}{WH^3}\left[X\left(\overline{\delta}^3 - s^3\right) + H_S s^3 + H_S\left(W - X - s\right)^3\right] \qquad (5.238)$$

Small X/W ratio ($X/W < 0.4$):

$$\pi_{\dot{V}} = 1 + \frac{\overline{\delta}X}{WH} - \frac{X}{W} - \frac{s}{W} + \frac{v_{Sz}}{v_{0z}}\left[\frac{HX + H_S\left(s + X\right)}{WH}\right]$$
$$- 2\frac{X}{W}\frac{H_S}{H}\frac{v_{Sz}}{v_{0z}}\left(1 - \frac{\rho_S}{\rho_M}\right)$$
$$- \pi_p' \frac{1}{WH^3}\left[X\left(\overline{\delta}^3 - s^3\right) + H_S s^3 + H^3\left(W - X - s\right)\right] \qquad (5.239)$$

with

$$\pi_p' = \frac{H^2}{6Kv_{0z}}\frac{dp}{dz} \qquad (5.240)$$

In contrast to the mere one-dimensional, Newtonian melt flow, a changed dimensionless volume flow with a dimensionless pressure gradient of $\pi_p = 0$ is yielded, depending on the solid bed geometry and velocity (Fig. 5.29). Furthermore, the slope of the created line changes, which can be explained by the changed velocity profile as becomes obvious in Fig. 5.30. The line for the mere melt transport and the throughput line intersect. On the right of the intersection point, the throughput in the melting section is greater; on the left it is less. For $X = 0$, $s = 0$ we obtain the Eq. for the mere melt transport. Fig. 5.31 shows the curves of a one-dimensional power law flow.

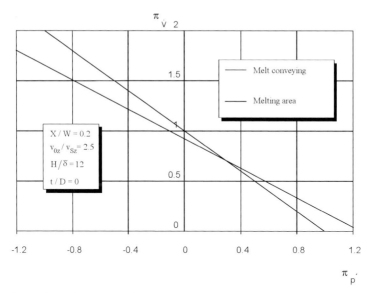

Figure 5.29   Comparison of the pressure/mass throughput behavior for a one-dimensional flow

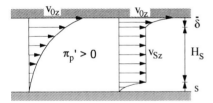

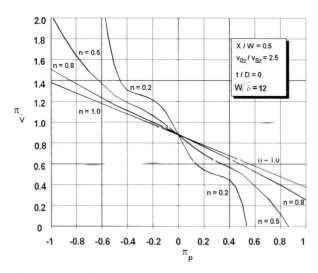

Figure 5.30   Comparison of the velocity profiles for a one-dimensional flow

Figure 5.31   Dimensionless volume throughput as a function of the dimensionless pressure gradient for one operating point during melting (one-dimensional)

In contrast to the channels that are exclusively filled with melt, this correlation between the pressure gradient and the throughput also depends on the arising solid bed velocity and the geometrical conditions (solid bed width, melt film thicknesses) within the melting section. Thus, different dimensionless characteristic curves are yielded for each operating mode.

Now, we will consider a multidimensional flow. In analogy to the dimensionless characteristic pressure/throughput curves with mere melt transport, the intersection point with the Newtonian line shifts at the transition from a one-dimensional flow to a multidimensional flow in direction of negative dimensionless pressure gradients. Furthermore, the additional parameters that describe the melting section, such as the solid bed geometry and velocity cause a change of the pressure/throughput behavior. The influence of the ratio of the solid bed width and the channel width can be recognized in Fig. 5.32.

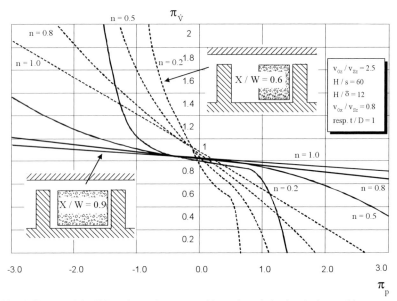

Figure 5.32   Influence of the X/B ratio on the pressure/throughput behavior during melting

An increase of the X/W ratio means a decrease of the melt pool width and thus a decrease of the volume flow with an identical negative pressure gradient; this is due to the changed velocity profile. The adding up of the drag flow and the pressure flow that increases the volume flow is replaced by a solid flow with a lower velocity within the solid bed so that the volume flow is reduced. For positive pressure gradients, the pressure flow counteracts to the drag flow; a greater X/W ratio thus causes an increase of the surface below the velocity profile.

Whereas a change of the dimensionless solid bed width considerably affects the pressure/throughput behavior in the melting section, even a great variation of the melt film thickness ratio at the barrel with regard to the channel height causes only little throughput change. Here, an increasing melt film thickness at the barrel goes along with an increasing volume flow with a constant pressure gradient; this is due to the newly created velocity profile.

An increase of the solid bed velocity and thus a reduction of the ratio of the rotational velocity component in z direction and the solid bed velocity causes an increase of the volume flow with a constant pressure gradient (Fig. 5.33).

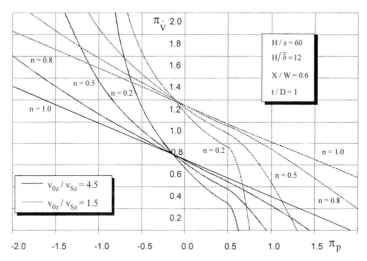

Figure 5.33   Influence of the velocity on the pressure/throughput behavior during melting

All these families of curves can be approximated by a polynomial of 5th order:

$$\pi_p = C_1 + C_2\pi_{\dot{V}} + C_3\pi_{\dot{V}}^2 + C_4\pi_{\dot{V}}^3 + C_5\pi_{\dot{V}}^4 + C_6\pi_{\dot{V}}^5 \tag{5.241}$$

where for the Newtonian case the coefficients C_3 to C_6 are zero. The coefficients have to be determined by a nonlinear regression, where certain boundary conditions must be met. A detailed description of the proceeding can be found in reference [46]. Eq.s for the computation of the power needed in the melting zone are listed in Table 5.7.

Due to the lack of space, it is not possible to deal with the break-up of the solid bed in detail. For this, please refer to [46].

Table 5.7   Power Equations for the Melting Zone

$$\pi_{Power} - X_1 + X_2\pi_{\dot{V}} + X_3\pi_{\dot{V}}^2$$

$$\pi_{\dot{V}} = \frac{2\dot{V}_M}{(W - X)v_{0z}H + (v_{Sz} + v_{0z})\overline{\delta}X}$$

$$\pi_{Power} = \frac{\Delta P H^n}{\Delta z W K(T)v_{0z}^{n+1}}$$

Table 5.7   Continued

$$X_1 = C_1 + Y_1 \frac{X}{W}$$

$$X_2 = C_2 + Y_2 \frac{X}{W} + K_2 \left(\frac{X}{W}\right)^2$$

$$X_3 = C_3 + Y_3 \frac{X}{W}$$

$$C_1 = \left[4 + 4\left(\frac{v_{0x}}{v_{0z}}\right)^2\right] \exp\left[\left(0.408\frac{v_{0x}}{v_{0z}} + 1\right)(n-1)\right]$$

$$C_2 = -3\left[1 + 0.61\frac{H}{W} + (1-n)\left(0.1665\frac{v_{0x}}{v_{0z}} - 0.34\frac{H}{W} - 0.534n\frac{v_{0x}}{v_{0z}} - 0.93n - 0.85\right)\right]$$

$$C_3 = (1-n)\left[0.18 \ln\left(\pi\frac{v_{0x}}{v_{0z}}\right) - 1.3n\right]$$

$$Y_1 = \exp\left\{\left[1.946 - \frac{6.082 + 19.421\frac{v_{0x}}{v_{0z}}}{W/H} + 66.19\frac{v_{0x}}{v_{0z}}\left(\frac{H}{W}\right)^2 - 0.806 \ln\left(\frac{W}{\overline{\delta}}\right)\right](1-n)\right.$$

$$\left. - 0.895(1-n)^2\left\{\left[\frac{1 + 3.889\left(v_{0x}/v_{0z}\right)^2}{\left(\overline{\delta}/H\right)} + \frac{0.11W/H - H/\overline{\delta}}{\left(v_{0z}/v_{Sz}\right)} - 3.358\right]\right.\right.$$

$$Y_2 = 1.6458\left(\frac{\overline{\delta}}{H}\right)^2 + 1.9133n - 4.7948\left(\frac{\overline{\delta}}{H}\right)^2 n^2$$

$$K_2 = -0.9054\frac{v_{Sz}}{v_{0z}} - 36.8785n\frac{\overline{\delta}}{W} + \left[1.231 - 8.8872\frac{\overline{\delta}}{H} - 0.00068\left(\frac{W}{H}\right)^2\right.$$

$$+ \left(16.1875 - 11.329\frac{v_{Sz}}{v_{0z}}\right)\left(\frac{\overline{\delta}}{H}\right)^2\right]n^2$$

$$Y_3 = \left(0.8704 + \frac{4.2292 - 4.015v_{0x}/v_{0z}}{v_{0z}/v_{Sz}} - 6.521\frac{\overline{\delta}}{H}\right)\left(n - n^2\right)$$

5.3   Solid Conveying

5.3.1   Smooth-Barrel Feed Zone

5.3.1.1   Throughput

The screw zones of an extruder are not able to convey more melt than is delivered by the feed zone. There are, however, reactions from these zones to the pressure at the end of the feed zone and thus to the conveying behavior of this zone [47–83].

For the mass flow let (Fig. 5.34),

$$\dot{m} = \rho_B A v_a f \tag{5.242}$$

with

ρ_B = packing depth with incompressible bulk materials considering the boundary influences of the screw channel; we will deal with this in detail in connection with the *grooved-barrel extruder*

A = cross-sectional area of the channel

v_a = axial velocity of the solid element

f = degree of filling

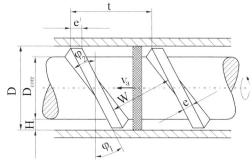

Figure 5.34   Scheme for solid conveying

For A,

$$A = \int_{R_K}^{R_z} \left(2\pi R - \frac{ie}{\sin \overline{\varphi}} \right) dR = \left[\frac{\pi}{4} \left(D_b^2 - D_{core}^2 \right) - \frac{ieH}{\sin \overline{\varphi}} \right] \tag{5.243}$$

With the factors v_a and f we have to distinguish three cases:

Case a: Forced conveying (nut/screw, this does not occur in smooth barrel extruders)

$$f = 1 \qquad v_a = tN = v_0 \tan \varphi_1 \tag{5.244}$$

t = Pitch
N = Speed

Case b: Archimedes conveying (Fig. 5.35)

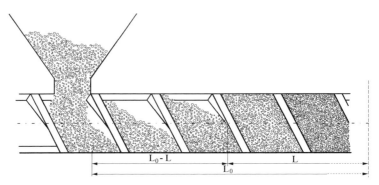

Figure 5.35   Archimedes conveying in the domain $L_0 - L$

$$f < 1 \qquad v_a = tN = v_0 \tan \varphi_1 \tag{5.245}$$

Case c: Frictionally engaged conveying (Fig. 5.36a)

$$f = 1$$

$$v_a = v_0 \frac{\sin \varphi_1 \sin \alpha}{\sin(\alpha + \varphi_1)} = v_0 \frac{\tan \alpha \tan \varphi_1}{\tan \alpha + \tan \varphi_1} \tag{5.246}$$

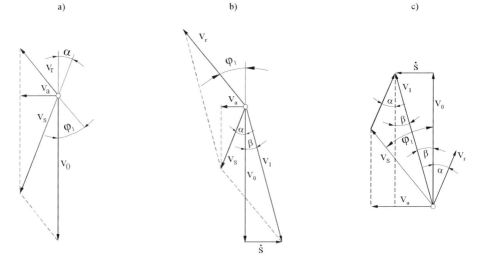

Figure 5.36   Velocity diagrams a) extruder; b) and c) injection molding machine.
Case a) and b): Fixed barrel, moving screw, Case c): Fixed screw, moving barrel

In injection molding machines, there is, in addition, an axial retraction velocity. With

$$\dot{m} = \rho_M \frac{\pi D^2}{4} \dot{s} = \rho_s A v_a \tag{5.247}$$

the following applies:

$$\dot{s} = \frac{4\dot{m}}{\rho_M \pi D^2} \tag{5.248}$$

and thus for the axial solid bed velocity,

$$v_a = \frac{\pi}{4} \frac{\rho_M}{\rho_S} \frac{D^2}{A} \dot{s} \tag{5.249}$$

i.e., the screw retraction velocity and the axial transport velocity of the solid are coupled.
    First, we consider a case where the coordinate system is fixed on the screw. Here, the velocity vector diagram in Fig. 5.35c is relevant. For the axial transport velocity we can say

$$v_a = v_0 \frac{\sin(\varphi_1 - \beta)\sin\alpha}{\sin(\alpha + \varphi_1)\cos\beta} + \dot{s} = v_0 \frac{\sin(\alpha + \beta)\sin\varphi}{\sin(\alpha + \varphi_1)\cos\beta} \tag{5.250}$$

with

$$\tan\beta = \frac{\dot{s}}{v_0}$$

    we get

$$v_a = v_0 \left(\frac{\tan\alpha + \tan\varphi_1}{\tan\alpha \tan\varphi_1} - \frac{4}{\pi} \frac{\rho_S}{\rho_M} \frac{A}{D^2} \frac{1}{\tan\alpha} \right)^{-1} \tag{5.251}$$

for the axial velocity. It is higher than in extrusion
    For the axial solid velocity in a fixed coordinate system, represented in the form of a velocity vector diagram in Fig. 5.35b, we obtain

$$v_{a1} = v_a - \dot{s} = v_0 \frac{\sin(\varphi_1 - \beta)\sin\alpha}{\sin(\alpha + \varphi_1)\cos\beta} \tag{5.252}$$

i.e.

$$v_{a1} = v_0 \left(\frac{\tan\alpha + \tan\varphi_1}{\tan\alpha \tan\varphi_1} + \frac{4}{\pi} \frac{\rho_S}{\rho_M} \frac{A}{D^2} \frac{1}{\tan\alpha} \right)^{-1} \tag{5.253}$$

This Equation is more of academic interest.

Fig. 5.37 shows the normalized axial solid transport velocity v_a/v_0 depending on the angle α for the extrusion and the injection molding, assuming a common screw geometry. There are no serious differences.

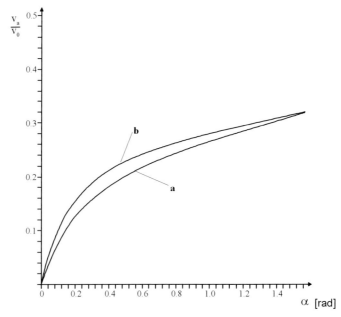

Figure 5.37   Normalized axial solid bed velocity depending on the conveying angle a) extrusion, b) injection molding

In the equations that have been written so far, the angle α is an unknown factor. To determine it, three different models are available. Essentially, they differ in the definition of the additional force F^* (Fig. 5.38). Only Schöppner [78, 80] does without any additional force. He proceeds from the assumption that a force transmission in the system is possible only via the pressure in the bulk material and arguments like this: "Due to the additional force in the known models, the load of the leading edge of flight is greater than that of the trailing edge of flight without the necessary pressure increase having any effect on the assumed force at the root surface and at the barrel wall. Thus, the average value of the pressure distribution does no longer correspond with the pressure at these walls." In a first approach, Schöppner proceeds from the assumption that the trailing edge of flight is relieved to the same degree as the additional load on the leading edge of flight. The average pressure distribution in vertical channel direction is then equal to the average pressure at the root surface and at the barrel wall leaving out the pressure anisotropies.

Schneider :

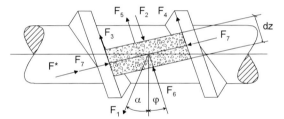

Campbell:

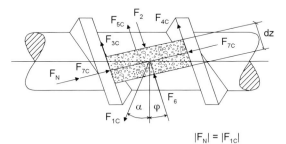

$$|F_N| = |F_{1C}|$$

Hyun, Spalding:

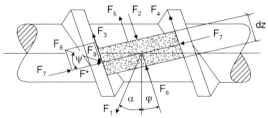

Figure 5.38   Solid conveying models

From the force balance in axial direction

$$\sum F_{ai} = 0 \tag{5.254}$$

and the momentum balance in tangential direction

$$\sum M_{ti} = 0 \tag{5.255}$$

we obtain the differential Eq.[*]

$$\frac{dp}{dz} - \beta p = 0 \tag{5.256}$$

---

[*]   Note: Models by Schneider and Schöppner: $z = \overline{Z}$,models by Campbell and Hyun: $z = z_b$

This indicates that the pressure profile in the channel direction is formally identical for all models. There are, however, differences in the mathematical expression for the coefficient β. For the models by Schneider (1968), Schöppner (1995), and Hyun (1997) the Eq.s for the determination of β are listed in Table 5.8. They differ only in the K expression, at which K is identical in the Schöppner and Hyun model. The Campbell model (1995) has been modified once again by Campbell in 1998. It is not possible, however, to deal with this here.

Table 5.8   Equations for Determining the Coefficient β to Compute the Pressure Profile and the Throughput in the Solid Conveying Zone

$$\beta = \frac{1 - K\Psi}{M_2\sqrt{1+\Psi^2}} - \frac{M_1}{M_2} \qquad \Psi = \tan\alpha = \frac{\pi_{\dot{m}}}{\rho_B{}^*(p) - \dfrac{\pi_{\dot{m}}}{\tan\varphi_1}}$$

$$\pi_{\dot{m}} = \frac{\dot{m}}{\rho_B(p_0)v_0 A} \qquad \rho_B{}^* = \frac{\rho_B(p)}{\rho_B(p_0)}$$

$$M_1 = 2\frac{\mu_s}{\mu_b}\frac{k_2}{k_1}\frac{HiE}{t - \dfrac{e}{\cos\overline{\varphi}}i}(K\tan\overline{\varphi} + E) + \frac{\mu_s}{\mu_b}\frac{k_3}{k_1}C\cos\varphi_2(K\tan\varphi_2 + C)$$

$$M_2 = \frac{1}{\mu_b}\frac{E}{k_1}H(K\sin\overline{\varphi} + E\cos\overline{\varphi})$$

$$C = 1 - 2\frac{H}{D} \qquad E = 1 - \frac{H}{D} \qquad \tan\varphi = \frac{t}{\pi D} \qquad \tan\overline{\varphi} = \frac{t}{\pi DE} \qquad \tan\varphi_2 = \frac{t}{\pi DC}$$

$$K = \frac{E(\tan\overline{\varphi} + \mu_s)}{1 - \mu_b\tan\overline{\varphi}} \qquad k_1 = 1; \qquad \frac{k_3}{k_1} = 1; \qquad \frac{k_2}{k_1} = 0.5 \qquad \text{Schneider [48]}$$

$$K = E\tan\overline{\varphi} \qquad k_1 = 1; \qquad \frac{k_3}{k_1} = 1; \qquad \frac{k_2}{k_1} = 0.4 \qquad\qquad \text{Schöppner [44]}$$

$$K = E\tan\overline{\varphi} \qquad k_1 = k_2 = k_3 = k \qquad\qquad\qquad\qquad \text{Hyun et al. [81]}$$

For constant material data, i.e., constant friction coefficients μ_b and μ_s and constant densities ρ_B, the differential equation is linear, but for variable material data it is not linear and thus can only be solved numerically due to the pressure-dependent material data. The solution of the linear equation is

$$p = p_0 \exp\left\{\left(\frac{1 - K\psi}{M_2\sqrt{1+\psi^2}} - \frac{M_1}{M_2}\right)\cdot z\right\} \tag{5.257}$$

With

$$M = M_1 + \frac{M_2}{Z} \ln \frac{p_1}{p_0} \tag{5.258}$$

the angular function gives

$$\tan \alpha = \frac{K - M\sqrt{1 + K^2 - M^2}}{K^2 - M^2} \tag{5.259}$$

and thus for the dimensionless throughput in the case of extrusion for example:

$$\pi_{\dot{m}} = \left[\frac{1}{\tan \varphi} + \frac{\left(K^2 - M^2\right)}{K - M\sqrt{1 + K^2 - M^2}} \right]^{-1} \tag{5.260a}$$

with

$$\pi_{\dot{m}} = \frac{\dot{m}}{\rho_B A v_0} \tag{5.260b}$$

and

and

$$\left. \begin{array}{ll} K > \dfrac{t}{\pi D} & \text{for} \quad \mu_s \tan \overline{\varphi} < 1 \\[2mm] 0 \le M \le 1 & \text{for} \quad p_1 > p_0 \end{array} \right\} \tag{5.261}$$

For the angle α we may say

$$90° - \varphi \ge \alpha \ge 0 \tag{5.262}$$

Values between

$$90° \le \alpha \le 90° - \varphi \tag{5.263}$$

are possible, too. This is the case if $p_0 > p_1$. Then, M is in the domain

$$-K \le M \le 0 \tag{5.264}$$

We will now deal with the nonlinear differential equation. Here, the coefficient β is a function of the pressure. The pressure-dependent density has a crucial influence. It is dealt with in [46, 71, 72, 77–79]. We choose the following relation:

$$\rho_B(p) = \frac{\rho_{p0}}{e^{-k \Delta p} + \left(\rho_{p0}/\rho_{B0} - 1\right) e^{-\lambda \Delta p}} \tag{5.265}$$

where ρ_{p0} is the solid density (particle density) with pressure p_0, and ρ_{B0} is the bulk density that has been corrected by means of the screw channel dimensions. The first term in the denominator furthermore describes the compression of the compact solid body; the second term describes the cavity volume change of the bulk material as a consequence of the pressure. In smooth-barrel extruders, the first term can be one $(\exp(-k \, \Delta p) = 1)$. A detailed description of the k and λ coefficients can be found in reference [46]. The compressibility coefficient λ is indicated in [65] for different materials at temperatures between 20° C and 50 °C in domain of $0.0015 < \lambda < 0.022$ (bar^{-1}). The bulk density ρ_{B0} at pressure p_0, which has been influenced by the channel dimensions (Fig. 5.39), can be described by the following relation [70, 77, 79]:

$$\frac{\rho_{B0}}{\rho_{p\infty}} = \frac{1 - \frac{\left(2-\sqrt{2}\right)}{2}\left(\frac{d}{H} + \frac{d}{W}\right) + \frac{\left(3-2\sqrt{2}\right)}{2}\frac{d^2}{HW}}{1 + \frac{d^2}{HW}}$$

(5.266)

where d is the average solid diameter and $\rho_{p\infty}$ the bulk density measured by means of a standardized method. Eq. 5.266 is an approximation. The exact relation can be found in references [77, 79].

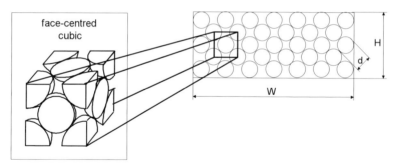

Figure 5.39   Packing of spheres within the screw channel

The factors k_1, k_2, k_3 are the anisotropic pressure coefficients. Schneider [48] gives the following values:

$$\frac{k_2}{k_1} = 0.5 \qquad \frac{k_3}{k_1} = 1$$

(5.267)

where k_1 is the anisotropic coefficient at the barrel, k_2 the coefficient at the flight, and k_3 the coefficient at the root surface. Jungmann [46] has made a detailed theoretical analysis of these coefficients. The relations are

$$k_1 = m_{Hb} \frac{1 - \frac{D_{core}}{D}}{\left(1 - \frac{D_{core}}{D}\right)^{\frac{m_{Hb}}{K_{wb}}}}$$

(5.268)

$$k_3 = m_{Hs} \frac{\dfrac{D}{D_{core}} - 1}{\left(\dfrac{D}{D_{core}}\right)^{m_{Hs}/K_{ws}} - 1}$$

(5.269)

$$m_{Hb} = \cos^2(\overline{\varphi}) + K_{wb} \sin^2(\overline{\varphi}) \qquad m_{Hs} = \cos^2(\overline{\varphi}) + K_{ws} \sin^2(\overline{\varphi})$$

(5.270)

$$\overline{\varphi} = \arctan\left(\frac{t}{\pi(D-H)}\right)$$

(5.271)

The coefficients K_{wb}, K_{ws}, and K_{wfl} are calculated according to the following equation:

$$K_w = \frac{\sqrt{1+\mu_i^2} - \mu_i \cos\left[-\arcsin\left(\dfrac{\mu_w\sqrt{1+\mu_i^2}}{\mu_i \cdot \sqrt{1+\mu_w^2}}\right) + \arctan \mu_w\right]}{\sqrt{1+\mu_i^2} + \mu_i \cos\left[-\arcsin\left(\dfrac{\mu_w\sqrt{1+\mu_i^2}}{\mu_i \sqrt{1+\mu_w^2}}\right) + \arctan \mu_w\right]}$$

(5.272)

in consideration of the corresponding external friction value μ_w (μ_{wb}, μ_{ws}, μ_{wfl}), assuming that the friction values at the flight and at the root surface are identical ($K_{ws} = K_{wfl}$). The anisotropic coefficient at the flight is [46]

$$k_2 = K_{wfl}$$

(5.273)

The greatest influence on the pressure-throughput calculations is exerted by the internal and external friction coefficients μ_i and μ_w. These are dealt with in detail in references [74–76, 83] with regard to pressure, temperature, and velocity. Especially if the friction coefficient at the screw does not differ much from that at the barrel, inexact friction coeffitions can entail serious computation faults because they take can be stated in the steep slope of the throughput curve shown in Fig. 5.40.

Fig. 5.41 shows experimental and theoretical results. The Hyun model and the improved Schneider model thus seem to differ only slightly with higher pressures if the pressure-dependent material data are taken into account for the solution of the differential Equation.

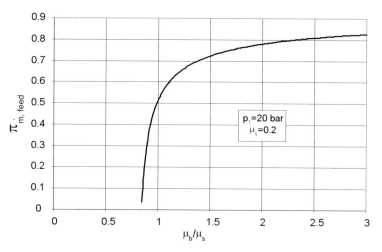

Figure 5.40   Dimensionless throughput as a function of the friction coefficient quotient

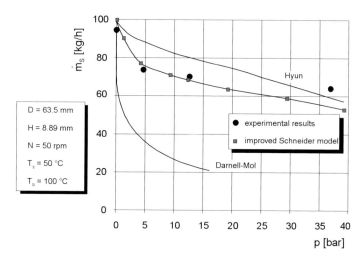

Figure 5.41   Comparison of different solid conveying models with experiments [46]

5.3.1.2   Power Consumption

The power

$$P = \tau_{0t} v_0 \pi D L \tag{5.274}$$

with

$$\tau_{0t} = \tau_0 \cos \alpha \tag{5.275}$$

and

$$\tau_0 = \frac{p_1 - p_0}{\ln p_1/p_0} \tag{5.276}$$

The dissipated energy per unit of time hence is

$$P_{diss} = P - \dot{V}\Delta p \qquad (5.277)$$

For the energy dissipated at the barrel wall we obtain

$$P_{b,diss} = v_r \tau_0 \pi DL$$
$$= v_0 \tau_0 \pi DL \frac{\sin \varphi_1}{\sin(\alpha + \varphi_1)} \qquad (5.278)$$

and thus for the energy dissipated at the screw per unit of time

$$P_{s,diss} = P_{diss} - P_{b,diss} \qquad (5.279)$$

5.3.2   Grooved Feed Section

5.3.2.1   Throughput

At first, our model is a rectangular channel across which a grooved plate is moving (Fig. 5.42). The material is conveyed in the form of a solid flow. Two cases may occur:

Case 1:   The material is conveyed in the grooves.
Case 2:   The material is not conveyed in the grooves.

These two cases can again be subdivided into two subgroups (Fig. 5.43). If the granules are smaller than the groove depth (case 2) a frictionally engaged solid flow predominates in the screw channel over the entire channel depth. If the granule diameter is greater than the groove depth (case 1), there is a reciprocal action between the grooved conveying and the channel conveying. If the channel is flatter than the double granule diameter, the maximum flow velocity can be stated in the entire channel due to the keyed granules (case 1a). If the channel is deeper, this is true only for the upper section of the channel (case 1b).

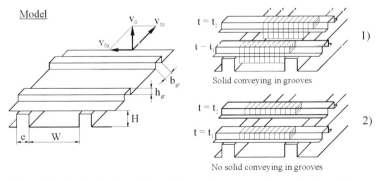

Figure 5.42   Model for the description of the solid conveying; conveying mechanism in the grooves

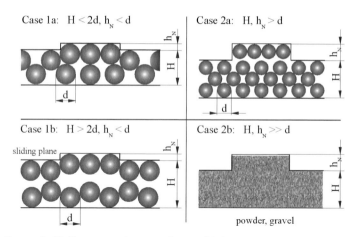

Figure 5.43   Geometrical influences on the conveying mechanism

If the throughput calculation is again based on Fig. 5.33 we obtain for case 1a

$$\dot{m} = \rho_B \left(A_{ch} + A_{gr} \right) tN \tag{5.280}$$

with
A_{ch} = cross-sectional channel area (Eq. 5.243)
$A_{gr} = i\, H_{gr}\, W_{gr}$ = grooved cross-sectional area

This is the maximum throughput possible.

In cases 2a and b we have a frictionally engaged conveying. Hence, we obtain for the throughput[*]

$$\dot{m} = \rho_B A_{ch} tN \, \frac{\tan \alpha}{\tan \varphi + \tan \alpha} \tag{5.281}$$

Case 1b is a mixture of the two conveying mechanisms. Hence, the first approximation:

$$\dot{m} = \rho_B tN \left\{ A_{gr} + A_{ch} \left[x + (1-x)\frac{\tan \alpha}{\tan \varphi + \tan \alpha} \right] \right\} \tag{5.282}$$

This is the most complex case because the variable x is unknown.

To determine the conveying angle α we now consider the pressure. The pressure p_1 at the end of the grooved barrel is determined by the extruder die pressure and the pressure difference along the screw zones up to the grooved barrel. Assuming that the grooved barrel

---

* Note: Principally, an additional grooved conveying may also occur: if the friction force in the boundary layer between the groove and the channel is equal to the sum of the friction forces at the grooved walls.

can reduce the pressure over a very short length, we can say that in a first approximation the pressure term in Eq. 5.258 can be neglected. Hence,

$$M = M_1 \tag{5.283}$$

For the determination of this factor, an average friction coefficient $\bar{\mu}_b$ at the barrel wall is necessary. It depends on the groove geometry and the number of grooves (Fig. 5.44) and can be described by the following approximation [60]:

$$\bar{\mu}_b = \mu_b + \left(\mu_i - \mu_b\right)\frac{N_{gr}W_{gr}}{\pi D}\left\{1 - \exp\left[-5\left(\frac{H_{gr}}{W_{gr}}\right)\right]\right\}^{0.9} \tag{5.284}$$

where μ_i is the friction coefficient polymer/polymer and μ_b is friction coefficient steel/polymer.

For the throughput we obtain

$$\pi_{\dot{m}} = \left[\frac{1}{\tan\varphi} + \frac{\left(K^2 - M_1^2\right)}{K - M_1\sqrt{1 + K^2 - M_1^2}}\right]^{-1} \tag{5.285}$$

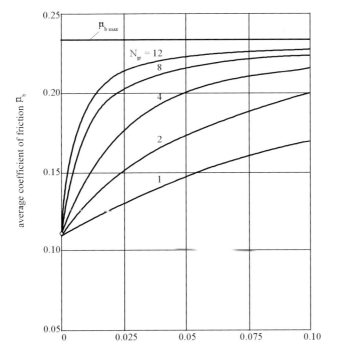

Figure 5.44   Average friction coefficient $\bar{\mu}_b$ at the barrel depending on the initial groove depth H_{gr}/D and the number of grooves N_{gr} (conical grooves) [58]

Useful correspondencies with experiments are yielded if

$$\frac{K_2}{K_1} = 0.5 \qquad \frac{K_3}{K_1} = 0.85 \tag{5.286}$$

is set (Fig. 5.45). But this must not obscure the fact that the grooved-barrel theory is imperfect. For the sake of completeness we will write the empirically determined throughput Equation by Peiffer [56]:

$$\frac{\dot{m}}{\rho_B N D^3} = 1.82 \left(\frac{\overline{\mu}_b}{\mu_b^{1.2}} - 1.3 \right)^{1/6} \left(\frac{H}{D} \right) \left(\frac{t}{D} \right) \left(1 + 0.9 \frac{H_{gr_{max}}}{D} \right) \left(1 - 2.3 \frac{e}{D} \right) \tag{5.287}$$

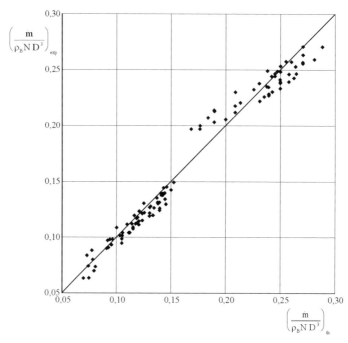

Figure 5.45   Comparison of theoretical and experimental throughput values

As a rule, a full solid friction, especially in larger machines, cannot be maintained. At the end of the grooved barrel it changes into a melt film friction. This point of change depends on the pressure p_{gr} at the end of the grooved barrel, of the rotational velocity and of the machine size. The pressure p_{gr} and the dimensionless throughput $\pi_{m1} = \dot{m}/\rho_B N H D^2$ will drop (Fig. 5.46). These critical values can be computed by means of the differential Equations and their solutions in Table 5.9. A melt film forms with

$$\xi = 0 \qquad \zeta = 1 \qquad \Theta = 1 \tag{5.288}$$

T_{max} is the crystallite melting temperature with semicrystalline polymers or the glass transition temperature with amorphous polymers respectively. T_{TM} is the temperature of the heat-cooling fluid, and T_0 the initial temperature of the solid.

Since only the temperature of the boundary layer at the point $\xi = 0$ is interesting, the complicated Eq. in Table 5.9 can be replaced by an approximation Equation. Under the conditions

$$\sqrt{\beta\zeta} \geq 2 \qquad \mathrm{Bi}\sqrt{\frac{\mathrm{Gz}}{\zeta}} \geq 2 \qquad\qquad (5.289)$$

it is

$$\Theta = \frac{\mathrm{Br}}{\mathrm{Bi} + \sqrt{\beta\mathrm{Gz}}} e^{\beta\zeta} + \Theta_{CM} \qquad\qquad (5.290)$$

Since we are interested only in the temperature at the end of the grooved barrel, the error occurring with smaller ζ values in the first condition is of no interest.

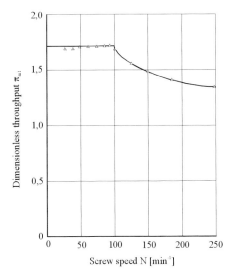

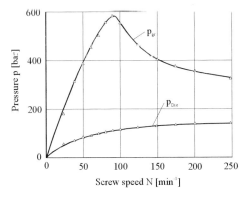

Figure 5.46 Pressure and dimensionless throughput depending on the speed

Table 5.9   Temperature Eq. for the Melt Film Formation in the Grooved Barrel

Differential Eq.	$$Gz\frac{\partial\Theta}{\partial\zeta} - \frac{\partial^2\Theta}{\partial\xi^2} = 0$$

Boundary condition $\dfrac{\partial\Theta}{\partial\xi} - Bi\Theta = -Bre^{\beta\zeta} - Bi\Theta_{CM}$ for $\xi = 0;\ \ \zeta \geq 0$

Initial condition $\Theta = 0$ for $\zeta = 0;\ \ \ \xi \geq 0$

$$\Theta = Br\left\{\frac{1}{2}\exp(\beta\zeta)\left[\frac{\exp\left(-\xi\sqrt{\beta Gz}\right)}{Bi + \sqrt{\beta Gz}}erfc\left(\frac{\xi}{2\sqrt{\dfrac{\zeta}{Gz}}} - \sqrt{\beta\zeta}\right)\right.\right.$$

$$\left.\left. + \frac{\exp\left(\xi\sqrt{\beta Gz}\right)}{Bi - \sqrt{\beta Gz}}erfc\left(\frac{\xi}{2\sqrt{\dfrac{\zeta}{Gz}}} + \sqrt{\beta\zeta}\right)\right] - \frac{Bi\exp\left(Bi\xi + \dfrac{Bi^2}{Gz}\zeta\right)}{Bi^2 - \beta Gz}erfc\left(\frac{\xi}{2\sqrt{\dfrac{\zeta}{Gz}}} + Bi\sqrt{\dfrac{\zeta}{Gz}}\right)\right\}$$

$$+ \Theta_{CM}\left\{erfc\left(\frac{\xi}{2\sqrt{\dfrac{\zeta}{Gz}}}\right) - \exp\left(Bi\xi + \frac{Bi^2}{Gz}\zeta\right)erfc\left(\frac{\xi}{2\sqrt{\dfrac{\zeta}{Gz}}} + Bi\sqrt{\dfrac{\zeta}{Gz}}\right)\right\}$$

$$\Theta = \frac{T - T_0}{T_{max} - T_0} \qquad \Theta_{CM} = \frac{T_{CM} - T_0}{T_{max} - T_0}$$

$$Br = \frac{\mu_b(T_{max})v_r p_0 H_E}{\lambda_S(T_{max} - T_0)}$$

$$Gz = \frac{\bar{v}_S H_E{}^2 \sin\varphi_1 \rho_S c_s}{\lambda_S L_{gr}}$$

$$Bi = \frac{\lambda_b}{\lambda_s}\frac{2H_E}{D\left[\ln\left(\dfrac{D_{Cch}}{D}\right) + \left(\dfrac{2\lambda_b}{\alpha_{cm}D_{Cch}}\right)\right]} \qquad \xi = \frac{x}{H_E} \qquad \zeta = \frac{L}{L_{gr}}$$

$$v_r = v_0\frac{\sin\varphi_1}{\sin(\alpha + \varphi_1)} \qquad \bar{v}_s = v_0\frac{\sin\alpha}{\sin(\alpha + \varphi_1)}$$

5.3.2.2   Power and Torque

For the calculation of the power, the pressure dependence of the material value, and thus the power dependence of the conveying angle can no longer be neglected, because, in general, the pressures in a grooved barrel are much higher than those in a smooth barrel. Hence,

$$P = \pi D v_0 k_1 \int_0^L \bar{\mu}_b(p) p(L) \cos[\alpha(p)] dL \qquad (5.291)$$

Strictly speaking, this Eq. can be applied only in case of a frictionally engaged conveying. For the torque we may thus say

$$M_d = \frac{P}{2\pi N} \qquad (5.292)$$

Peiffer [56] has determined following relation for powder by means of empiric experiments:

$$\frac{M_d}{D^3} = \left(0.382 \Delta p + 11 \cdot 10^5\right) \frac{\bar{\mu}_b^{0.3}}{\mu_s^{0.2}} \left(\frac{H}{D}\right)^{0.65} \left(\frac{t}{D}\right) \left(1 - 2.6 \frac{e}{D}\right) v^{0.137} \qquad (5.293)$$

where the pressure Δp has to be indicated in N/m^2. v is the Poisson number.
In case of additional shear forces, as can occur with granules, the Eq.s are no longer applicable.

5.3.3   High-Speed Conveying

With smooth-barrel extruders, the throughput is determined by the total system; with grooved-barrel extruders it is mainly determined by the grooved barrel and with high-speed extruders by the boundary layer of feed hopper/extruder feed zone. Depending on the speed, the following facts can be stated by means of experiments with regard to the feeding of the material: At a low screw speed, the material is fed on both sides of the screw channel through the feed hopper. With an increasing speed, this two-sided feeding changes into a one-sided feeding. At a very high speed, the material is fed only at the side in direction of the rotation. The material passes into the screw channels against the transport direction. The throughput is determined by two limiting cases [118]:

- *The throughput $\dot{m}_0$ for a completely filled screw length.* For low velocities, this is a linear function of the rotational velocity. The function is the tangential at the throughput curve for the high-speed conveying, which intersects the zero point:

$$c_0 = \frac{\dot{m}_0}{v_0} = \frac{d\dot{m}}{dv_0}\bigg|_{v_0=0} \qquad (5.294)$$

- The Eq.s are listed in Table 5.10.

- *The maximum throughput possible.* This is the throughput that can be supplied by the hopper. It is the vertical tangential at the throughput curve for the high-speed conveying. The dimension analysis yields

$$\dot{m}_{max} = c_1 \rho_B \sqrt{2gD} DH \tag{5.295}$$

- where g is the acceleration of gravity and c_1 a geometry- and material-dependent parameter.

Table 5.10 Equations for Computing the Throughput Gradient at the Point $v_0 = 0$, Y_1, Y_2 (see Table 5.12)

Smooth-barrel extruder
$c_0 = \dfrac{\dot{m}_0}{v_0} = \dfrac{1}{2} \rho W H \cos \varphi \left(Y_1 - Y_2 \pi_p \right)$

Injection molding machine
$c_0 = \dfrac{\dot{m}_0}{v_0} = \dfrac{1}{2} \rho W H \cos \varphi \left(1 + \tan \varphi \tan \beta \right) \left[Y_1 - Y_2 \pi_p \right] \qquad \tan \beta = \dfrac{\dot{s}}{v_0}$

Grooved-barrel extruder
$c_0 = \dfrac{\dot{m}_0}{v_0} = \rho_B A \dfrac{\tan \alpha \tan \varphi}{\tan \alpha + \tan \varphi}$

The differential equation that meets the two conditions is

$$\frac{d\dot{m}}{dv_0} = \frac{\dot{m}_0}{v_0} \left[1 - \left(\frac{\dot{m}}{\dot{m}_{max}} \right)^j \right] \tag{5.296}$$

with $j \geq 1$. The greater j is, the better is the smoothness of the differential Eq. solution to the two tangentials. Based on experiments $j = 2$ has been chosen. The throughput can be described by the function

$$\dot{m} = \dot{m}_{max} \tanh \left(c_2 \frac{v_0}{\sqrt{2gD}} \right) \tag{5.297}$$

The parameter c_2 is obtained from the Eqs. 5.294 and 5.296:

$$c_2 = \sqrt{\frac{2gD}{v_0^2} \frac{\dot{m}_0}{\dot{m}_{max}}} = c_0 \frac{\sqrt{2gD}}{\dot{m}_{max}} \tag{5.298}$$

and hence the following expression for Eq. 5.296:

$$\dot{m} = \dot{m}_{max} \tanh\left(\frac{c_0 v_0}{\dot{m}_{max}}\right)$$

(5.299)

where $\dot{m}_{max}$ must be determined by experiments. Fig. 5.47 shows the normalized representation of the function. The throughput $\dot{m}_{max}$ and thus the dimensionless parameter c_1 are extremely dependent on the channel depth, the particle geometry and the material. It can take values between 0 and 1:

$$0 \le c_1 \le 1$$

(5.300)

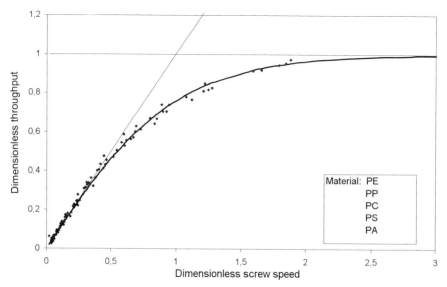

Figure 5.47   Throughput function

The value 0 occurs if the channel depth is smaller than the minimal particle diameter d_{min}, the value 1 occurs if the material reaches the velocity $v = \sqrt{2gD}$ in vertical direction in the screw channel after passing the height $h = D$. Hence follows the function

$$c_1 = \frac{\dot{m}_{max}}{\rho_B \sqrt{2gD}DH} = f\left(\frac{H - d_{min}}{d_{max}}\right)$$

(5.301)

A conveying that may be due to granular shearing has been neglected. Empirically, we can say

$$c_1 = \left[a\left(\frac{H}{d_{max}} - \frac{d_{min}}{d_{max}}\right)\right]^b$$

(5.302)

In Table 5.11 some values are given for the parameters a and b.

Table 5.11   Experimental Values a and b for Determining the Dimensionless Parameter c_1

Material	d_{max} (mm)	$\dfrac{d_{min}}{d_{max}}$	a	b
PC	3.1	0.61	0.275	0.9
PMMA	2.5	0.88	0.255	0.9
PA	2.6	0.846	0.295	0.9
SB	2.7	0.6296	0.25	0.9
PS	4.0	0.50	0.38	0.8
PE	3.5	1.0	0.195	0.52
PP	3.4	0.882	0.20	0.65
Note	4.25 mm $\leq$ H $\leq$ 9 mm			
	18.00 mm $\leq$ D $\leq$ 50 mm			
	$D_{hopper} \mathrel{\hat=} D_{screw}$			

With Eq. 5.301 and the prerequisite $c_1 = 1$ we can calculate a value H, which determines the maximum feeding speed for the conveying from the full feed hopper, which cannot be increased any more. However, this has not yet been proved in experiments.

Eq. (5.298) describes the increasing branch of the curve. However, the throughput has a theoretical maximum because at a very high speed the screw flights below the feed hopper impede the material supply. In this case, the throughput decreases again.

At correspondingly high speeds, the feed zone is never completely filled. The degree of filling can be determined by the relation

$$f = \frac{\dot m}{\dot m_0} = \frac{\dot m_{max}}{c_0 v_0} \tanh\left(\frac{c_0 v_0}{\dot m_{max}}\right) \tag{5.303}$$

5.4   Composite Extruder Models

5.4.1   Integral Treatment

5.4.1.1   Throughput Behavior of Smooth-Barrel Extruders with Melt-Dominated Conveying

When representing the throughput behavior of a smooth-barrel plasticating extruder as a parameter, dependent on the pressure difference Δp with the screw speed N, we obtain the family of the characteristic screw curves (Fig. 5.48). With the characteristic numbers

$$\pi_{\dot m} = \frac{2\dot m}{i\rho WHv_{0z}} \tag{5.304}$$

and

$$\pi_p = \frac{\Delta p H^{1+n}}{6K(T_M)v_{0_z}{}^n Z}$$

(5.305)

the dimensionless mass throughput $\pi_{\dot m}$ (Fig. 5.49) can be determined from the dimensional representation. By such a way of plotting, the families of curves from Fig. 5.47 can be combined in a quite approximate characteristic curve, the so-called characteristic extruder curve, where in the ideal case the single characteristic die curves appear as one point of the characteristic extruder curve.

All material data ρ, K, n in these Eq.s are referred to the melt temperature in front of the screw tip. Besides, H is the channel depth of the reference zone, $Z = L/\sin\varphi$ is the length of

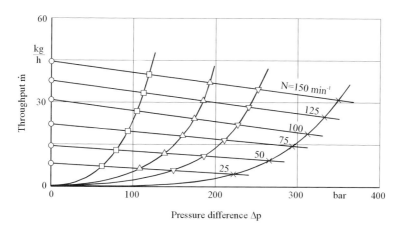

Figure 5.48   Characteristic curves of a smooth barrel extruder

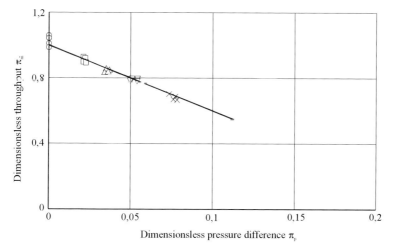

Figure 5.49   Characteristic dimensionless extruder curve for a smooth barrel extruder

the complete unwound screw channel, W is the channel width, i the number of flights, v_{0z} the velocity in screw channel direction, and Δp the pressure difference across the complete screw length.

To compute the throughput in a smooth-barrel extruder we assume a melt film and melt pool that forms very early in the screw channel; i.e., the throughput is melt-dominated. Fig. 5.50 shows a scheme of the geometrical dimensions, temperatures, and pressure that allow the computation of the throughput. The temperature profile is predefined in a first approximation as a linear increasing curve from the front of the hopper, where T_1 is the glass transition temperature T_g with amorphous or the crystallization temperature T_{cry} with semicrystalline thermoplastics, respectively, up to the melt temperature in front of the screw tip T_4.

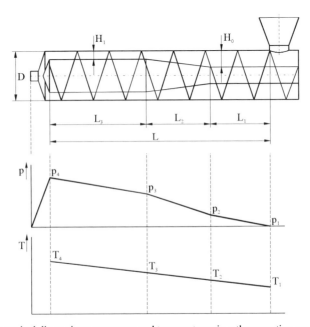

Figure 5.50   Geometrical dimensions, pressures and temperatures in a three-section screw (schematic)

Just like with a melt extruder the pressure differences of the single zones are summed up. The resulting Eq. is solved for the throughput and is normalized with the drag flow. We get the dimensionless throughput

$$\pi_{\dot{m}} = f\left(\pi_p; n; \text{temperature; geometry}\right) \tag{5.306}$$

This is a linear function of π_p:

$$\pi_{\dot{m}} = Y_1 - Y_2 \pi_p \tag{5.307}$$

with

$$Y_1 = \phi_1 X_{1,iso} \tag{5.308}$$

and

$$Y_2 = \frac{\phi_2}{2}\left(\frac{1}{X_{2,iso}} + \frac{1}{X_{2,n-iso}}\right) \qquad (5.309)$$

The values ϕ and X for a three-section screw are listed in Table 5.12. The values describe the screw geometry and the material behavior. The index "iso" means isothermal and "n-iso" means nonisothermal.

Table 5.12   Throughput Equation of a Smooth-Barrel Plasticating Extruder (three-section screw)

$$\boxed{\pi_{\dot{m}} = Y_1 - Y_2\pi_p}$$

$$Y_1 = C\phi_1 X_{1,iso} \qquad Y_2 = \frac{\phi_2}{2}\left(\frac{1}{X_{2,iso}} + \frac{1}{X_{2,n-iso}}\right)$$

$$\pi_{\dot{m}} = \frac{2\dot{m}}{i\rho v_{0z}W_1H_1} \qquad \pi_p = \frac{(p_4 - p_1)H_1^{1+n}}{6K(T_4)v_{0z}^n Z}$$

$$\Phi_1 = \frac{n^A}{\cos^{Bn}\varphi} \qquad \Phi_2 = C\frac{\cos^{Dn}\varphi}{n^{E\sin\varphi + F\cos\varphi}} \qquad \text{(for coefficients A – F see Table 5.2)}$$

$$X_{1,iso} = \frac{\dfrac{L_1}{L}\left(\dfrac{H_1}{H_0}\right)^{1+n} + \dfrac{1}{n}\left(\dfrac{L_2}{L}\right)\left(\dfrac{H_1}{H_0}\right)^n \dfrac{(H_0/H_1)^n - 1}{(H_0/H_1)-1} + \dfrac{L_3}{L}}{\dfrac{L_1}{L}\dfrac{W_1}{W_0}\left(\dfrac{H_1}{H_0}\right)^{2+n} + \dfrac{1}{1+n}\left(\dfrac{L_2}{L}\right)\dfrac{2}{W_0/W_1+1}\left(\dfrac{H_1}{H_0}\right)^{1+n}\dfrac{(H_0/H_1)^{1+n}-1}{(H_0/H_1)-1} + \dfrac{L_3}{L}}$$

$$X_{2,iso} = \frac{L_1}{L}\frac{W_1}{W_0}\left(\frac{H_1}{H_0}\right)^{2+n} + \frac{1}{1+n}\left(\frac{L_2}{L}\right)\frac{2}{W_0/W_1+1}\left(\frac{H_1}{H_0}\right)^{1+n}\frac{(H_0/H_1)^{1+n}-1}{(H_0/H_1)-1} + \frac{L_3}{L}$$

$$X_{2,n-iso} = \frac{L_1}{L}\frac{W_1}{W_0}\left(\frac{H_1}{H_0}\right)^{2+n}\Omega_1 + \frac{1}{1+n}\frac{2}{\frac{W_0}{W_1}+1}\left(\frac{L_2}{L}\right)\left(\frac{H_1}{H_0}\right)^{1+n}\frac{(H_0/H_1)^{1+n}-1}{(H_0/H_1)-1}\Omega_2 + \frac{L_3}{L}\Omega_3$$

$$\Omega_1 = \frac{\beta(T_2 - T_1)}{e^{-\beta(T_4-T_2)} - e^{-\beta(T_4-T_1)}} \qquad \Omega_2 = \frac{\beta(T_3 - T_2)}{e^{-\beta(T_4-T_3)} - e^{-\beta(T_4-T_2)}} \qquad \Omega_3 = \frac{\beta(T_4 - T_3)}{1 - e^{-\beta(T_4-T_3)}}$$

$$T_1 = T_{Fl} \qquad T_2 = T_1 + (T_4 - T_1)\frac{L_1}{L} \qquad T_3 = T_4 - (T_4 - T_1)\frac{L_3}{L} \qquad T_4 = T_M$$

For extremely short plasticating units with a ratio of $L/D < 20$, $D < 45$ mm, and short metering zones $L_M/D < 3$, Y_1 has to be multiplied with the correction factor

$$C = 1 - \exp\left[-0.011\left(\frac{A_0}{A_1}\frac{1}{1-A_1/A_0}\frac{Z}{D}\right)\right] \tag{5.310}$$

where A_0 is the cross-sectional screw channel area of the feed zone and A_1 that of the metering zone. Under the conditions,

$$\frac{A_0}{A_1}\frac{1}{1-A_1/A_0}\frac{Z}{D} > 300 \qquad \frac{L}{D} \geq 20 \tag{5.311}$$

$C = 1$. Three-section screws are relatively easy to handle. Complicated screws are too time-consuming for a manual computation. In this case, a corresponding computer program should be used. The equations are no longer applicable if the granule diameter is similar to the channel depth of the feed zone.

5.4.1.2   Throughput Behavior of Grooved Barrel Extruders

In case of solid friction, the throughput is described with the equations indicated in Section 5.3.2. As has already been discussed there, the solid friction changes into a melt film friction at the end of the grooved barrel, depending on the pressure, the rotational velocity, and the geometry. To derive the throughput equation for this case, we proceed from Fig. 5.51, and the throughput and the rotational velocity are referred to the variables in the point of change:

$$y = \frac{\dot{m}}{\dot{m}_{li}} \qquad x = \frac{v_0}{v_{0,li}} \tag{5.312}$$

where $\dot{m}_{li}$ and $v_{0,li}$ are the variables in the point of change.

We furthermore presume that for the melt film friction the throughput-speed characteristic of the grooved-barrel extruder corresponds with that of the smooth barrel extruder, i.e., that the curves a and b in Fig. 5.50 are parallel. In consideration of the coordinate transformation we obtain

$$\psi = y - 1 \qquad \zeta = x - 1 \tag{5.113}$$

For the throughput we get the following equation with

$$a = \frac{\dot{m}_{sm}}{\dot{m}_{li}x} \tag{5.314}$$

where $\dot{m}_{sm}$ is the throughput of the smooth-barrel extruder and $\dot{m}_{li}$ that of the grooved-barrel extruder in case of solid friction for $y = 1$.

$$\dot{m} = \dot{m}_{li} + \dot{m}_{sm}\left(1 - \frac{v_{0,li}}{v_0}\right)$$ (5.315)

or

$$\pi_{\dot{m}} = \pi_{\dot{m}_{li}}\frac{v_{0,li}}{v_0} + \pi_{\dot{m}_{sm}}\left(1 - \frac{v_{0,li}}{v_0}\right)$$ (5.316)

respectively, for $x > 1$ and

$$\pi_{\dot{m}} = \pi_{\dot{m}_{li}}$$ (5.317)

for $x < 1$. Thus, the throughput of the grooved-barrel extruder with melt film friction is greater than in the smooth-barrel extruder; it is equal only for $v_0 \to \infty$.

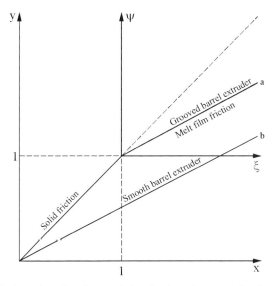

Figure 5.51   Schematic throughput functions to derive the throughput equation for melt film conveying in grooved barrel extruders

5.4.1.3   Throughput Behavior of Venting Extruders

In venting single screw extruders several screws with different functions are arranged in tandem. Generally, these are two systems, or three for a gradual venting, which are mostly set up in the same barrel and are operated together. Fig. 5.52 shows a scheme of the typical basic structure of a venting system.

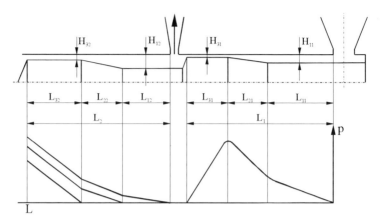

Figure 5.52   Scheme of a venting plasticating unit

The following requirements are important for a smooth operation:

- The channels in the middle section of the venting zones may be only partially filled with melt, because if they were completely filled, there would be no free surface for the escape of gas and the melt could escape through the vent at low pressure increases.
- The whole length of the metering zone should be permanently filled, because a pulsating output can be expected in case of a partial filling.

Fig. 5.53 shows the throughput characteristics of the two stages. Since at a constant melt temperature and constant material data for the screw the throughput depends only on the speed N, the dimensionless throughput of stage 1 is independent of the back pressure to be produced in front of the screw tip. Only the throughput behavior of the stage 2 depends hereupon. But since a partial filling is always necessary in the venting section so that no material emanates from the barrel opening and a free surface allows the escape of the gas,

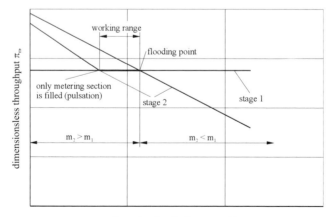

Figure 5.53   Throughput characteristics of the two stages of a venting screw

the conveying rate of stage 2 must be greater than that of stage 1. The upper characteristic extruder curve of stage 2 (Fig. 5.53) shows the throughput behavior depending on the dimensionless back pressure for a completely filled screw stage 2 (Fig. 5.52). The intersection point with the characteristic curve of stage 1 represents the operating mode that must not be reached. This is granted if the dimensionless back pressure, and thus the pressure in front of the screw tip, is lower than the flooding pressure.

The lower characteristic curve of stage 2 shows the throughput behavior of the metering zone depending on the dimensionless back pressure. Here, only the metering zone is filled with melt. The intersection point of the characteristic curve and the characteristic throughput curve of stage 1 yields the minimal back pressure. If this pressure is not achieved, a pulsation may occur. The operative range of the machine lies between this pressure and the flooding pressure. For a better venting and homogenization, a restriction ring is installed at the end of the first screw stage.

5.4.1.4   Power Input

In a thermodynamical sense, the extrusion process is an open process; i.e., there is an exchange of energy and mass. When defining the inside of the barrel as balance area, we obtain

$$P \pm \dot{Q} = \dot{m}\Delta h \tag{5.318}$$

The injection molding process, on the contrary, is a partially open and partially closed process. We speak of a closed process if there is no mass exchange. This is the case in injection molding during the downtime phase. Neglecting the injection phase, the total residence time of the melt in the barrel is composed of the residence time during the rotation phases $\bar{t}_{R1}$ and of the residence time during the downtime phases $\bar{t}_{R2}$:

$$\bar{t}_R = \bar{t}_{R1} + \bar{t}_{R2} \tag{5.319}$$

with

$$\bar{t}_{R1} = \frac{m}{\dot{m}} = it_{dos} \tag{5.320}$$

$$\bar{t}_{R2} = it_{st} \tag{5.321}$$

where i is the number of shots in the machine, t_{dos} is the mold filling time, and t_{st} is the downtime between the mold filling phases. The energy balance thus yields

$$\Delta Pit_{dos} \pm \Delta\overline{\dot{Q}}_{dos}it_{dos} \pm \Delta\overline{\dot{Q}}_{st}it_{st} = \dot{m}t_{dos}(\Delta h_1 + \Delta h_2) \tag{5.322}$$

When introducing an average heat flow $\Delta\overline{\dot{Q}}$ over a the cycle time

$$t_{cyc} = t_{dos} + t_{st} \tag{5.323}$$

we get

$$P \pm \overline{Q}\, \frac{t_{cyc}}{t_{dos}} = \dot{m}\Delta h \tag{5.324}$$

Fig. 5.54 shows the measuring values for different polymers. The quotient $P/\dot{m}\Delta h$ can take high values at about 2. This is the case if the material has a high viscosity and a low enthalpy. Then, much heat must be carried off.

5.4.2   Coupling the Models of the Function Zones

For a correct treatment of the extrusion process, the models of the single function zones have to be combined in one overall model. This coupling is mainly achieved via the pressure and the temperature as well as the material data, which are functions of the temperature and of the pressure. For the pressures and temperatures we obtain

$$\Delta p = \sum_{i=1}^{j} \Delta p_{i,Sc} + \sum_{i=1}^{k} \Delta p_{i,Mel} + \sum_{i=1}^{l} \Delta p_{i,Mc} \tag{5.325}$$

$$T - T_0 = \sum_{i=1}^{j} \Delta T_{i,Sc} + \sum_{i=1}^{k} \Delta T_{i,Mel} + \sum_{i=1}^{l} \Delta T_{i,Mc} \tag{5.326}$$

Fig. 5.55 shows the flow diagram for the entire simulation. For this, the total plasticating unit is subdivided into small intervals so that the geometry, the material data, and the melting-specific variables can be considered constant within these subdivisions.

The throughput can be predefined or calculated within the scope of the simulation. If it shall be computed, a distinction can be made between smooth-barrel and grooved-barrel extruders. To grooved-barrel extruders, a rigid conveying model or a model of the solid, melting, and melt conveying zone, which are coupled by the pressure and the throughput, can be applied. For smooth-barrel extruders, the throughput can be determined on the basis of a mere melt-dominated model or via a coupling of the solid, melting, and melt conveying function zones. As soon as the throughput is known, further important variables, such as the pressure profile or the temperature development across the screw length, can be computed.

Fig. 5.56 shows the pressure profile with a melt-dominated computation and the pressure profile with a correct coupling of the function zones. The experimental values correspond much better with the latter pressure profile. But it also becomes obvious that the throughput is not considerably influenced.

The following Fig.s show some typical curves that have been determined by means of simulation computations. Fig. 5.57 shows the typical differences in the pressure profiles of a smooth-barrel extruder and a grooved-barrel extruder. Fig. 5.58 shows the pressure profile of a grooved-barrel extruder if the solid friction has changed into a melt film friction. This reduces the throughput, too. Fig. 5.59 shows the influence of slip and no slip at the wall. Slip at the wall causes a pressure reduction. Fig. 5.60 shows the pressure profile of a

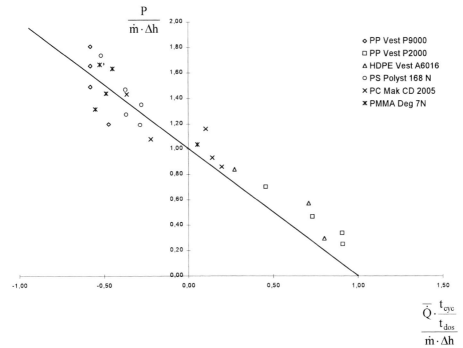

Figure 5.54   Characteristic power diagram

smooth-barrel extruder with a barrier screw. Fig. 5.61 shows the melting profile (solid bed profile) and the melted material flow for a well- and badly- designed screw. Fig. 5.62 illustrates the effect of slip at the wall on the melting process. Slip at the wall causes an obviously worse melting behavior. Slip at the wall also considerably reduces the melt temperature (Fig. 5.43), but the quality of such simulation computations ultimately depends on the quality of the material data and material laws.

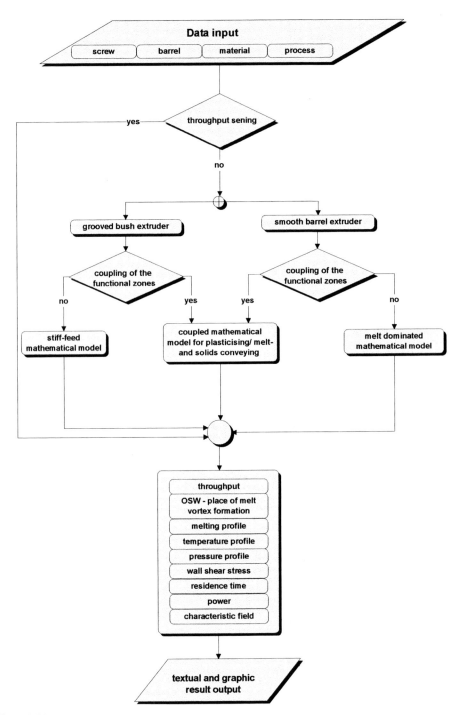

Figure 5.55   Flow diagram for the process simulation

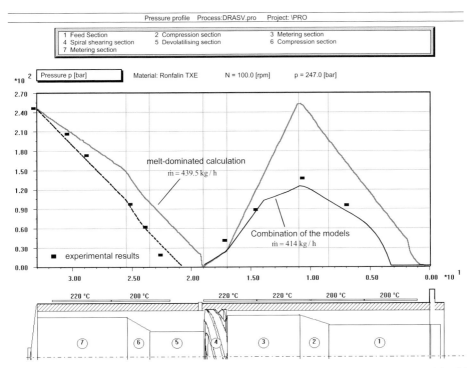

Figure 5.56   Pressure profiles for a melt-dominated conputation and a computation where the models of the function zones have been considered correctly (REX simulation)

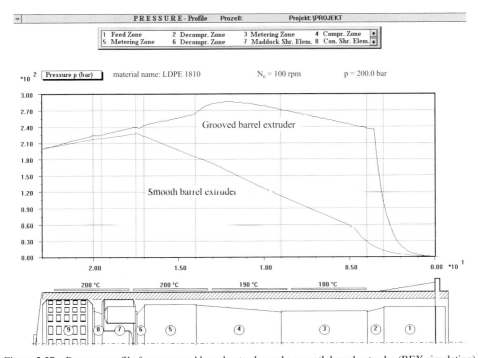

Figure 5.57   Pressure profile for a grooved barrel extruder and a smooth barrel extruder (REX simulation)

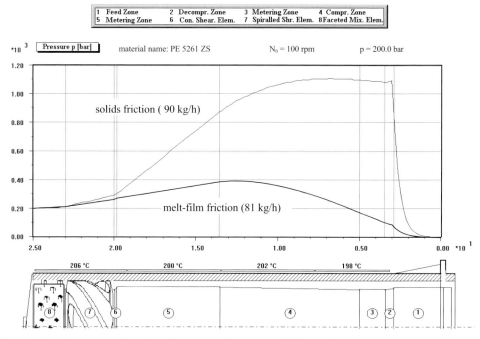

Figure 5.58 Pressure profile with solid and melt film friction (REX simulation)

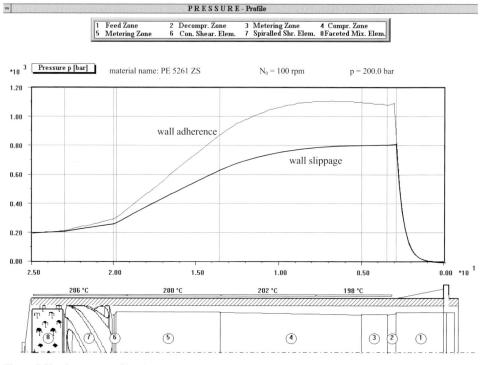

Figure 5.59 Pressure profile with no slip at the wall and with slip at the wall (REX simulation)

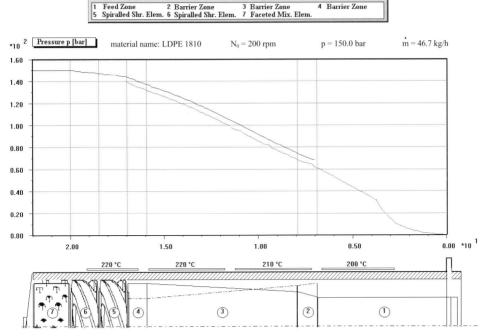

Figure 5.60   Pressure profile of a barrier screw (REX simulation)

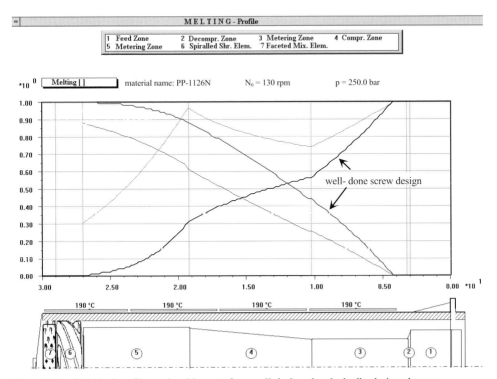

Figure 5.61   Solid bed profiles and melting rate for a well-designed and a badly-designed screw

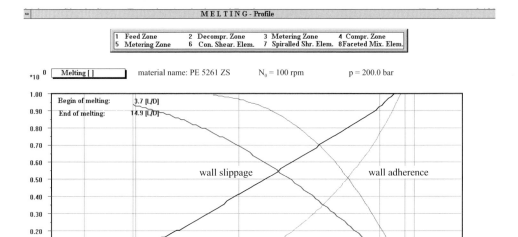

Figure 5.62   Solid bed profiles and melting rates for the conditions of slip and no slip at the wall

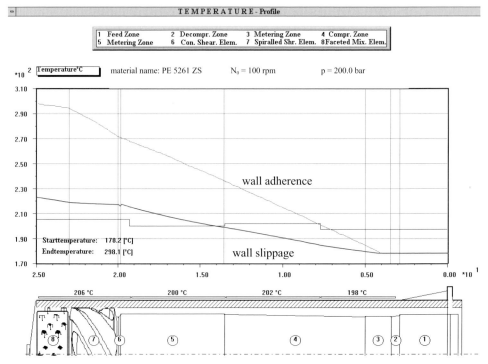

Figure 5.63   Temperature profiles for the conditions of no slip ant slip at the wall

5.5   Scale-Up

5.5.1   Principle of Similarity

By means of scale-up rules, operating and design data of known machines – so-called models – can be applied to larger and smaller machines, the so-called pilot plants. Scale-up rules are based on the principle of similarity. According to this principle, all physical and technical facts are similar that are described by the same dimensionless characteristic values (dimensionless products), also called π values, if they are allocated to the same state of the dimensionless space of characteristic numbers [84]. An overview of the development of the scale-up rules can be found in references [84–117].

5.5.2   General Formulation of the Scale-Up Rules

In global writing, we may say for the extrusion process (Section 5.4.1.4)

$$\pi_P = 1 - \pi_{\dot{Q}} \tag{5.327}$$

and for the injection molding process

$$\pi_P P = 1 - \pi_{\dot{Q}} \frac{t_{cyc}}{t_{dos}} \tag{5.328}$$

with

$$\pi_P = \frac{P}{\dot{m}\Delta h} \qquad \pi_{\dot{Q}} = \frac{\dot{Q}}{\dot{m}\Delta h} \tag{5.329}$$

where

$$\Delta h = c_v \Delta T + \left(\frac{p_2}{\rho_2} - \frac{p_1}{\rho_1} \right) \tag{5.330}$$

We demand an invariant enthalpy difference and input and output temperature (T_1, T_2). Then, the output pressure p_2 must be constant in case of a constant input pressure p_1. The same applies to the material data. The processes of the model and of the pilot plant will be similar with regard to the energy if

$$\pi_P = idem \qquad \pi_{\dot{Q}} = idem \qquad \frac{t_{cyc}}{t_{dos}} = idem \tag{5.331}$$

The dimensionless time value is only relevant for injection molding. Due to different moldings – here, especially, the wall thickness is decisive – this value cannot be kept invariant so that a complete similarity in injection molding cannot be achieved.

The invariancy of π_P and $\pi_{\dot{Q}}$ yields

$$\frac{P}{\dot{m}} = \frac{P_0}{\dot{m}_0} \tag{5.332}$$

$$\frac{\dot{Q}}{\dot{m}} = \frac{\dot{Q}_0}{\dot{m}_0} \tag{5.333}$$

where the index 0 describes the model machine. For the throughput let

$$\pi_{\dot{m}} = \frac{\dot{m}}{\rho DNWH} = idem \tag{5.334}$$

With

$$\frac{H}{H_0} = \left(\frac{D}{D_0}\right)^{\psi} \qquad \frac{N}{N_0} = \left(\frac{D}{D_0}\right)^{-x} \tag{5.335}$$

it follows that

$$\frac{P}{P_0} = \frac{\dot{Q}}{\dot{Q}_0} = \frac{\dot{m}}{\dot{m}_0} = \left(\frac{D}{D_0}\right)^{2+\psi-x}$$

Arguments for the exponents ψ and x will be given later on. Due to the assumption $\Delta h = const$, the temperatures and pressures have to follow the following scale-up rules:

$$\frac{T_1}{T_{1,0}} = \frac{T_2}{T_{2,0}} = \left(\frac{D}{D_0}\right)^0 \qquad \frac{p_1}{p_{1,0}} = \frac{p_2}{p_{2,0}} = \left(\frac{D}{D_0}\right)^0 \tag{5.336}$$

From

$$M_T \sim \frac{P}{N} \tag{5.337}$$

it follows for the torque

$$\frac{M_T}{M_{T,0}} = \left(\frac{D}{D_0}\right)^{2+\psi} \tag{5.338}$$

The essential scale-up rules for the design are thus available in their general form.

5.5.3   Treatment by Zones

5.5.3.1   Melt Conveying Zone

We proceed from the differential equation of conservation of energy. In its reduced form it is

$$\rho c \left(v_x \frac{\partial T}{\partial x} + v_z \frac{\partial T}{\partial z} \right) = -\frac{\partial \dot{q}}{\partial y} + \tau_{yx} \frac{\partial v_x}{\partial y} + \tau_{yz} \frac{\partial v_z}{\partial y} \tag{5.339}$$

It will be transformed into a dimensionless equation with the following variables:

$$y^* = \frac{y}{H} \qquad x^* = \frac{y}{W} \qquad z^* = \frac{z}{Z} \tag{5.340}$$

$$v_x^* = \frac{v_x}{v_0} \qquad v_z^* = \frac{v_z}{v_0} \tag{5.341}$$

$$\tau_{yx}^* = \frac{\tau_{yx}}{K_0} \left(\frac{H}{v_0} \right)^n \qquad \tau_{yx}^* = \frac{\tau_{yz}^*}{K_0} \left(\frac{H}{v_0} \right)^n \tag{5.342}$$

$$\theta = \frac{T - T_0}{T_1 - T_0} \qquad \dot{q}^* = \frac{\dot{q}}{\dot{q}_0} \tag{5.343}$$

where T_1 is a representative melt temperature, e.g., in front of the screw tip, and $\dot{q}_0$ a representative heat flow at the barrel wall. Thus, the dimensionless form of the DE is

$$\frac{L}{W} v_x^* \frac{\partial \theta}{\partial x^*} + v_z^* \frac{\partial \theta}{\partial z^*} = -\frac{1}{Gz_1} \frac{\partial \dot{q}^*}{\partial y^*} + \frac{Br_1}{Gz_1} \left(\tau_{yx}^* \frac{\partial v_x^*}{\partial y^*} + \tau_{yz}^* \frac{\partial v_z^*}{\partial y^*} \right) \tag{5.344}$$

With an invariant φ, let for the characteristic numbers

$$\frac{Z}{W} \sim \frac{L}{D} = idem \tag{5.345}$$

$$Gz_1 = \frac{c \rho v_0 H (T_1 - T_0)}{\dot{q}_0 Z} = idem \tag{5.346}$$

$$\frac{Br_1}{Gz_1} = \frac{K v_0^{\,n} Z}{\rho c H^{1+n} (T_1 - T_0)} = idem \tag{5.347}$$

With invariant material data, invariant melt temperature T_1, and the heat flow density $\dot{q}_0$ as well as $Z = L/\sin \varphi$, we have

$$\psi = 1 - \frac{n}{1+n} x \tag{5.348}$$

and

$$\psi = x = \frac{1+n}{1+2n} \tag{5.349}$$

for

$$\frac{L}{D} = \frac{L_0}{D_0} \tag{5.350}$$

By substituting $\dot{q}_0 = idem$ with the barrel wall temperature $T_b = idem$, and with

$$\theta = \frac{T - T_0}{T_b - T_0} \tag{5.351}$$

and

$$\dot{q}^* = -\frac{\lambda(T_b - T_0)}{\dot{q}_0 H} \frac{\partial \theta}{\partial y^*} \tag{5.352}$$

we get the following characteristic numbers from the differential equation

$$\frac{L}{D} = idem \tag{5.353}$$

$$Gz_2 = \frac{v_0 H^2}{aZ} = idem \tag{5.354}$$

$$\frac{Br_2}{Gz_2} = \frac{K v_0{}^n Z}{\rho c H^{1+n}} = idem \tag{5.355}$$

and hence

$$\psi = 1 - \frac{n}{1+n} x \tag{5.356}$$

and

$$\psi = \frac{x}{2} = \frac{1+n}{1+3n} \tag{5.357}$$

for

$$\frac{L}{D} = \frac{L_0}{D_0}$$ (5.358)

By introducing

$$\frac{\dot{q}_0}{\dot{q}_{0,0}} = \left(\frac{D}{D_0}\right)^{-\alpha\psi}$$ (5.359)

for the heat flow density at the barrel wall,[*] the Graetz number for the exponent α gives the relation

$$\alpha = \frac{x}{\psi} - 1$$ (5.360)

i.e., for

$$\dot{q}_0 = const \qquad \alpha = 0$$ (5.361)

and for

$$T_b = const \qquad \alpha = 1$$ (5.362)

For

$$\alpha < 0$$ (5.363)

the heat influx and efflux, respectively, increase with the scaling-up, and they decrease for

$$\alpha > 0$$ (5.364)

In a general formulation, the ψ and x values are

$$\psi = \frac{1+n}{1+(2+\alpha)n}$$ (5.365)

$$x = \frac{(1+n)(1+\alpha)}{1+(2+\alpha)n}$$ (5.366)

for

$$\frac{L}{D} = \frac{L_0}{D_0}$$ (5.367)

---

[*] Note: In former works $\dot{q} \sim D^{-\alpha x}$ has been set. Under the condition $L \sim D^{1+\varpi}$ for $T_b = const$ this leads to a contradiction.

The $(L/D = idem)$ is superfluous if presuming average values for the velocity and the dissipated energy in the differential equation

$$\bar{v}^* \frac{\partial \theta}{\partial z^*} = -\frac{1}{Gz_1} \frac{\partial \dot{q}^*}{\partial y^*} + \frac{Br_1}{Gz_1} \left(\overline{\tau\dot{\gamma}}\right)^* e^{-\beta(T-T_0)} \tag{5.368}$$

Then, the L/D ratio can be variable, and we have the following expressions with

$$\frac{L}{L_0} = \left(\frac{D}{D_0}\right)^{1+\omega} \tag{5.369}$$

for the exponents

$$\psi = \frac{(1+\omega)(1+n)}{1+(2+\alpha)n}$$

$$x = \frac{(1+\omega)(1+n)(1+\alpha)}{1+(2+\alpha)n} - \omega \tag{5.370}$$

5.5.3.2   Plasticating Zone

As becomes obvious for the melting in Section 5.2, the melting process is predominated by the following characteristic numbers:

$$\pi_0 = Gz_0 = \frac{k_1 \rho v_{0x} \Delta h \bar{\delta}_0^2}{\lambda(T_b - T_{Fl})W} \tag{5.371}$$

$$Br_0 = \frac{k_2 K(T_{Fl}) v_r^{1+n} \bar{\delta}_0^{1-n}}{\lambda(T_b - T_{Fl})} \tag{5.372}$$

$$\pi_1 \zeta = \frac{k_1 \rho v_0 \bar{\delta}_0 L_S}{2\dot{m}} \tag{5.373}$$

$$k_1 = f\left[\frac{\beta}{n}(T_b - T_{Fl})\right] \tag{5.374}$$

$$k_2 = f\left[n, \frac{\beta}{n}(T_b - T_{Fl})\right] \tag{5.375}$$

For constant material data we obtain

$$\frac{\beta}{n}\left(T_b - T_{Fl}\right) = idem \tag{5.376}$$

and thus

$$T_b = T_{b0} \tag{5.377}$$

Thus, k_1 and k_2 are invariant for the model and the pilot plant. With

$$v_{0x} \sim v_0 \tag{5.378}$$

$$v_r \sim v_0 \tag{5.379}$$

we obtain

$$Gz_0 \sim \frac{v_0 \overline{\delta}_0{}^2}{W} = idem \tag{5.380}$$

and hence with

$$\overline{\delta}_0 \sim D^{\psi_\delta} \tag{5.381}$$

the relation

$$\psi_\delta = \frac{x}{2} \tag{5.382}$$

From

$$\frac{Br_0}{Gz_0} = idem \tag{5.383}$$

we have

$$\psi_\delta - 1 - \frac{n}{1+n}x \tag{5.384}$$

The elimination of x gives

$$\psi_\delta = \frac{1+n}{1+3n} \tag{5.385}$$

i.e.,

$$\psi_\delta = \psi \tag{5.386}$$

for constant barrel temperature in the model and in the pilot plant.

We now consider the general case. If $\psi = y = 1$, then

$$\frac{\pi_0}{Br_0} = 1 + \frac{2}{Br_0} \tag{5.387}$$

according to Section 5.2.2.2. As can be seen in Fig. 5.64, the curve approaches two limiting cases:

$$\left.\begin{array}{lll} \dfrac{\pi_0}{Br_0} = 1 & \text{for} & Br_0 \Rightarrow \infty \\[2ex] \pi_0 = 2 & \text{for} & Br_0 \Rightarrow 0 \end{array}\right\} \tag{5.388}$$

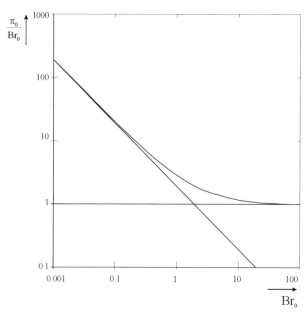

Figure 5.64   Characteristic number function of the melting process

If the channel depth is added to these characteristic numbers, we obtain

$$\frac{\pi_{oH}}{Br_{0H}}\left(\frac{H}{\overline{\delta}_0}\right)^{-(1+n)} = idem \tag{5.389}$$

$$\pi_{0H}\left(\frac{\overline{\delta}_0}{H}\right)^{2} = idem \tag{5.390}$$

Furthermore, we have to consider the characteristic number $\pi_1 \zeta$. For the throughput we introduce into this number

$$\dot{m} = \rho W H v_0 \pi_{\dot{m}} \tag{5.391}$$

and for the solid bed length

$$L_S \sim D^{1 + \omega_S} \tag{5.392}$$

The invariancy of this characteristic number for the model and pilot plant yields

$$\frac{\bar{\delta}_0}{H} \sim \frac{D}{L_S} \sim D^{-\omega_S} \tag{5.393}$$

and thus from the characteristic number (Eq. 5.389):

$$\omega_S = 0 \tag{5.394}$$

i.e., with great Brinkman numbers,

$$\frac{L_S}{D} = const \qquad for \qquad T_b \neq T_{b0} \tag{5.395}$$

From the characteristic number (Eq. 5.390) we obtain, however,

$$\omega_S = \frac{1}{2} \frac{(1+n)(1-\alpha)}{1+(2+\alpha)n} \tag{5.396}$$

i.e., in case of an enlargement of the machine, L_S/D increases according to the scale-up rule

$$\frac{L_S}{L_{S0}} = \left(\frac{D}{D_0} \right)^{1+\omega_S} \tag{5.397}$$

if $T_b \neq T_{b0}$

The curve betwen these two limits (Fig. 5.64) can be approximated domainwise by the following equation (Fig. 5.65):

$$\frac{\pi_0}{Br_0} = a Br_0^{-b} \tag{5.398}$$

Hence, we have with

$$\frac{\pi_0}{Br_0^{1-b}} = idem \qquad \frac{L}{D} = idem \tag{5.399}$$

the following expression for ω_S:

$$\omega_s = \frac{1+n}{1+(2+\alpha)n} - \frac{(1+n)(1-b)}{(1+n)+b(1-n)}$$
$$+ \frac{n-b(1+n)}{(1+n)+b(1-n)}\left[\frac{(1+n)(1+\alpha)}{1+(2+\alpha)n}\right] \qquad (5.400)$$

where

$$0 \le b \le 1 \qquad 0 \le \alpha \le 1 \qquad (5.401)$$

Figure 5.66a shows the dependency on α and b for $n = 0.5$. The exponent ω_S is zero for $\alpha = 1$, $0 \le b \le 1$ and for $b = 0$, $0 \le \alpha \le 1$. The corresponding ψ_δ values of the melt film are shown in Fig. 5.66b. With a variable L/D, i.e., $L \sim D^{1+\omega}$, we obtain

$$\omega_S = \frac{(1+\omega)(1+n)}{1+(2+\alpha)n} - \frac{(1+n)(1-b)}{(1+n)+b(1-n)}$$
$$+ \frac{n-b(1+n)}{(1+n)+b(1-n)}\left[\frac{(1+\omega)(1+n)(1+\alpha)}{1+(2+\alpha)n} - \omega\right] \qquad (5.402)$$

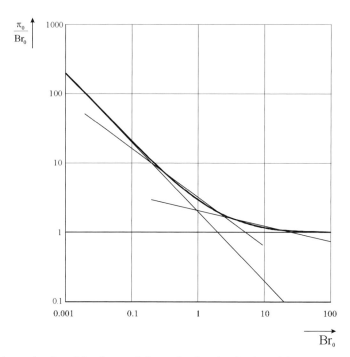

Figure 5.65   Approximation of the characteristic number function for the melting process

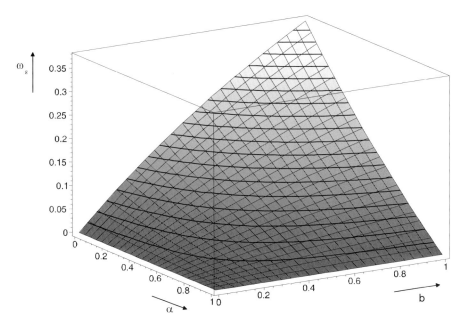

Figure 5.66a   Solid bed length exponent as a function of the exponents α and b

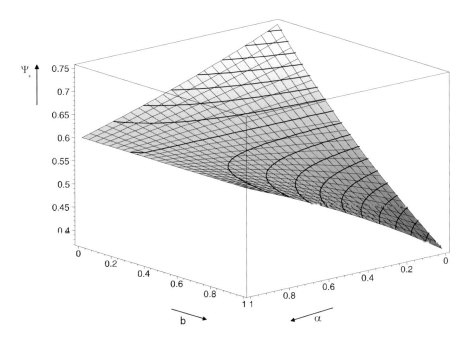

Figure 5.66b   Melt film thickness exponent as a function of the exponents α and b

5.5.3.3   Solid Conveying

5.5.3.3.1   Smooth-Barrel Extruder

In the smooth-barrel extruder we have to distinguish two cases:

- Partially filled screw channels
- Filled screw channels

In the first case we speak of an Archimedes conveying, and in the second case of a frictionally engaged conveying. The functions of the characteristic numbers are

- Archimedes conveying

$$\frac{\dot{m}}{\rho_B HDtN} = f\left(\frac{H}{D}; \frac{t}{D}; \frac{e'}{D}; f\right) \tag{5.403}$$

where f is the degree of filling.
- Frictionally engaged conveying

$$\frac{\dot{m}}{\rho_B HDtN} = f\left(\frac{H}{D}; \frac{H}{L}; \frac{t}{D}; \frac{e'}{D}; \mu_s; \mu_b; k_i; \frac{p_1}{p_0}\right) \tag{5.404}$$

where k_i are the anisotropic pressure coefficients.

If we consider only the solid conveying zone, a complete geometrical similarity has to be required for the scaling-up. However, this entails a changing channel depth ratio with an increasing machine size according to

$$\frac{H_E}{H_i} = \frac{H_{E0}}{H_{i0}}\left(\frac{D}{D_0}\right)^{1-\psi} \tag{5.405}$$

This affects the compression and thus the throughput characteristics. Therefore, the scale-up rule of the channel depths is adjusted to the following zones:

$$\frac{H_E}{H_{E0}} = \left(\frac{D}{D_0}\right)^{\psi} \tag{5.406}$$

For the throughput we obtain for $t \sim D$

$$\frac{\dot{m}}{\dot{m}_0} = \frac{\rho_B}{\rho_{B0}}\left(\frac{D}{D_0}\right)^{2+\psi-x} \tag{5.407}$$

The bulk density has been left deliberately in the transfer rule because it changes with the channel geometry (Section 5.3). In the simplest (one-dimensional) form, we may say

$$\frac{\rho_B}{\rho_{B0}} = \frac{\dfrac{H_0}{d} - \left(1 - \dfrac{1}{\sqrt{2}}\right)\left(\dfrac{D_0}{D}\right)^{\psi}}{\dfrac{H_0}{d} - \left(1 - \dfrac{1}{\sqrt{2}}\right)} \qquad (5.408)$$

where d is the average granule diameter. This influence always has to be taken into account if d is about the channel depth H_{E0}. In this case, the feed zone dominates the throughput.

5.5.3.3.2 Grooved-Barrel Extruder

In the grooved-barrel extruder we have to distinguish

- Inherently engaged conveying
- Frictionally engaged conveying
- Conveying through melt film friction

With inherently engaged conveying (nut/screw) we obtain

$$\frac{\dot{m}}{\rho_B H D t N} = idem \qquad (5.409)$$

With frictionally engaged conveying we obtain formally

$$\frac{\dot{m}}{\rho_B H D t N} = f\left(\frac{H}{D}; \frac{t}{D}; \frac{ie'}{D}; \mu_s; \bar{\mu}_b; k_i; \frac{H_{gr}}{D}; \frac{W_{gr}}{D}; \frac{N_{gr}}{D}\right) \qquad (5.410)$$

where H_{gr} is the groove depth, W_{gr} the groove width, and N_{gr} the number of grooves. The average friction coefficient $\bar{\mu}_b$ can be determined via the relation

$$\bar{\mu}_b = \mu_b + (\mu_i - \mu_b)\frac{N_{gr}W_{gr}}{\pi D}\left\{1 - \exp\left[-5\left(\frac{H_{gr}}{W_{gr}}\right)^{0.9}\right]\right\} \qquad (5.411)$$

with
μ_b = external friction value (plastics/steel)
μ_i = internal friction value (plastics/plastics)

Since the groove geometry has been considered in the average friction value, it may at first be skipped in the function of the characteristic number. Furthermore, the characteristic numbers H/D and t/D have a negligible influence (Figs. 5.67 and 5.68). Hence,

$$\frac{\dot{m}}{\rho_B H D t N} = f\left(\frac{ie'}{D}; \mu_s; \bar{\mu}_b; k_i\right) \qquad (5.412)$$

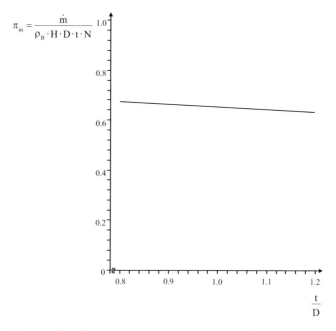

Figure 5.67   Dimensionless throughput as a function of the normalized pitch

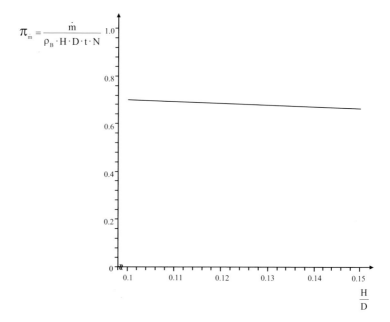

Figure 5.68   Dimensionless throughput as a function of the normalized channel depth

This corresponds with the experiments executed by Peiffer [56], where he still stated a slight influence of the groove depth. With invariant material values and an invariant characteristic number for the channel depth, we thus have

$$\frac{\dot{m}}{\rho_B HDtN} = idem \tag{5.413}$$

i.e., for $t \sim D$

$$\frac{\dot{m}}{\dot{m}_0} = \frac{\rho_B}{\rho_{B0}} \left(\frac{D}{D_0}\right)^{2+\psi-x} \tag{5.414}$$

This applies to inherently engaged and frictionally engaged conveying. The condition

$$\bar{\mu}_b = idem \tag{5.415}$$

yields for the groove geometry:

$$\frac{N_{gr} W_{gr}}{D} = idem \qquad \frac{H_{gr}}{W_{gr}} = idem \tag{5.416}$$

The scale-up rule (Eq. 5.414) also applies to the melt film friction if the characteristic throughput number (Eq. 5.413) is invariant for the model and the pilot plant. It is no longer applicable if there is solid friction in the model and melt film friction in the pilot plant or if the characteristic throughput number $\pi_{\dot{m}}$ depends on the rotational velocity v_0 (Fig. 5.69). This can be explained by the fact that the velocity $v_{0,li}$ (curve a) decreases with an

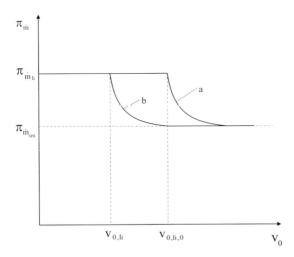

Figure 5.69   Shift of the dimensionless throughput curve at scaling up,a: model,b: pilot plant

increasing machine size. For this velocity the following characteristic number quotients are relevant according to Eq. 5.290:

$$\frac{Bi}{Br} \sim \frac{1}{v_{0,li}D} = idem \qquad \frac{\beta Gz}{Br^2} \sim \frac{1}{v_{0,li}L} = idem \qquad (5.417)$$

With a constant L/D ration, the following scale-up rule is yielded:

$$\frac{v_{0,li}}{v_{0,li,0}} = \left(\frac{D}{D_0}\right)^{-1} \qquad (5.418)$$

In case of a model transfer, curve a becomes curve b. The classic model theory is no longer be applicable. Therefore, in [113, 115] we have tried to solve this problem of a non transferability by an approximation of the form $\dot{m} \sim D^{2+\psi-x-\zeta_k}$, where ζ_k is an empirically determined correction factor. However, a correct solution shall be given here. For this we need, as in the melting case, a model transfer for variable characteristic numbers.

The starting point is the characteristic throughput curve of the model machine (Fig. 5,70). It gives the characteristic numbers

$$\pi_{\dot{m}_{li}} = \frac{\dot{m}_{li,0}}{\rho_{B0}H_0W_0v_{0,li,0}} \qquad \pi_{\dot{m}_{sm}} = \frac{\dot{m}_{sm,0}}{\rho_{B0}H_0W_0v_{0,0}} \qquad (5.419)$$

For the model and the pilot plant we may thus say according to Eq. 5.316:

Model:

$$\pi_{\dot{m},0} = \pi_{\dot{m}_{li}} \frac{v_{0,li,0}}{v_{0,0}} + \pi_{\dot{m}_{sm}} \left(1 - \frac{v_{0,li,0}}{v_{0,0}}\right) \qquad (5.420)$$

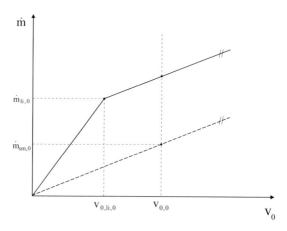

Figure 5.70   Throughput of a grooved barrel extruder and a smooth barrel extruder depending on the speed, solid friction: $v_0 \le v_{0,li,0}$, melt film friction at the end of the grooved barrel: $v_0 > v_{0,li,0}$

Pilot plant:

$$\pi_{\dot{m}} = \pi_{\dot{m}_{li}} \frac{v_{0,li}}{v_0} + \pi_{\dot{m}_{sm}} \left(1 - \frac{v_{0,li}}{v_0}\right)$$

With the scale-up rule for $v_{0,li}$, $\pi_{\dot{m}}$ becomes dependent on the diameter ratio:

$$\pi_{\dot{m}} = \pi_{\dot{m}_{li}} \frac{v_{0,li,0}}{v_{0,0}} \left(\frac{D}{D_0}\right)^{x-2} + \pi_{\dot{m}_{sm}} \left[1 - \frac{v_{0,li,0}}{v_{0,0}} \left(\frac{D}{D_0}\right)^{x-2}\right] \qquad (5.421)$$

Fig. 5.68 shows a scheme of this function, and Fig. 5.71 shows a three-dimensional respresentation. Only if

$$x = 1 \qquad \text{i.e.} \quad \frac{v_0}{v_{0,0}} = 1 \qquad\qquad (5.422)$$

is Eq. 5.421 exact, because only then does $\pi_{\dot{m}}$ start at point $v_{0,li}$ (Fig. 5.72). For

$$x \neq 1 \qquad\qquad (5.423)$$

the initial values are less or greater than $v_{0,li}$ (Fig. 5.72). In this case, Eq. (5.421) is no longer exact.

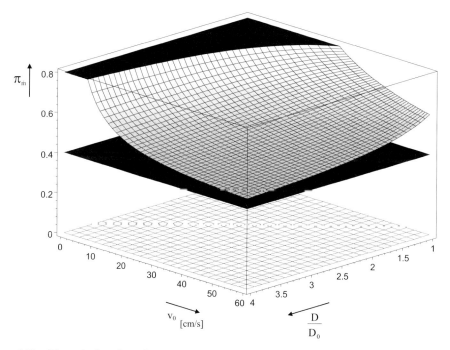

Figure 5.71   Dimensionless throughput depending on the rotational velocity and the diameter ratio

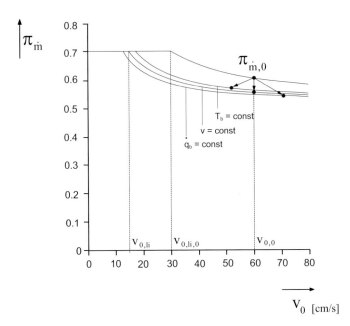

Figure 5.72   Dimensionless throughput for different scale-up conditions

In the further discussion we have to distinguish two cases:

Case 1: $$\frac{v_{0,0}}{v_{0,li,0}} \geq 1$$ (5.424)

Case 2: $$\frac{v_{0,0}}{v_{0,li,0}} \leq 1$$ (5.425)

In the first case, $v_{0,0}$ lies in the melt film friction domain, and in the second case it lies in the solid friction domain.

Case 1
With

$$\pi_{\dot{m}} = \frac{\dot{m}}{\rho_B W H v_0}$$ (5.425)

and the scale-up rules for *W, H,* and v_0 we have

$$\dot{m} = \rho_B W_0 H_0 v_{0,0} \pi_{\dot{m}} \left(\frac{D}{D_0}\right)^{2+\psi-x}$$ (5.426)

Divided by $\dot{m}_0$, the new scale-up rule for the throughput is obtained with Eq. 5.421:

$$\frac{\dot{m}}{\dot{m}_0} = \frac{1 + \dfrac{\pi_{\dot{m}_{sm}}}{\pi_{\dot{m}_{li}}}\left[\dfrac{v_{0,0}}{v_{0,li,0}}\left(\dfrac{D}{D_0}\right)^{2-x} - 1\right]}{1 + \dfrac{\pi_{\dot{m}_{sm}}}{\pi_{\dot{m}_{li}}}\left(\dfrac{v_{0,0}}{v_{0,li,0}} - 1\right)}\left(\frac{D}{D_0}\right)^{\psi} \tag{5.427}$$

where again the solution is exact only with $x = 1$. The channel depth exponent thus is

$$\psi = \frac{1}{1+n} \tag{5.428}$$

By means of Eq. 5.427 we can now explain why with experimental data for grooved-barrel extruders, the scale-up rule for the throughput

$$\frac{\dot{m}}{\dot{m}_0} = \left(\frac{D}{D_0}\right)^{\chi} \qquad 1.4 \le \chi \le 1.65 \tag{5.429}$$

occurs. As becomes obvious in the double logarithmic plotting of Eq. 5.427 (Fig. 5.73), it can be well approximated by an exponential approach (Eq. 5.429), where we may say for χ

$$\chi = f\left(n, \frac{v_{0,0}}{v_{0,li,0}}, \frac{\pi_{\dot{m}_{sm}}}{\pi_{\dot{m}_{li}}}\right) \tag{5.430}$$

This functional relation is represented in Fig. 5.74. For a predefined material, the variables $v_{0,li,0}$, $\pi_{\dot{m}_{sm}}$ and $\pi_{\dot{m}_{li}}$ depend only on the screw geometry. Thus, the above χ values can be explained by the fact that $\pi_{\dot{m}}$ does not remain invariant in case of an enlargement of the machine.

Case 2
If $v_{0,0}$ is less than $v_{0,li,0}$, we again have to distinguish two cases. For the model transfer, the point

$$v_{0,0} = v_{0,li,1} \tag{5.431}$$

is decisive (Fig. 5.75). Due to the scale-up rule (Eq. 5.418) we may say

$$\frac{D_1}{D_0} = \frac{v_{0,li,0}}{v_{0,0}} = \frac{v_{0,li,0}}{v_{0,li,1}} \tag{5.432}$$

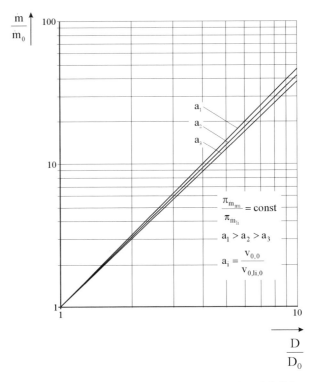

Figure 5.73   Double logarithmic representation of the throughput function (5.5.101)

If $D < D_1$ (Case 2a), the dimensionless throughput is

$$\pi_{\dot{m}_{li}} = idem \tag{5.433}$$

and thus the scale-up rule

$$\frac{\dot{m}}{\dot{m}_0} = \left(\frac{D}{D_0}\right)^{2+\psi-x} \tag{5.434}$$

If, however, $D > D_1$ (Case 2b), Eq. 5.427 yields in consideration of D_1

$$\frac{\dot{m}}{\dot{m}_1} = \left\{1 + \frac{\pi_{\dot{m}_{sm}}}{\pi_{\dot{m}_{li}}}\left[\left(\frac{D}{D_1}\right)^{2-x} - 1\right]\right\}\left(\frac{D}{D_1}\right)^{\psi} \tag{5.435}$$

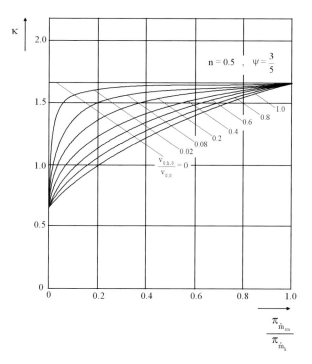

Figure 5.74 Exponent of the throughput model law as a function of the quotient of the characteristic throughput number, with the velocity quotient as parameter

With the scale-up rule for D_1 and

$$\frac{\dot{m}_1}{\dot{m}_0} = \left(\frac{D_1}{D_0}\right)^{2+\psi-x} = \left(\frac{v_{0,li,0}}{v_{0,0}}\right)^{2+\psi-x}$$ (5.436)

we obtain

$$\frac{\dot{m}}{\dot{m}_0} = \left\{1 + \frac{\pi_{\dot{m}_{sm}}}{\pi_{\dot{m}_{li}}}\left[\left(\frac{D}{D_0}\frac{v_{0,0}}{v_{0,li,0}}\right)^{2-x} - 1\right]\right\}\left(\frac{D}{D_0}\right)^{\psi}\left(\frac{v_{0,li,0}}{v_{0,0}}\right)$$ (5.437)

for

$$\frac{D}{D_0} \geq \frac{D_1}{D_0} = \frac{v_{0,li,0}}{v_{0,0}}$$ (5.438)

and

$$x = 1 \qquad \psi = \frac{1}{1+n} \tag{5.439}$$

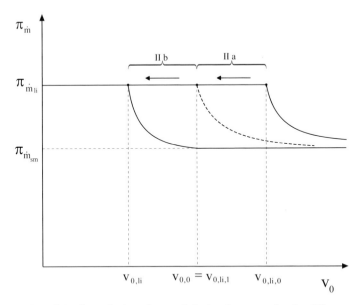

Figure 5.75   Function of the dimensionless characteristic throuhgput number for different scale-up conditions, Case 2a: $\pi_{\dot{m}_{li}} = idem$; Case 2b: $\pi_{\dot{m}} \neq idem$

Power, torque, and specific throughput will then follow the known rules:

$$\frac{P}{P_0} = \frac{\dot{m}}{\dot{m}_0} \tag{5.440}$$

$$\frac{M_t}{M_{t0}} = \frac{\dot{m}/N}{\dot{m}_0/N_0} \tag{5.441}$$

Thus, there are also consistent scale-up rules for grooved-barrel extruders.

5.5.4   Summary

5.5.4.1   Scale-Up Rules for Melt and Smooth-Barrel Plasticating Machines

The generally formulated scale-up rules are listed in Table 5.13 and the corresponding exponents ψ and x are given in Table 5.14. The scale-up rules for variable L/D rations should not be applied to plasticating extruders, because, here, the melting length can increase considerably (great ω_S).

Table 5.13   Scale-Up Rules for Smooth-Barrel Extruders and for Grooved-Barrel Extruders on the Condition $\pi_{\dot{m}}$ = idem

	$\dfrac{L}{D}$ = const	$\dfrac{L}{D}$ = variable
Geometry		
$\dfrac{i}{i_0}$	$\left(\dfrac{D}{D_0}\right)^0$	$\left(\dfrac{D}{D_0}\right)^0$
$\dfrac{L}{L_0} = \dfrac{L_i}{L_{i0}}$	$\left(\dfrac{D}{D_0}\right)^1$	$\left(\dfrac{D}{D_0}\right)^{1+\omega}$
$\dfrac{t}{t_0}$	$\left(\dfrac{D}{D_0}\right)^1$	$\left(\dfrac{D}{D_0}\right)^1$
$\dfrac{W}{W_0} = \dfrac{e}{e_0}$	$\left(\dfrac{D}{D_0}\right)^1$	$\left(\dfrac{D}{D_0}\right)^1$
$\dfrac{H_i}{H_{i0}}$	$\left(\dfrac{D}{D_0}\right)^{\psi}$	$\left(\dfrac{D}{D_0}\right)^{\psi_\omega}$
Process parameters		
$\dfrac{T_2}{T_{2,0}} = \dfrac{T_1}{T_{1,0}}$	$\left(\dfrac{D}{D_0}\right)^0$	$\left(\dfrac{D}{D_0}\right)^0$
$\dfrac{p_i}{p_{i,0}} = \dfrac{p_1}{p_{1,0}} = \dfrac{p_2}{p_{2,0}}$	$\left(\dfrac{D}{D_0}\right)^0$	$\left(\dfrac{D}{D_0}\right)^0$
$\dfrac{N}{N_0}$	$\left(\dfrac{D}{D_0}\right)^{-x}$	$\left(\dfrac{D}{D_0}\right)^{-x_\omega}$
$\dfrac{\dot{m}}{\dot{m}_0}$	$\left(\dfrac{D}{D_0}\right)^{2+\psi-x}$	$\left(\dfrac{D}{D_0}\right)^{2+\psi_\omega-x_\omega}$
$\dfrac{P}{P_0}$	$\left(\dfrac{D}{D_0}\right)^{2+\psi-x}$	$\left(\dfrac{D}{D_0}\right)^{2+\psi_\omega-x_\omega}$

Table 5.13   Continued

	$\dfrac{L}{D}$ = const	$\dfrac{L}{D}$ = variable
$\dfrac{\dot{Q}}{\dot{Q}_0}$	$\left(\dfrac{D}{D_0}\right)^{2+\psi-x}$	$\left(\dfrac{D}{D_0}\right)^{2+\psi_\omega-x_\omega}$
$\dfrac{M_t}{M_{t0}}$	$\left(\dfrac{D}{D_0}\right)^{2+\psi}$	$\left(\dfrac{D}{D_0}\right)^{2+\psi_\omega}$
$\dfrac{\dot{m}/N}{\dot{m}/N_0}$	$\left(\dfrac{D}{D_0}\right)^{2+\psi}$	$\left(\dfrac{D}{D_0}\right)^{2+\psi_\omega}$
$\dfrac{t_R}{t_{R0}}$	$\left(\dfrac{D}{D_0}\right)^{x}$	$\left(\dfrac{D}{D_0}\right)^{x_\omega}$

Table 5.14   General Formulation of the Exponents of the Channel Depth and the Speed Scale-Up Rules

$\dfrac{L}{D}$ = const	$\dfrac{L}{D}$ = variable
$\psi = \dfrac{1+n}{1+(2+\alpha)n}$	$\psi_\omega = \dfrac{(1+\omega)(1+n)}{1+(2+\alpha)n}$
$x = \dfrac{(1+n)(1+\alpha)}{1+(2+\alpha)n}$	$x_\omega = \dfrac{(1+\omega)(1+n)(1+\alpha)}{1+(2+\alpha)n} - \omega$
Note $0 \le \alpha \le 1$	

Principally, the scale-up rules are used for invariant L/D ratios. The corresponding ψ and x exponents are represented in Fig. 5.76. The usual area lies between the limits $\dot{q}_0 = const$, i.e., $\alpha = 0$ and $T_b = const$, i.e., $\alpha = 1$. It is cross-hatched.

Smooth-barrel extruders are situated at the upper limit $(\dot{q}_0 = const)$, i.e.,

$$\psi = x = \frac{1+n}{1+2n} \tag{5.442}$$

Then, the throughput and the power follow the scale-up rule

$$\frac{\dot{m}}{\dot{m}_0} = \frac{P}{P_0} = \left(\frac{D}{D_0}\right)^2 \tag{5.443}$$

The corresponding range data of some European extruder manufacturers are shown in the Fig. 5.77 and 5.78.

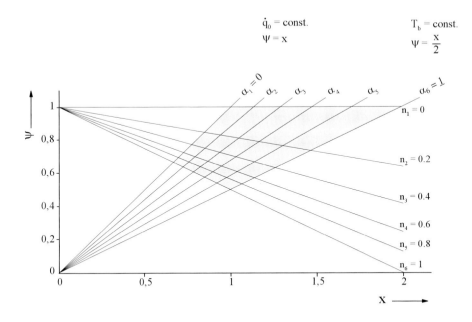

Figure 5.76   Diagram of the speed and channel depth exponents for L/D = const

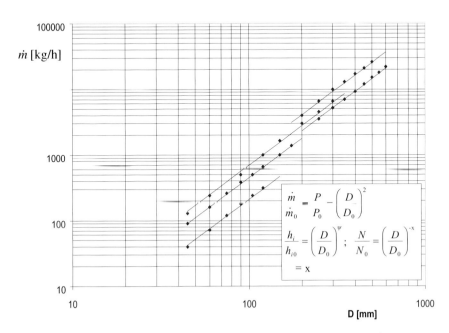

Figure 5.77   Throughput as a function of the diameter of different machine manufacturers

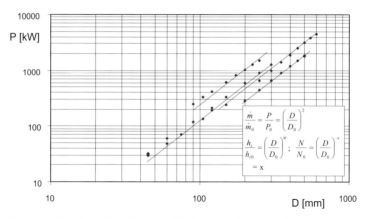

Figure 5.78   Power as a function of the diameter of different machine manufacturers

In the model law for the maximal speed, the speed exponent x as preset by theory is not applied (Fig. 5.79). A greater number of extruder manufacturers seem to follow the model law

$$\frac{N_{\text{max}}}{N_{\text{max},0}} = \left(\frac{D}{D_0}\right)^{-0.7} \tag{5.444}$$

[103]. Since, however, $x = 0.7$ is less than most of the values in the cross-hatched area, the theoretically correct value for x can be followed in the pilot plant.

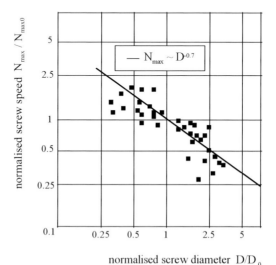

Figure 5.79   Normalised screw speed versus normalised screw diameter

With injection molding machines, the ψ-x exponents cover almost the entire cross-hatched area. Here, the exponent of the throughput model law lies between [116, 117]

$$1.45 \leq 2 + \psi - x \leq 2 \tag{5.445}$$

i.e., the upper values are at about

$$\psi = x \quad \text{d.h.} \quad \dot{q}_0 = const \tag{5.446}$$

and the lower values are at

$$\psi = \frac{x}{2} \quad \text{d.h.} \quad T_b = const \tag{5.447}$$

The latter are used for general-purpose screws [117].

5.5.4.2   Scale-Up Rules for Grooved-Barrel Extruders

For the model transfer of a grooved-barrel extruder, an exact distinction of the cases is necessary. In Case 1,

$$v_{0,0} \geq v_{0,li,0} \tag{5.448}$$

the generally formulated model laws listed in Table 5.15 apply, where

$$x = 1 \quad \psi = \frac{1}{1+n} \tag{5.449}$$

has to be set if the theory of similarity is applied in its narrow sense. If $v_{0,0}$ is distant enough from $v_{0,li,0}$ (Fig. 5.71), the ψ and x exponents of Table 5.14 can be used for $L/D = const$, too. The error is negligible because the $\pi_{\dot{m}}$ values do not differ much (Fig. 5.71). In Case 2,

$$v_{0,0} \leq v_{0,li,0} \tag{5.450}$$

we again have to distinguish two cases. To

$$\frac{D}{D_0} \leq \frac{D_1}{D_0} = \frac{v_{0,li,0}}{v_{0,0}} \tag{5.451}$$

the scale-up rules of Tables 5.13 and 5.14 apply for

$$\frac{L}{D} = const \tag{5.452}$$

Table 5.15   Scale-Up Rules for Grooved-Barrel Extruders for $\pi_{\dot{m}} \neq const$, $v_{0,0}$, $v_{0,li,0} \geq 1$, $L/D =$ **const**

	$\dfrac{v_{0,0}}{v_{0,li,0}} \geq 1 \; ; \; \dfrac{L}{D} = \textbf{const}$
Geometry	
$\dfrac{i}{i_0}$	$\left(\dfrac{D}{D_0}\right)^0$
$\dfrac{L}{L_0} = \dfrac{L_i}{L_{i0}}$	$\left(\dfrac{D}{D_0}\right)^1$
$\dfrac{t}{t_0}$	$\left(\dfrac{D}{D_0}\right)^1$
$\dfrac{W}{W_0}$	$\left(\dfrac{D}{D_0}\right)^1$
$\dfrac{H_i}{H_{i0}}$	$\left(\dfrac{D}{D_0}\right)^{\psi}$
Process parameters	
$\dfrac{T_2}{T_{2,0}} = \dfrac{T_1}{T_{1,0}}$	$\left(\dfrac{D}{D_0}\right)^0$
$\dfrac{p_i}{p_{i,0}} = \dfrac{p_1}{p_{1,0}} = \dfrac{p_2}{p_{2,0}}$	$\left(\dfrac{D}{D_0}\right)^0$
$\dfrac{N}{N_0}$	$\left(\dfrac{D}{D_0}\right)^{-x}$
$\dfrac{\dot{m}}{\dot{m}_0} = \dfrac{P}{P_0} = \dfrac{\dot{Q}}{\dot{Q}_0}$	$\dfrac{1 + \dfrac{\pi_{\dot{m}_{sm}}}{\pi_{\dot{m}_{li}}}\left[\dfrac{v_{0,0}}{v_{0,li,0}}\left(\dfrac{D}{D_0}\right)^{2-x} - 1\right]}{1 + \dfrac{\pi_{\dot{m}_{sm}}}{\pi_{\dot{m}_{li}}}\left(\dfrac{v_{0,0}}{v_{0,li,0}} - 1\right)} \left(\dfrac{D}{D_0}\right)^{\psi}$

Table 5.15   Continued

	$\dfrac{v_{0,0}}{v_{0,li,0}} \geq 1$; $\dfrac{L}{D} = \text{const}$
$\dfrac{M_t}{M_{t0}} = \dfrac{\dot{m}/N}{\dot{m}/N_0}$	$\dfrac{1 + \dfrac{\pi_{\dot{m}_{sm}}}{\pi_{\dot{m}_{li}}}\left[\dfrac{v_{0,0}}{v_{0,li,0}}\left(\dfrac{D}{D_0}\right)^{2-x} - 1\right]}{1 + \dfrac{\pi_{\dot{m}_{sm}}}{\pi_{\dot{m}_{li}}}\left(\dfrac{v_{0,0}}{v_{0,li,0}} - 1\right)}\left(\dfrac{D}{D_0}\right)^{\psi + x}$
Exp.	$x = 1, \qquad \psi = \dfrac{1}{1+n}$

If, however,

$$\frac{D}{D_0} > \frac{D_1}{D_0} = \frac{v_{0,li,0}}{v_{0,0}} \qquad (5.453)$$

the scale-up rules of Table 5.16 are applicable, where

$$x = 1 \qquad \psi = \frac{1}{1+n} \qquad (5.454)$$

must be set again if the theory of similarity is applied in its narrow sense.

Table 5.16   Scale-Up Rules for Grooved-Barrel Extruders for $\pi_{\dot{m}} \neq const$,

	$\dfrac{D}{D_0} \geq \dfrac{v_{0,li,0}}{v_{0,0}}$; $\dfrac{v_{0,0}}{v_{0,li,0}} < 1;$ $\dfrac{L}{D} = \text{const}$
Geometry	
$\dfrac{i}{i_0}$	$\left(\dfrac{D}{D_0}\right)^0$
$\dfrac{L}{L_0} = \dfrac{L_i}{L_{i0}}$	$\left(\dfrac{D}{D_0}\right)^1$
$\dfrac{t}{t_0}$	$\left(\dfrac{D}{D_0}\right)^1$

Table 5.16   Continued

	$\dfrac{D}{D_0} \geq \dfrac{v_{0,li,0}}{v_{0,0}}; \quad \dfrac{v_{0,0}}{v_{0,li,0}} < 1; \quad \dfrac{L}{D} = \text{const}$
$\dfrac{W}{W_0}$	$\left(\dfrac{D}{D_0}\right)^1$
$\dfrac{H_i}{H_{i0}}$	$\left(\dfrac{D}{D_0}\right)^{\psi}$
Process parameters	
$\dfrac{T_2}{T_{2,0}} = \dfrac{T_1}{T_{1,0}}$	$\left(\dfrac{D}{D_0}\right)^0$
$\dfrac{p_i}{p_{i,0}} = \dfrac{p_1}{p_{1,0}} = \dfrac{p_2}{p_{2,0}}$	$\left(\dfrac{D}{D_0}\right)^0$
$\dfrac{N}{N_0}$	$\left(\dfrac{D}{D_0}\right)^{-x}$
$\dfrac{\dot{m}}{\dot{m}_0} = \dfrac{P}{P_0} = \dfrac{\dot{Q}}{\dot{Q}_0}$	$\left\{1 + \dfrac{\pi_{\dot{m}_{zm}}}{\pi_{\dot{m}_{li}}}\left[\left(\dfrac{D}{D_0}\dfrac{v_{0,0}}{v_{0,li,0}}\right)^{2-x} - 1\right]\right\}\left(\dfrac{D}{D_0}\right)^{\psi}\left(\dfrac{v_{0,li,0}}{v_{0,0}}\right)^{2-x}$
$\dfrac{M_t}{M_{t0}} = \dfrac{\dot{m}/N}{\dot{m}/N_0}$	$\left\{1 + \dfrac{\pi_{\dot{m}_{zm}}}{\pi_{\dot{m}_{li}}}\left[\left(\dfrac{D}{D_0}\dfrac{v_{0,0}}{v_{0,li,0}}\right)^{2-x} - 1\right]\right\}\left(\dfrac{D}{D_0}\right)^{\psi+x}\left(\dfrac{v_{0,li,0}}{v_{0,0}}\right)^{2-x}$
Exp.	$x = 1; \qquad \psi = \dfrac{1}{1+n}$

Nomenclature

Symbols

A	surface area
A, B, C	Carreau parameters
a	center line distance
$a = \lambda/\rho\, c_v$	thermal diffusivity
c_p, c_v	specific heat capacity
D	screw diameter

d, d_{ij}	rate of deformation tensor, components
d	pellet diameter
E	Young modulus of elastic
E_A, R	arrhenius parameters
e	flight thickness
f	filling level
G	shear modulus
H	enthalpy
H	screw channel depth
h	specific enthalpy
i	number of flights
K	power law consistency parameter
L	screw length
M	torque
$\dot{m}$	mass throughput
N	screw speed (rpm)
n	power law exponent parameter
P	power
p	pressure
$\dot{q}$	specific heat flux
$\dot{Q}$	heat flux
$Q = \dot{V}$	volumetric throughput
r, x_i	distance vector components
T	temperature
T_B	reference temperature
T_S	standard temperature
t	screw pitch
t	time
U	internal energy
u	specific internal energie
V	volume
$\dot{V} = Q$	volumetric throughput
v	specific volume
v, v_i	velocity vector, components
W	width of screw channel
x, y, z	location coordinates
r	exponent of the screw speed scale-up rule
Z	screw length in z-direction
α	heat transfer coefficient
α	exponent of scale-up rules
β	coefficient of thermal expansion
γ	shear strain
$\dot{\gamma}$	shear rate
$\acute{o}$	clearance between screw flight and barrel
$\dot{\varepsilon}$	rate of extension
η	shear viscosity

η_D	extension viscosity
η_0	zero shear viscosity
λ	thermal conductivity
μ	friction coefficient
ν	Poisson´s ratio
ϱ	density
ϱ_B	bulk density
σ, σ_{ij} $(i,j=1,\,2,\,3)$	stress tensor, component
τ	relaxation time
τ	shear stress
φ	screw helix angle
ψ_1, ψ_2	principle and second normal stress coefficients
ψ	exponent of the channel scale-up rule
ω	angular velocity
$I,\ II,\ III$	second-order tensor invariants
$I_d,\ II_d,\ III_d$	rate of deformation tensor invariants

Dimensionless Parameters

Bi	Biot number
Br	Brinkman number
Fr	Froude number
Gr	Grashof number
Gz	Graetz number
Nu	Nusselt number
Pe	Peclet number
Pr	Prandtl number
Re	Reynolds number
Θ	dimensionless temperature
π	dimensionless parameter
π_p	dimensionless pressure Gradient
$\pi_{\dot{V}}$	dimensionless volume throughput

Indices

a	axial
B	bulk
b	barrel
CM	cooling medium
Cry	crystalline (melting point)
c	critical

ch	channel
cl	clearance
core	core
cyc	cycle
dos	mold filling
Fl	flow
g	glass transition
gr	grooved bush
in	in
iso	isothermal
li	limit
M	melt
Mc	melt conveying bzw. metering
Mel	melting
n-iso	non-isothermal
out	out
P	pore
PM	premelting
pool	pool
R	residence
r	relativ
S	solid
Sc	solid conveying
s	screw
sl	slide
sm	smooth barrel
st	stop

References

1. Tadmor, Z., Gogos, C.G., *Principles of Polymer Processing* (1979) John Wiley & Sons, New York
2. Potente, H., *Rheol. Acta* (1983) 22, pp. 387–395
3. Effen, N., (1996) Dr.-Ing. dissertation, University-GH Paderborn
4. Tadmor, Z., Klein, I., *Engineering Principles of Plasticating Extrusion* (1978) Robert F. Krieger, Publishing Company, Huntington, NY
5. Potente, H., Effen, N., Liu, J., *Polym. Eng. Sci.* (1996) 36, pp. 1557–1564
6. Potente, H., Obermann, Ch., *Intern. Polym. Proc. XIV* (1999) 1, pp. 21–27
7. Middleman, S., *Fundamentals of Polymer Processing* (1977) McGraw-Hill, New York
8. Potente, H., *Rheol. Acta* (1983) 22, pp. 387–395
9. Hensen, F., Knappe, W., Potente, H., *Handbuch der Kunststoff-Extrusionstechnik,* I: *Grundlagen.* (1989) Carl Hanser Verlag, Munich
10. Prüß, J., (1991) Non published manuscript. University-GH Paderborn
11. Tautz, H., *Wärmeleitung und Temperaturausgleich* (1971)Verlag Chemie GmbH, Weinheim/Bergstr.
12. Worth, R.A., Parnaby, J., *Polym. Eng. Sci.* (1977) 17, pp. 257/25
13. Uhland, E., *Rheol. Acta* (1979) 18, pp. 1–24
14. Ingen Housz, I.F., *Z. Lebensmitteltechnologie* (1983) 2, pp. 88–93
15. Mennig, G., *Kunststoffe* (1984) 74, pp. 296–298

16. Meijer, H.E.H., Verwraak, C.P.I.M., *Polym. Eng. Sci.* (1988) 28, (11), p. 758–772
17. Adewale, K.E.P., Olabisi, O., Oyediran, A.A., Olnuloyo, V.O.S., *Intern. Polym. Proc. VI* (1991) 3, pp. 195–198
18. White, J., Han, M.J., Nakajima, N., Brzoskowski, R., *J. Rheol.* (1991) 35, p. 167
19. Ji, Z., Gotsis, A.D., *Intern. Polym. Proc. VII* (1992) 2, pp. 132–139
20. Hatzikiriakos, S.G., *Intern. Polym. Proc. VIII* (1993) 2, pp. 135–142
21. Lawal, A., Kalyon, D.M., *Polym. Eng. Sci.* (1994) 34, (19) pp. 1471–1479
22. Potente, H., Schöppner, V., Ujma, A., *J. Polym. Eng.* (1997) 19, (2), pp. 153–169
23. Maddock, B.H., *SPE J.* (1959) 15, (5), pp. 383–389
24. Tadmor, Z., *Polym. Eng. Sci.* (1966) 6, p. 185
25. Klenk, K.P., (1969) Dr.-Ing. dissertation, RWTH Aachen
26. Lindt, J.T., *Polym. Eng. Sci.* (1976) 16, (4) pp. 284–291
27. Ingen Housz, I.F., Meijer, E.H., *Polym. Eng. Sci.* (1981) 21, (6) pp. 232–39; (17), pp. 1156–1161
28. Mount, E.M., Watson, J.G., Chung, C.I., *Polym. Eng. Sci.* (1982) 22, (12), pp. 729–737
29. Fukase, H., Takeshiki, Shinya, S., Nomura, A., *Polym. Eng. Sci.* (1982) 22, (9), pp. 578–586
30. Lindt, J.T., Elbirli, B., *Polym. Eng. Sci.* (1983) 25, (9), pp. 412–418
31. Lindt, J.T., *Polym. Eng. Sci.* (1985) 25, (10), pp. 585–588
32. Rauwendaal, C., *Polymer Extrusion* (1986) Carl Hanser Verlag, Munich
33. Potente, H., Koch, M., *Plastverarbeiter* (1987) 38, (11), pp. 112–116; 166–172
34. Koch, M., (1987) Dr.-Ing. dissertation, University-GH Paderborn
35. McClelland, D.E., Chung, C.I., *Polym. Eng. Sci.* (1988) 23, (2), pp. 180–184
36. Bruker, J., Balek, G.S., *Polym. Eng. Sci.* (1989) 29, (9), pp. 258–267
37. Grünschloß, E., *Der Extruder im Extrusionsprozeß*, (1989) VDI-Verlag, Düsseldorf, pp. 155–199
38. Lee, K.Y., Han, Ch.D., *Polym. Eng. Sci.* (1990) 30, (11), pp. 665–676
39. Potente, H., *Intern. Polym. Proc.* (1991) 6, (4), pp. 297–303
40. Potente, H., Stenzel, H., *Intern. Polym. Proc.* (1991) 6, (2), pp. 126–135
41. Zhu, F., Chen, L., *Polym. Eng. Sci.* (1991) 31, (15), pp. 1113–1116
42. Schulte, H., (1992) Dr.-Ing. dissertation, University-GH Paderborn
43. Stenzel, H., (1992) Dr.-Ing. dissertation, University-GH Paderborn
44. Schöppner, V., (1994) Dr.-Ing. dissertation, University Paderborn
45. Lipshitz, S.D., Lavie, R., Tadmor, Z., *Polym. Eng. Sci.* (1974) 14, (8), pp. 553–559
46. Jungemann, J., (1998) Dr.-Ing. dissertation, University-GH Paderborn
47. Darnell, W.H., Mol, E.A. J., *SPE J.* (1956) 20, (12), pp. 20–29
48. Schneider, K., (1968) Dr.-Ing. dissertation, RWTH Aachen
49. Tadmor, Z., Breyer, E., *Polym. Eng. Sci.* (1972) 19, (5), pp. 378–386
50. Menges, G., Hegele, R., Langecker, G., *Plastverarbeiter* (1972) 23, (7), pp. 455–460
51. Hegele, R., (1972) Dr.-Ing. dissertation, RWTH Aachen
52. Lovegrove, I.G.A., Williams, J.G., *Polym. Eng. Sci.* (1974) 14, (9), pp. 589 ff.
53. Langecker, G., (1977) Dr.-Ing. dissertation, RWTH Aachen
54. Ingen Housz, J.F., In *Kunststoff-Fortschrittsberichte*, Band. 1 (1976) Carl Hanser Verlag, Munich
55. Fritz, H.G., In *Kunststoff-Fortschrittsberichte*, Band 1 (1976) Carl Hanser Verlag, Munich
56. Peiffer, H., (1981) Dr.-Ing. dissertation, RWTH Aachen
57. Rautenbach, R., Peiffer, H., *Kunststoffe* (1982) 72, pp. 137–143; 262–266; 696–700
58. Grünschloß, E., *Kunststoffe* (1985) 75, pp. 405–409
59. Feistkorn, W., (1985) Dr.-Ing. dissertation, RWTH Aachen
60. Potente, H., *Kunststoffe* (1985), 75, pp. 439–441
61. Peng, Y., Cheng, J., Wu, Y., *ANTEC, Technical Papers* (1985) 31, pp. 73–75
62. Potente, H., *Kunststoffe* (1988) 78 (4), pp. 355–363
63. Koch, M., (1987) Dr.-Ing. dissertation, University-GH Paderborn
64. Rauwendaal, Ch., *Adv. Polym. Technol.* (1989) 9 (4), pp. 301–308
65. Hwang, C., McKelvey, J.M., *Adv. Polym. Technol.* (1989) 9, (3), pp. 227–251
66. Potente, H., Koch, M., *Adv. Polym. Technol.* (1989) 9, (2), pp. 119–127
67. Potente, H., Koch, M., *Intern. Polym. Proc. IV* (1989) 4, pp. 208–218
68. Potente, H., *Kunststoffe* (1990) 80, (1), pp. 80–84
69. Potente, H., Stenzel, H., Bergedieck, J., *Adv. Polym. Technol.* (1990) 10 (4), pp. 285–295

70. Potente, H., (Hrsg.) 1992) Kunststofftechnisches Seminar – Rechnergestützte Extruderauslegung, Universität-GH Paderborn

71. Yousuff, M., Page, N.W., *Powder Technol.* (1993) 76, pp. 299–307

72. Chen, W., Malghan, S.G., *Powder Technol.* (1994), 81, pp. 75–81

73. Campbell, G.A., Dontula, N., *Intern. Polym. Proc. X* (1995) 1, pp. 30–35

74. Spalding, M.A., Kirkpatrick, D.E., Hyun, K.S., *Polym. Eng. Sci.* (1993) 33, (7), pp. 423–430

75. Spalding, M.A., Hyun, K.S., *Polym. Eng. Sci.* (1995) 35, (7), pp. 557–563

76. Spalding, M.A., Hyun, K.S., Jenkins, S.R., Kirkpatrick, D.E., *Polym. Eng. Sci.* (1995) 35, (23), pp. 1907–1916

77. Potente, H., Schöppner, V., *Intern. Polym. Proc. X* (1995) 1, pp. 1014

78. Potente, H., Schöppner, V., *Intern. Polym. Proc. X* (1995) 4, pp. 289–295

79. Spalding, M.A., Hyun, K.S., Hughes, K.R., *SPE-ANTEC Technical Papers* (1996)

80. Schöppner, V., (1996) Dr.-Ing. dissertation, University-GH Paderborn

81. Hyun, K.S., Spalding, M.A., Hinton, C., *Composites* (1997) 16, (13), pp. 1210–1219

82. Spalding, M.A., *SPE-ANTEC Technical Papers* (1997)

83. Yamamuro, J., Penumadu, D., Campbell, G.A., *Intern. Polym. Proc. XIII* (1998) 1, pp. 3–8

84. Pawlowski, J., *Ähnlichkeitsprinzip in der physikalisch technischen Forschung* (in German) (1971) Springer Verlag, Berlin

85. Strub. R.A., In *Proceedings of the Second Midwestern Conference on Fluid Mechanics,* The Ohio State University (1952) p. 481

86. Carley, J.F., McKelvey, J.M., *Ind. Eng. Chem.* (1953) 45, p. 989

87. Schenkel, G., *Kunststoffe* (1959), 49, pp. 15, 63, 122

88. Maddock, B.H., *SPE J.* (1959) Nov., p. 983

89. Schenkel, G., *Industrie-Anzeiger* (1972) 94, p. 113

90. Maddock, B.H., *Polym. Eng. Sci.* (1974), 4, p. 853

91. Pearson, J.R.A., *Plastics and Rubber: Processing* (1976), Sept., p. 113

92. Yi, B., Fenner, R.T., *Plastics and Rubber: Processing* (1976), Sept., p. 119

93. Fischer, P., (1976) Dr.-Ing. dissertation, RWTH Aachen

94. Potente, H., Fischer, P., *Kunststoffe* (1977) 67, p. 242

95. Potente, H., *Rheol. Acta* (1978) 17, p. 406; *Handbuch des 9. Kunststofftechnischen Kolloquiums des IKV* (1978) Aachen

96. Schenkel, G., *Kunststoffe* (1978) 68, p. 155

97. Potente, H., *A comprehensive Model Theorie for Conventional Single-Screw Machines in Plasticating Processing: Handbook of the Polymer Extrusion Conference I.* (1979) The Plastics and Rubber Institute, London, pp. 3.1–3.11

98. Potente, H., *Auslegung von Schneckenmaschinenbaureihe: Kunststoff-Fortschrittsberichte,* Band 6 (1981) Carl Hanser Verlag, Munich

99. Peiffer, H., (1981) Dr.-Ing. dissertation, RWTH Aachen

100. Rautenbach, R., Peiffer, H., *Kunststoffe* (1982), 72, pp. 137, 262, 616

101. Ingen Housz, J., *Handbook of the Polymer Extrusion Conference II (* (1982). The Plastics and Rubber Institute, London,Paper 8

102. Menges, G., Wortberg, J., Mayer, A., *Adv. Polym. Technol.* (1983), 3, p. 157

103. Mayer, A., (1984) Dr.-Ing. dissertation, RWTH Aachen

104. Limper, A., (1984) Dr.-Ing. dissertation, RWTH Aachen

105. Feistkorn, W., (1985) Dr.-Ing. dissertation, RWTH Aachen

106. Chung, Ch. I., *Polym. Eng. Sci.* (1984), 24. p. 626

107. Fritz, H.G., In *Kunststoff-Extrusionstechnik,* Band 2, Hensen, F., Knappe, W., Potente, H., (Eds.) (1986) Carl Hanser Verlag, Munich pp. 365–381

108. Rauwendaal, Ch., (1987) *Polym. Eng. Sci.* (1987), 27, p. 1059

109. Potente, H. In *Kunststoff-Extrusionstechnik,* Band 1, Hensen, F., Knappe, W., Potente, H. (Eds.), (1989) Carl Hanser Verlag, Munich, pp. 227–278

110. Grünschloß, E., In *Der Extruder im Extrusionsprozeß – Grundlagen für Qualität und Wirtschaftlichkeit* (1989) VDI-Verlag, Düsseldorf, pp. 155–199

111. Potente, H., *Kunststoffe* (1990), 80, p. 80

112. Potente, H., *Kunststoffe* (1990), 80, p. 208

113. Potente, H., *Plastverarbeiter* (1980), 71, p. 474

114. Potente, H., *Intern. Polym. Process. VI* (1991), p. 297
115. Potente, H., *Intern. Polym. Process. VI* (1991) 4, pp. 267–278
116. Bürkle, E., (1989) Dr.-Ing. dissertation, RWTH Aachen
117. Johannaber, F., *Injection Molding Machines* (1994) Hanser Publishers, Munich
118. Obermann, Ch., Dr.-Ing. dissertation, University, Paderborn 2000

6 Twin and Multiscrew Extrusion

James L. White

6.1 Introduction

While single screw extruders are the dominant type of screw extrusion machines, twin screw extruders have become of increasing importance since the 1950s. In this chapter we describe (1) the development of the technology of twin screw extruders, (2) experimental studies of their behavior, and (3) models of their behavior based on fluid mechanics and energy balances, which may be used for design. There are many different classes of twin screw machines. These differ from each other in terms of directions of rotation of the screws and types of interactions between the screws. They also vary in their industrial applications. We organize this chapter into separate sections dealing with self-wiping, co-rotating machines; intermeshing, counter-rotating machines, tangential, counter-rotating machines; and continuous mixers.

6.2   Intermeshing Co-Rotating Twin Screw Extrusion

6.2.1   Technology

The first self-wiping, co-rotating twin screw extruder was described by Adolph Wunsche [1] in a 1901 German patent. Similar self-wiping twin screw machines were subsequently described by Easton [2, 3]. These early machines are shown in Fig. 6.1. They were invented because of their self-cleaning features. The first commercial co-rotating twin screw extruder was due to R. Colombo and *Lavoratorie Materie Plastische* [4–9] in Turin who applied them to profile extrusion. The modern modular machine grew out of activities of R. Erdmenger and his associates with *Farbenfabriken Bayer AG* (now *Bayer AG)* in Leverkusen beginning in the late 1940s [9–12].

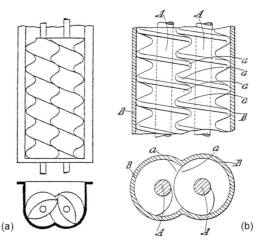

Figure 6.1   Early self-wiping co-rotating twin screw machine: (a) Wunsche [1];   (b) Easton [2, 3]

Self-wiping co-rotating twin- and triple-screw machines for mixing and devolatilization were devised and applied by Erdmenger and W. Meskat [13–15] at the *IG Farbenindustrie* Wolfen Works during the early 1940s. With the end of the war and the advance of the Soviet army, Erdmenger and Meskat went into western Germany and found employment with the newly independent *Farbenfabriken Bayer AG,* which had previously been part of the *IG Farbenindustrie.* Here Erdmenger in Leverkusen and Meskat in Dormagen continued their activities. Erdmenger developed new mixing machines based on the principle of self-wiping, co-rotating pumping machines [16–19]. These included "knetscheiben" or kneading disk block machines in which the rotors consist of pairs of blocks of self-wiping single-, double-, or triple-tipped staggered disks (see Fig. 6.2). [*] Riess and Erdmenger [22] published in 1951 the first paper describing kneading disk block machines in the open literature.

---

[*] American patent examiners found the 1932 patent [20] of W.K. Nelson of *Universal Gypsum and Lime,* which described a double-tipped "self-wiping" paddle system. The American patent system of Erdmenger [21] was then limited to exclude double-tipped kneading disk blocks, which had been included in the German patent [17].

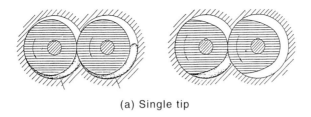

(a) Single tip

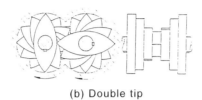

(b) Double tip

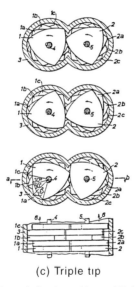

(c) Triple tip

Figure 6.2   Kneading disk block (Knetscheiben) machines of Erdmenger [17, 19]:
(a) single tip;   (b) double tip;   (c) triple tip

The first modular, co-rotating twin screw extruder was described in a 1950 German patent application of Meskat and Pawlowski [23] of *Farbenfabriken Bayer AG*. Their machine, which is shown in Fig. 6.3, consisted of combinations of right-handed and left-handed screw modules. This machine is described in a 1951 paper of Riess and Meskat [24]. In a 1955 paper by Riess [25], a system of six Meskat-Pawlowski machines in series are described as involved in the esterification of cellulose (Fig. 6.4).

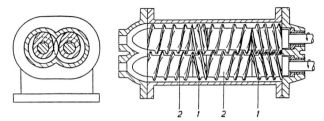

Figure 6.3   Modular, self-wiping, co-roatating twin screw extruder of Meskat and Self-Pawlowski [23]

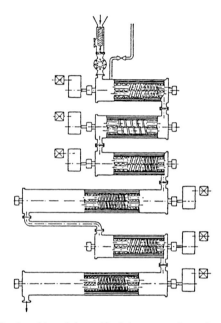

Figure 6.4   Use of Meskat–Pawlowski modular, self-wiping co-rotating twin screw extruders for the esterification of cellulose after Riess

Subsequently, Erdmenger [10, 12, 26, 27] developed modular, co-rotating twin screw machines with right- and left-handed kneading disk blocks as well as screws (Fig. 6.5). This machine proved far more versatile because it combined the modularity of the Meskat–Pawlowski machine and the mixing and masticating capability of Erdmenger's kneading disk blocks.

In the mid 1950s, the machine design of Erdmenger was licensed by *Farbenfabriken Bayer AG* to *Werner and Pfleiderer GmbH* (later *Krupp Werner and Pfleiderer* and now *Coperion Werner and Pfleiderer*) of Stuttgart for commercial development. The commercial machine would be called the "ZSK (Zwei Schnecken Kneten) System Erdmenger." The new commercial compounding machine was first described in the open literature in a 1959 paper by Fritsch and Fahr [28] of *Werner and Pfleiderer* (Fig. 6.6). Subsequent papers were published by Herrmann [29, 30] of the same firm in 1964 and 1966, which describe machines of varying scale including 160-mm screws. Applications to

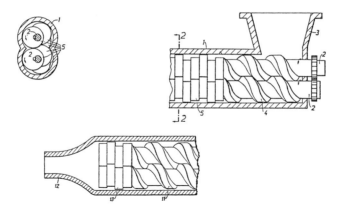

Figure 6.5   Modular, self-wiping, co-rotating twin screw extruder of Erdmenger [26]

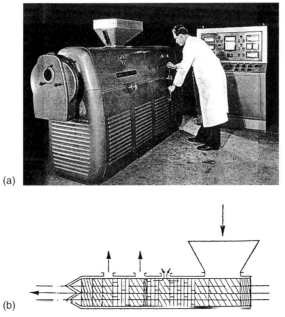

(a)

(b)

Figure 6.6   Early Werner and Pfleiderer modular, co-rotating twin screw extruders: (a) photograph of a 1957 ZSK 83700 machine (courtesy of Werner and Pfleiderer);   (b) diagram showing applications of machine from Fritsch and Fahr

compounding, devolatilization, and homogenization are noted as well as applications to many commercial polymers. Herrmann featured the ZSK machine prominently in his 1972 monograph *Schneckenmaschinen in der Verfahrenstechnik* [11]. The modern modular co-rotating twin screw machine is starve fed receiving material from various feeders and containing storved as well as fully fed regins along the screw axes.

Werner and Pfleiderer engineers made many improvements on the original Erdmenger machine. The early Erdmenger machines had the two self-wiping screws arranged vertically. They changed this to a horizontal arrangement, which greatly improves the feeding. *Werner and Pfleiderer* engineers obtained a number of patents [31–38] on machine improvements and applications to polymer processing and chemical and food technology, including the manufacture of chocolate [36], compounding thermosetting resins [38], and continuous soap production [37].

In the United States, Loomans and Brennan of the Pennsylvania-based *Readco Machine Company* filed patent applications [39, 40] for a special modular, co-rotating twin screw extruder. These were granted by the U.S. Patent Office. By the time the patents issued, they had been licensed to *Baker Perkins* (later *APV Chemical Machinery*), an English firm that had engineering and manufacturing facilities in Saginaw, Michigan. Loomans joined *Baker Perkins*.

The first patent application using a self wiping co-rotating twin screw extruder for polymerization had been a 1943 *I.G. Farbenindustrie* German patent application [42] using a series of Colombo LMP machines that were applied to sodium polymerization of butadiene and to polyurethanes. In 1964, Illing and Zahradnik of the University of Erlangen filed a patent [43] for the continuous anionic polymerization of caprolactam in a modular, co-rotating twin screw extruder arguing the superiority over single screw extruders. Illing also described this process in a subsequent paper [44]. Illing in a later patent application [45] proposed the free radical polymerization of vinyl monomers in the same machine. These patents were licensed to *Werner and Pfleiderer GmbH*. A *Baker Perkins* patent application in 1969 by Wheeler, Irving, and Todd [46] treats the continuous polycondensation of polyester and polyamide from prepolymer in a modular, co-rotating twin screw extruder. A 1971 *Bayer AG* patent application of Erdmenger, Ullrich, and coworkers describe a co-rotating twin screw extrusion process to produce cross-linked lacquers [47]. A 1976 patent application by Ullrich, Meisert, and Eitel [48] describes the polymerization of polyurethane elastomers in this type of machine. Various polymerization processes now came to be investigated by different chemical companies using twin screw extruders.

In the 1970s, the patent system of Bayer AG on self-wiping, co-rotating twin screw extruder systems expired and other European companies began to manufacture modular, intermeshing, co-rotating twin screw extruders. Among the early German firms to enter the market place with similar machines were *Hermann Berststorff Machinenbau* of Hannover and *Maschinenfabrik Paul Leistritz* of Nuremburg. In Japan, *Ikegai Corporation* of Tokyo, *Japan Steel Works* of Hiroshima, and *Toshiba Machine Company* of Tokyo began to produce these machines, as did *Clextral* in France. Today there are over 60 manufacturers of modular, co-rotating twin screw machines around the world.

There has been continued interest since the introduction of the modular, co-rotating twin screw extruder in the development of new elements. In 1962/63 new, more complex designs of screw elements were proposed by Erdmenger [49] of *Bayer AG* and in 1973 by Loomans [50] of *Baker Perkins*. These are shown in Fig. 6.7a. The purpose of the Erdmenger design was to develop elements with geometric complexity transverse to the screw axis and the direction of flow so that the thickness of layers of material being extruded can be better controlled. This element has a groove cut in the screw channel near the loading flight. The Loomans screw element of a decade later has a hump in the profile. It is recommend for handling thermally sensitive materials. The Loomans design screw was introduced by *Baker Perkins* as a module.

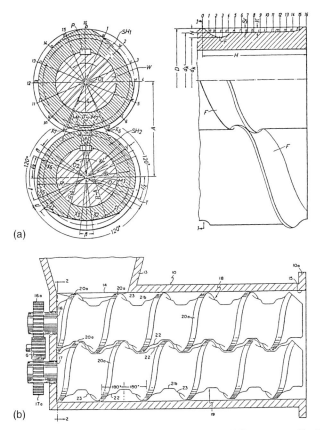

Figure 6.7   New self-wiping screw profiles: (a) Erdmenger 1962/3 screw profile [49];   (b) Loomans 1973 screw profile [50]

Another type of element that has been introduced into modular, co-rotating twin screw extruders is valves. The first valve sections were introduced into modular, co-rotating twin screw extruders in 1961 by Colombo [51] of *LMP*. They were intended for devolatilization applications and were Venturi-shaped elements (Fig. 6.8a). Todd [52] of *Baker Perkins* described an adjustable saddle section in a 1977 patent application (Fig. 6.8b). Such valves are able to introduce large resistances to flow and in their wake regions of starvation.

Many different modules in addition to kneading disk blocks have been proposed for the purpose of enhancing mixing and homogenization in co-rotating twin screw extruders. The earliest design was in 1950 patent application [53, 54] by Kraffe de Laubarede, an independent French inventor. This design, which was intended to be machined onto a screw shaft, is shown in Fig. 6.9a. It consists of intermeshing disks containing slots for material passage. A second self-wiping, co-rotating design intended for mixing and homogenization was a screw flighted rotor with a very large helix angle (Fig. 6.9b). This was proposed by Ellermann [55] in 1953 for a continuous mixing machine where this element and normal screws would be machined onto a shaft. Ellermann's screw rotor design was incorporated into the *Krauss-Maffei* and *Japan Steel Works* DSM (Doppel Schnecken Mischer) [56, 57].

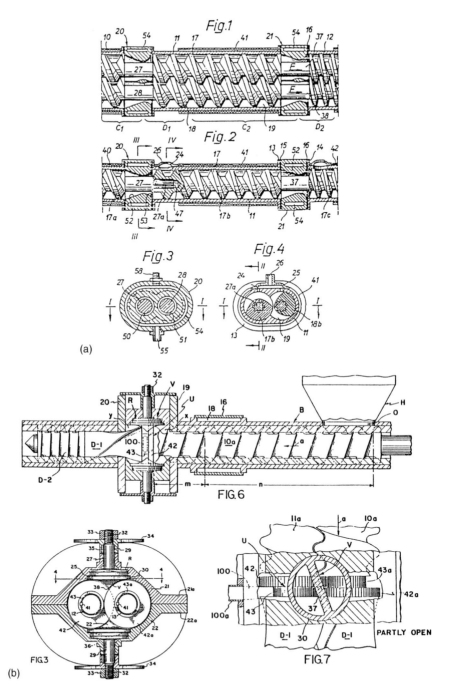

Figure 6.8   Valve sections in a modular, self-wiping twin screw section: (a) Colombo 1961 Venturi section [51];   (b) Todd 1979 [52]

Another early mixing element for self-wiping co-rotating twin screw extruders was introduced by Colombo [58] in 1964. This consisted of left-handed screw elements with slices (Fig. 6.9c). In a 1981 patent application, Rathjen and Ullrich [59] of *Bayer AG* described screw elements containing transverse cuts for handling the extrusion of crystallizing melts (Fig. 6.9d).

In the 1980s, *Berstorff* introduced the use of gears as mild mixing elements in twin screw extruders. *Werner and Pfleiderer* subsequently introduced Turbine Mixing Elements (TME) [60] and also the Zahn Misch Elemente (ZME), which is a left-handed screw with slices in its flights [61] (Fig. 6.10).

Another trend is the introduction of high flight angle rotors in the place of kneading disk blocks. Inspiration for this design seems not to be the Ellermann machine [55] shown in Fig. 6.9b, but rather the Farrel Continuous Mixer (FCM) [62], a nonintermeshing, counter-rotating machine with (see Section 6.5). A self-wiping, co-rotating twin screw extruder with high flight angle rotors was first introduced by *Kobe Steel* in the late 1980s. Similar designs have been developed in more recent years by *Rockstedt, Farrel Corp.,* and *Pomini SpA*. The *Farrel* machine has been described in a paper by Valsamis and Canedo [63]. Rotor elements are shown in Fig. 6.11.

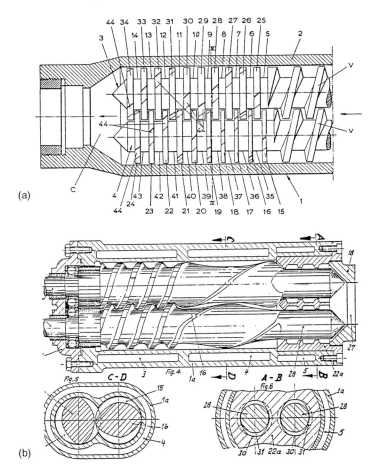

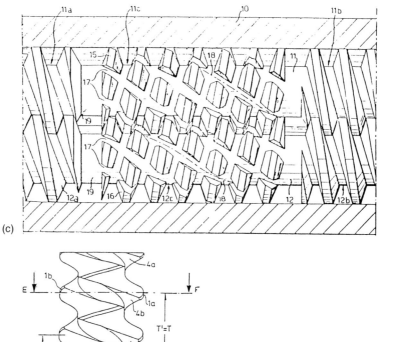

Figure 6.9   Mixing homogenization sections for modular, self-wiping, co-rotating twin screw extruders: (a) Kraffe de Laubarede 1950 homogenizing section [53, 54];   (b) Ellermann 1953 high helix angle screw rotor [55];   (c) Colombo 1964 Mixing element [58];   (d) Rathjen–Ullrich element for crystallizing melts [59]

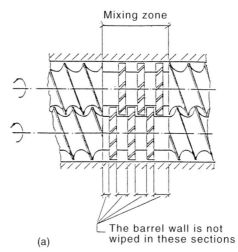

Figure 6.10   (a) Newer mixing/homogenization sections for modular, self-wiping, co-rotating twin screw extruders: Berstorff gear element

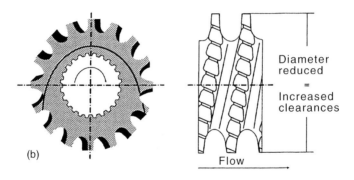

Figure 6.10   (b) Werner and Pfleiderer ZME element [61]

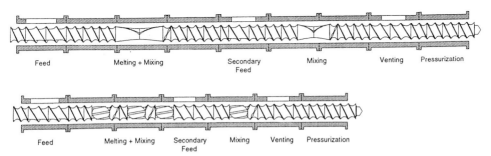

Figure 6.11   Rotor elements for modular, co-rotating twin screw extruder [63]

6.2.2   Geometry

For two screws that rotate in the same direction to wipe each other clean, the screws must have special shapes. The problem is first discussed and modeled in a 1901 patent of Wunsche [1] and subsequently by later investigators [64, 65]. Erdmenger [12] states his associate A. Geberg developed the solution for the shape of the profile. In the scientific literature, the first published solution is by Booy [65] of the *DuPont Company*. With reference to Fig. 6.12, the solution involves the form $H(0)$ the depth of cut in the shaft as a function of angle around the self-wiping screw cylinder. The form of $H(\theta)$ will be the same for screw or kneading disk blocks, as they are both self-wiping. The primary variables to define $H(\theta)$ are the screw radius R_s, and the distance C_L between the center-lines of the self-wiping shafts. Clearly, the maximum value of $H(\theta)$ is $H(0) = H$:

$$H(0) = H = 2R_s - C_L \tag{6.1}$$

If α is the angle of the screw tip, then for a screw with m flights one may write

$$m(2\alpha + 4\psi) = 2\pi \tag{6.2}$$

where there are m screw channels (e.g., m would be two in a double-flighted screw channel). Booy argues that

$$\psi = \cos^{-1}\left(\frac{C_L}{2R_s}\right)$$

(6.3)

which leads to

$$\alpha = \frac{\pi}{m} - 2\cos^{-1}\left(\frac{C_L}{2R_S}\right)$$

(6.4)

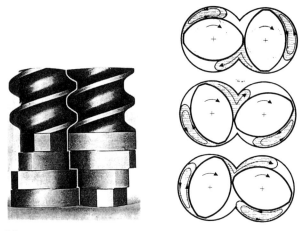

Figure 6.12   Self-wiping screws

Eq. 6.4 indicates that the value of α is determined by C_L and R_s. For a double-flighted screw ($m = 2$), where the value of C_L/R_s is $\sqrt{2}$, the value of α will be zero.

Booy determines that the shape of $H(\theta)$ as a function of position around the shaft should be

$$H(\theta) = R_S(1 + \cos\theta) - \sqrt{C_L^2 - R_S^2 \sin^2\theta}$$

(6.5)

where angle θ is measured from the position of maximum channel cut. When θ equals zero, Eq. 6.5 becomes equivalent to Eq. 6.1.

6.2.3   Experimental Studies

6.2.3.1   Early Studies to 1975

The earliest experimental studies of flow in a modular, co-rotating twin screw extruder are those of Erdmenger [10, 12] with *Farbenfabriken Bayer AG,* which showed that material moved forward in the machine in a roughly helical Figure-eight motion (Fig. 6.13) in these machines and that there were regions of fill and starvation along its axis. Erdmenger also presented the first investigations of material flow in kneading disk blocks (Fig. 6.14).

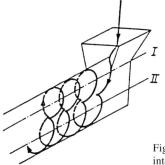

Figure 6.13   Erdmenger's [10] observations of flow in screw regions in intermeshing, co-rotating twin screw extruder

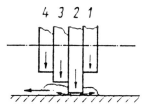

Figure 6.14   Erdmenger's [10] consideration of flow in a kneading disk block

There is a second generation of key experimental studies, which are associated with the commercialization of co-rotating twin screw extruders by *Werner and Pfleiderer* and *Baker Perkins*. Studies published in the period 1965–1975 emphasize the measurement of residence time distribution. This may be seen in the work of Herrmann [30], Todd and Irving [66], and Todd [67]. Residence times were found to be of order 100 seconds and decreased with screw speed.

Another early published study was by Armstroff and Zettler [68] of *BASF* in 1973. They fashioned a transparent barrel and introduced modular *Werner and Pfleiderer* ZSK screws into them. Using a liquid polyisobutylene (Oppanol B3/B10 mixture), they studied the fluid motions and found alternating fully filled and starved regions. Generally, kneading disk blocks and left-handed screws were filled and right-handed screw elements were starved. Fig. 6.15, taken from their paper, which shows surmised pressure profiles together with screw geometry, illustrates their conclusion that there were pressure buildups in the screw regions and pressure losses in the kneading disk blocks. Filled lengths and pressures were also measured, but they were strangely correlated with fluid Reynolds numbers.

6.2.3.2   Flow Visualization Including Solid Conveying and Melting

There have been many subsequent flow visualization studies for the modular, co-rotating twin screw extruder. They tend to supplement and extend the investigations of Erdmenger [10] and Armstroff and Zettler [68]. Transparent barrel experiments for moderate viscosity oils containing markers have been reported by Sakai [69] of *Japan Steel Works*. *Werner and Pfleiderer GmbH* in a commercial videotape describing their ZSK machine showed flow

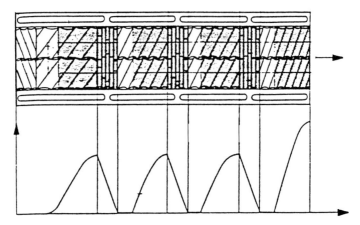

Figure 6.15   Armstroff and Zettler's [68] observations presumption of pressure profiles in a modular, co-rotating twin screw extruder

visualization investigations. Meijer and Elemans [70], and P. J. Kim and White [71] subsequently made flow visualization investigations. In these various studies the Fig.-eight motions of the advancing fluid moving through screw elements was observed. The occurrence of filled kneading disk blocks and starved screws were noted. The *Werner and Pfleiderer* videotapes and subsequently more explicitly Kim and White [71] indicate backward flows between the tips of the individual staggered kneading disks. Pulsating flows were also observed.

The early observation [68] of starved screws and fully filled regions near kneading disk blocks and left-handed screws are confirmed by later investigators who have carried out experiments with polymer systems and subsequently examined the material distributions along the screws [72–75]. In more recent years, attention has given to the motion of solid pellets in the feed region of a modular co-rotating twin screw extruder by Bawiskar and White [76] and by Potente and Melisch [77]. In a videotape of the former authors [76] the existence of two solid conveying regions is shown. One conveying region is above and between the horizontal screws and a second below the screws where pellets are dragged into the region of intermesh (Fig. 6.16).

Figure 6.16   Regions of solid conveying in a self-wiping, co-rotating twin screw extruder

Observations have also been made on the melting of thermoplastics in modular twin screw extrusion. These have been based on operating under steady-state conditions, then cooling down the machine, removing the modular screw shaft and inspecting them. Such efforts have been made by Todd [78], Bawiskar and White [76, 79], Potente and Melisch [77], and Curry [80]. Bawiskar and White [76, 79] observe on sectioned carcasses when the barrel is hotter than the polymer melting temperature a melted region along the barrel with a solid bed along the screw surface (Fig. 6.17). Potente and Melisch [77] see pellets dispersed in melt matrix, which seems to suggest a later stage of melting. More recently Jung and White [80b] have found that depending upon operating conditions, melting can begin at the barrel surtace, the screw survace and the bulk. Barrel melting occurs when the barrel temperature is above the melting temperature. The other mechanisms occur when the barrel temperature is below the crystalline melting temperature.

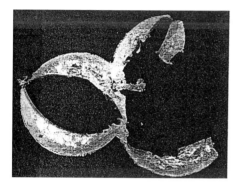

Figure 6.17   Carcass cross sections indicating melting initiated at barrel [78]

Bawiskar and White [79] note the importance of the modular screw design used in determining the occurrence of melting. Generally, in starve-fed, co-rotating twin screw extruders, melting occurs only in fully filled pressurized regions. If there are only screw elements present, melting is delayed until the pressurized region just before the die. Introduction of nonpumping or poorly pumping elements even in a kneading disk block leads at this location to the machine becoming completely filled and pressurized. The rate of movement is slowed and heat conduction into the material greatly enhanced. This is where melting begins. Generally, the greater the resistance to pumping, the more rapidly melting proceeds. Placing additional left-handed kneading disk blocks or screws to the back of a right-handed kneading disk block increases the extent of fill not only in the kneading disk block but in the right-handed screw elements prior to it. This leads to a more rapid and earlier melting, with melting occurring even in these right-handed screw elements.

6.2.3.3   Residence Time Distributions

There have been numerous investigations of residence time distributions in twin screw extruders during the past generation. Different experimental techniques have been used and the influence of various process conditions studied. The first studies were by Herrmann [30] and Todd and Irving [66, 67]. There were subsequent investigations by Sakai [69], Hornsby [81], Bur and Gallant [82], Kim and White [71], Kye and White [83] and Shon et al. [83a] among others. Various experimental techniques have ranged from tracer particles [30, 66, 72, 83] to radioactive tracers [68, 81] and fluorescence [82].

Certain results are clear. Increasing screw speed in any screw configuration reduces the residence time [30, 71, 83, 83a]. Increasing throughput reduces residence time [71, 83a]. Replacing right-handed screw elements with left-handed screws or with kneading disk blocks increases the residence time and broadens the residence time distribution [71, 83] (Fig. 6.18).

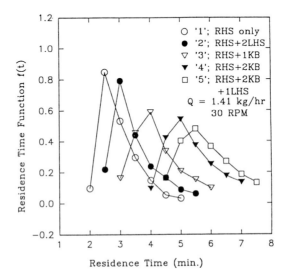

Figure 6.18   Residence time distributions for a modular, co-rotating twin screw extruder

6.2.3.4   Pumping Characteristics

Certain points need to be borne in mind to understand pumping characteristics. First, commercial modular machines are generally operated in a starved manner with the throughput controlled by the metering actions at the various input hoppers. The throughput of the machine is determined by the programmed input. In machines that are flood fed, such as single screw extruders, the output varies with screw speed and the changed levels of throughput occurs in all of the elements. In a starved-feed screw extruder with metered input, variations in screw speed can change only levels of fill in elements but not the throughput, which is maintained constant.

The pumping characteristics of a modular, co-rotating twin screw extruder are generally expressed in terms of screw characteristic curves, which relate throughput to pressure rise and loss at specific screw speeds. Studies of the pumping characteristics of twin screw extruders were first reported by Armstroff and Zettler [68] as part of their previously cited investigations. Experimental studies of the pumping characteristics of various screw and paddle (kneading disk) elements were presented in 1989 by Todd [84] of *APV Chemical Machinery* based on work carried out with H.G. Karian. The pumping was represented by

$$Q = \alpha N - \frac{\beta'}{\eta L} \Delta p \qquad (6.6)$$

where α and β' are experimental parameters, Q is flow rate, and Δp is pressure rise. Values of α and β' were determined.

Experimental studies of Y. Wang et al. [74] and White et al. [75] published in 1989–90 determined the filled lengths L_f in modular, co-rotating twin screw extruders, which were mentioned in Section 6.2.3.2. These were used as the basis of determining pumping characteristics of polymer melts. The Q–Δp relationship of an element may be represented by an expression of the type of Eq. 6.6, where L_f would be proportional to Δp. These authors used these characteristics to test flow simulations of their model [74].

In 1994, Gogos et al. [85] at the Polymer Processing Institute described a device for experimentally determining the pumping characteristics of elements in self-wiping, co-rotating screw extruders. Most recently N.H. Wang et al. [86] of *Japan Steel Works* have experimentally determined screw characteristic curves of various elements in a modular, co-rotating twin screw extruder.

6.2.3.5   Heat Transfer

The only study of heat transfer in a self-wiping, co-rotating twin screw extruder has been that of Todd [87] of *APV Chemical Machinery* published in 1988 (Figs. 6.19 and 6.20). The heat

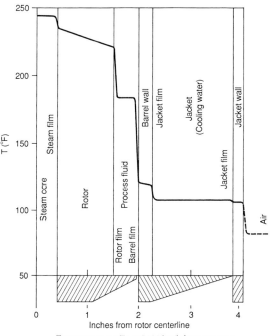

Temperature profile at 5000 btu/h from rotors to
barrel, and 500 btu/h from jacket to ambient

Figure 6.19   Temperature profile in a twin screw extruder after Todd [87]

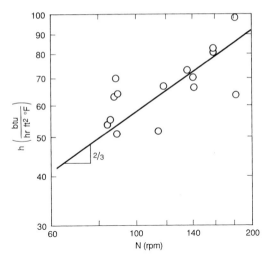

Figure 6.20   Internal heat transfer coefficient in intermeshing co-rotating twin screw extruder after Todd [87]

transfer rate to fluid moving through a two-tipped kneading disk block (paddle) with a 90°
configuration in a 100-mm-diameter screw configuration was determined. The process fluids
were glucose, polyvinyl chloride, polyethylene terephthalate, and polyamide-6. Various heat
transfer media were used in the screws and barrel. Todd gives a correlation of form

$$\frac{hD}{k} = 0.94 \left(\frac{D^2 N \rho}{\eta} \right)^{0.28} \left(\frac{c \eta}{k} \right)^{0.33} \tag{6.7}$$

where h is the heat transfer coefficient, D is the screw diameter, N is the screw speed, k is
the thermal conductivity, η is the viscosity, and ρ is the density of the process fluid. A
fundamental basis for this Eq. (6.7) was suggested by White [9] and later given in detail by
White et al. [87a] using a convective heat transfer analysis.

6.2.3.6   Mixing

The first basic study of mixing in a modular, intermeshing, co-rotating twin screw extruder
was in 1985 by Bigio and Erwin [88]. They studied the striation thickness of black and
white silicone layers along the length of a modular, co-rotating twin screw extruder. The
apparatus of Bigio and Erwin [88] is shown in Fig. 6.21. They found that in the screw
section the number of striations and the associated interfacial area between black and white
layers increased linearly with position along the length of the screw (Fig. 6.22). In the
kneading disk block, they observed a rapid increase in number of striations and graying
(Fig. 6.23).

  During the period 1985–1990 various investigations, notably Ess and Hornsby [89, 90]
and Kalyon and his coworkers [91, 92], have published experimental studies on the distribution of solid particulates in modular, intermeshing, co-rotating twin screw extruders.
Optical microscopy [89–91] and nuclear magnetic resonance [92] imaging methods have

been used. Generally, the thrust of these studies is that enhanced mixing improves solid particle dispersion. Little consideration is given to extruder design variables.

From the late 1980s, various investigators, beginning with Plochocki et al. [93], have studied the development of phase morphology of polymer blends in modular, co-rotating twin screw extruders [93–100]. These papers have had different emphases. Sundaraj et al. [95] and Bordereau et al. [96] were interested in the mechanism of development of phase morphology with special concern for the role of the interfacial tension as the pellets of the minor phase evolve to their final dispersed state. This mechanistic concern is also found in

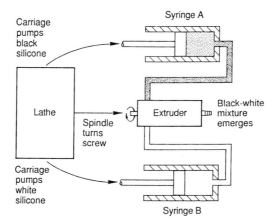

Figure 6.21 Apparatus of Bigio and Erwin [87] for studying flow in an intermeshing, co-rotating twin screw extruder

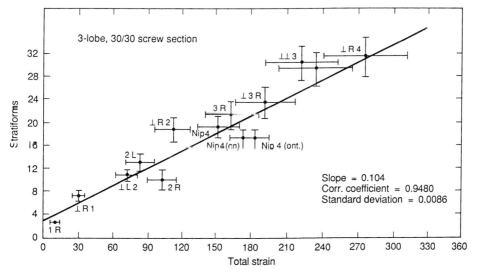

Figure 6.22 Bigio and Erwin's [88] observation of the number of striations versus total strain in screw channel

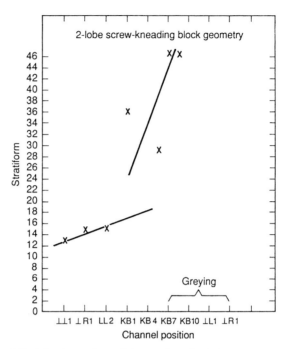

Figure 6.23   Bigio and Erwin's observations of the number of striations versus position in channel for a screw kneading disk block combination [88]

the papers of Lim, Cho, Lee, and White [94, 97, 99, 100], but from different viewpoints. Lim and White [97] show that introducing (or, better, in situ-creating) a compatibilizing agent leads to a local rapid increase in the rate of mixing and a significantly lowered disperse phase size. Lee and White [100] explore the role of viscosity ratio in mixing and show that a low-viscosity phase acts as a lubricant and greatly reduces the rate of mixing.

The influence of modular screw design on the development of phase morphology in a 20/80 polyamide-6/polyethylene blend system has been described by Lim and White [93]. These authors compare mixing being carried out in modular screws of varying design. The screws were removed following the preparation of the blends and inspected. Optical and scanning electron microscopy observations were used to determine the dimensions of the dispersed (polyamide-6) phase. The modular screw designs investigated and the number average polyamide-6 disperse phase size variation along the screw axis are shown in Fig. 6.24. The results may be summarized as follows. No mixing occurs until both phases are melted. When a screw has only right-handed screw elements, melting of the polyamide phase is delayed until the polymer reaches the region in front of the die. Generally, however, if both polymers are added into the first hopper, melting and mixing will occur in the first set of kneading disk blocks. A very finely dispersed blend may be obtained by having additional sets of kneading disk blocks along the screw.

When one of the materials to be mixed is a low-viscosity liquid, it is best that the more viscous polymer be added into the first feed port and melted before the low-viscosity phase is added.

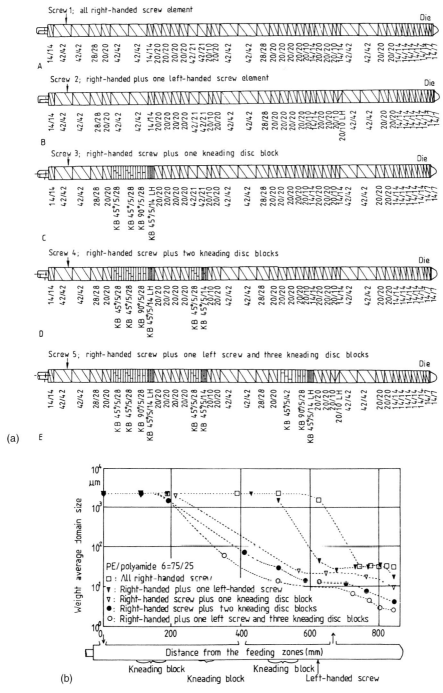

Figure 6.24   (a) Modular screw designs investigated by Lim and White to determine the influence of screw design on mixing [94];   (b) Variation of weight average polyamide-6 phase size with positions along the screw axis for different modular screw design [94]

6.2.4   Flow Modeling

6.2.4.1   General

The earliest efforts at modeling flow in self-wiping, co-rotating twin screw extruders tend to presume that the machine behaves as a fully filled screw pump and to simulate the forward drag and backward pressure as in a single screw extruder. The papers of Herrmann and Burkhardt [101] and Denson and Hwang [102] published from 1978 to 1980 are very much in this spirit. In this same period, studies by Armstroff and Zettler [69] and Werner [72], among others, who had done flow visualization on disassembled machines, pointed out that there were fully filled regions and starved regions along the machine axis. It was clear from the work of the former authors that occurrence of fill and starvation was associated with the pumping capacity of elements. Poorly pumping elements such as kneading blocks tended to be filled and screw elements starved except in front of dies. In 1987–88, papers by White and Szydlowski [103] and by Meijer and Elemans [70] described procedures for making calculations of pressure profiles and positions of fill along the axis of modular screws. This has since been extended and put into quantitative form by the University of Akron group [74, 75, 104–106] and by researchers at the University of Paderborn [107, 107a] among others.

The results described above consider pumping, global fluid mechanics, and heat transfer but did not consider the mechanisms of mixing in modular, co-rotating twin screw extruders. Of primary concern here is flow in kneading disk blocks. The earliest efforts at modeling flow in kneading disk blocks was by H. Werner [72], but this was done without fluid mechanics. Papers in 1987/8 by Szydlowski et al. [108–109] used a hydrodynamic lubrication theory formulation for Newtonian fluids to compute flow fluxes in kneading disk blocks which lead to mixing. In 1989, Gotsis and Kalyon [110] presented the first simulation of flow in kneading disk blocks which used the total Navier-Stokes Eq.s. It was carried out by a finite element numerical simulation method.

The problems cited above – (1) the fluid mechanics/pumping of screw and kneading disk blocks, (2) the global nonisothermal fluid mechanics, including melting in modular twin screw machines, (3) the phenomena of mixing, and (4) reactive extrusion – have been the four research thrusts of modeling activities on these machines. We have organized this chapter to consider each of these in turn.

6.2.4.2   Flow Pumping in Modules

6.2.4.2.1   Screw Elements

The flow has been analyzed by considering the fluid to move from screw 1 to screw 2 and back to screw 1 as it advances. This follows Erdmenger's [10] mechanism of Fig. 6.13 schematically for flattened out screws and is shown in Fig. 6.25. Generally, two types of flows are observed to occur in the screw element sections of modular, co-rotating twin screw extruders. The screw may be fully filled with fluid or polymer melt or the screw sections may be partially empty, i.e., starved, with fluid occurring only along the leading flight (see Fig. 6.26). We shall first consider the starved-flow case and then in greater detail the fully filled screw channel.

For flow in a starved screw channel, the equations of motion need be considered only within the viscous fluid phase. If we establish a coordinate system embedded in the

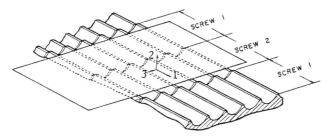

Figure 6.25   Screw channel layout for 'flattened out approximation' in screw regions of self-wiping, co-rotating twin screw extruder

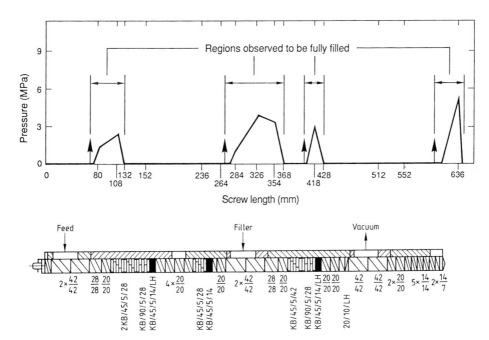

Figure 6.26   Fully filled and starved extruder screws

unwrapped screw and directed along the screw channel so that 1 is the direction of flow, 2 is perpendicular to the screw axis, and 3 is the transverse direction, then only two stress components, σ_{12} and σ_{32}, need be considered to a first approximation. If each screw has m thread starts, the composite self-wiping, co-rotating twin screw machine (except for a single-flighted screw) may be represented as having $(2m-1)$ parallel channels for flow [65]. The equations of motion in cartesian coordinates in the hydrodynamic lubrication approximation are

$$0 = -\frac{\partial p}{\partial x_1} + \frac{\partial \sigma_{12}}{\partial x_2} \qquad (6.8a)$$

$$0 = -\frac{\partial p}{\partial x_3} + \frac{\partial \sigma_{32}}{\partial x_2} \qquad (6.8b)$$

Since the viscous fluid in starved flow has an interface contacting the air along the 1 direction, the pressure gradient in direction of slow must be zero, i.e.,

$$\frac{\partial p}{\partial x_1} = 0 \tag{6.9}$$

Thus, we must have

$$\frac{\partial \sigma_{12}}{\partial x_2} = 0 \qquad \sigma_{12}(x_2) = \text{constant} \tag{6.10}$$

This leads to a velocity field of simple shear flow. If we neglect transverse drag flow, this is

$$v_1(x_1, x_2) = U_1\left(\frac{x_2}{H}\right) = \pi DN \cos\phi\left(\frac{x_2}{H}\right) \tag{6.11}$$

where we have introduced $U_1 = \pi DN \cos\phi$ (Eq. 2.15). The flow rate in the starved flow region is Q, which is given by

$$\begin{aligned}
Q &= \int v_1(x_2, x_3)dx_2 dx_3 \\
&= \pi DN \cos\phi\int\left(\frac{x_2}{H}\right)dx_2 dx_3 \\
&= \left[(1/2)\pi DN \cos\phi\right]A_c\Phi
\end{aligned} \tag{6.12}$$

where A_c is the cross section area of the channel and Φ the fill factor of the cross section. For m-flighted screws, the total flow Q_T may be shown to be

$$Q_T = (2m - 1)Q \tag{6.13}$$

For fully filled channels, there is an extensive literature on flow simulation for both Newtonian and non-Newtonian fluids. The earliest studies were by Herrmann and Burkhardt [101] who considered one-dimensional shearing between the screw and barrel for a Newtonian fluid. Here the velocity field is

$$v_1(x_2) = \pi DN \cos\phi\left(\frac{x_2}{H}\right) - \frac{H^2}{2h}\left(-\frac{\partial p}{\partial x_1}\right)\left[\frac{x_2}{H} - \left(\frac{x_2}{H}\right)^2\right] \tag{6.14}$$

If this is integrated from the screw root to the barrel, one obtains

$$q_1 = \int_0^H v_1 dx_2 = \frac{1}{2}\pi DHN \cos\phi - \frac{H^3}{12\eta}\left(-\frac{\partial p}{\partial x_1}\right) \tag{6.15}$$

The flow through the cross section of a self-wiping screw channel is

$$Q = \int q_1 dx_3 = \frac{1}{2} \pi D \overline{HW_{ch}} N \cos\phi - \frac{\overline{H^3 W_{ch}}}{12\eta}\left(-\frac{\partial p}{\partial x_1}\right) \tag{6.16}$$

Here W_{ch} is the characteristic channel width where

$$\overline{HW_{ch}} = \int_0^{W_{ch}} H(x_3)dx_3 \qquad \overline{H^3 W_{ch}} = \int_0^{W} H^3(x_3)dx_3 \tag{6.17}$$

In a 1980 paper, Denson and Hwang [102] gave a more general solution for flow of a Newtonian fluid through the cross section of a self-wiping screw channel. The velocity field in the flow direction was taken as

$$v_1 = v_1(x_2, x_3) \tag{6.18}$$

The equation of motion for the force balance for such a flow is

$$0 = -\frac{\partial p}{\partial x_1} + \frac{\partial \sigma_{12}}{\partial x_2} + \frac{\partial \sigma_{13}}{\partial x_3} \tag{6.19}$$

For a Newtonian fluid

$$\sigma_{12} = \eta \frac{\partial v_1}{\partial x_2} \qquad \sigma_{13} = \eta \frac{\partial v_1}{\partial x_3} \tag{6.20}$$

where η is a constant shear viscosity. This leads to

$$0 = -\frac{\partial p}{\partial x_1} + \eta\left(\frac{\partial^2 v_1}{\partial x_2} + \frac{\partial^2 v_1}{\partial x_3^2}\right) \tag{6.21}$$

Eq. 6.21 was solved by Denson and Hwang [102] using finite element methods for the velocity field $v_1(x_2, x_3)$ using Eq. 6.5 for the screw channel shape. Their result for the total flow Q is

$$Q = \iint v_1(x_2, x_3)dx_2 dx_3 \tag{6.22}$$

$$Q = \frac{1}{2}\pi DH_{ch}W_{ch}N\cos\phi F_D - \frac{H_{ch}^3 W_{ch}}{12\eta}\left(\frac{\partial p}{\partial x_1}\right)F_P \tag{6.23}$$

where F_D and F_P are "screw channel" shape factors and H_{ch} and W_{ch} are characteristic channel depth and width. The intermeshing region was ignored.

In a 1987 paper, Szydlowski and White [111] describe another approach to the screw pumping problem. They take into consideration the role of the screw crests in re-directing the flow from screw to screw (Fig. 6.27) and raising the pressure losses along the screw channel. This was done by taking the equations of motion to be those of hydrodynamic lubrication theory, i.e. (compare Eq. 6.8).

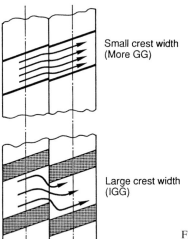

Small crest width
(More GG)

Large crest width
(IGG)

Figure 6.27   Screw crest effect on flow in intermeshing region in self-wiping, co-rotating twin screw extruder [111]

$$0 = -\frac{\partial p}{\partial x_1} + \frac{\partial \sigma_{12}}{\partial x_2} \quad \text{and} \quad 0 = -\frac{\partial p}{\partial x_3} + \frac{\partial \sigma_{32}}{\partial x_2} \tag{6.24}$$

or for a Newtonian fluid

$$0 = -\frac{\partial p}{\partial x_1} + \eta \frac{\partial^2 v_1}{\partial x_2^2} \quad \text{and} \quad 0 = -\frac{\partial p}{\partial x_3} + \eta \frac{\partial^2 v_3}{\partial x_2^2} \tag{6.25}$$

The solution of Eq. 6.25 for the fluxes is

$$q_1 = \int_0^H v_1 dx_2 = \frac{1}{2} U_1 H - \frac{H^3}{12\eta} \frac{\partial p}{\partial x_1} \tag{6.26a}$$

$$q_3 = \int_0^H v_3 dx_2 = \frac{1}{2} U_3 H - \frac{H^3}{12\eta} \frac{\partial p}{\partial x_3} \tag{6.26b}$$

These fluxes are balanced in a finite difference mesh as

$$q_1(x_1 + \Delta x_1, x_3)\Delta x_3 + q_3(x_1, x_3 + \Delta x_3)\Delta x_3 - q_1(x_1, x_3)\Delta x_3 - q_3(x_1, x_3)\Delta x_1 = 0 \tag{6.27}$$

and solved. This method generalizes the FAN method of Tadmor et al. [112] devised for simulation of die extrusion and injection molding. Szydlowski and White [111] calculated screw characteristic curves for various self-wiping screw designs and considered the intermeshing region. The effect of screw flight crest dimension at the intermesh on pressure development from their calculations is shown in Fig. 6.28.

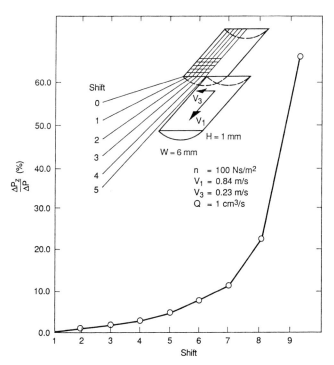

Figure 6.28   Effect of screw flight crest thickness on pressure development [112]

From the calculations of both Denson and Hwang [102] and Szydlowski and White [111] we may calculate screw characteristic curves, which have the form of Eq. 6.16 and 6.22. It is useful if these can be put in dimensionless form. If we divide the latter equation by $2\pi R_s^3 N \cos\phi$, which has dimensions of flow rate, we obtain

$$\frac{Q}{2\pi R_S^3 N \cos\phi} = \frac{H_{ch}W_{ch}}{4R_S^2}F_D - \frac{H_{ch}^3 W_{ch}}{24\pi\eta R_S^3 N \cos\phi}\frac{\Delta p}{x_1}F_P$$

$$= \frac{1}{4}\frac{H_{ch}W_{ch}}{R_S^2}F_D - \frac{H_{ch}^3 W}{12R_S^4}F_P\left(\frac{R_S}{2\pi\eta N \cos\phi}\frac{\Delta p}{x_1}\right)$$ (6.28)

Fig. 6.29 presents a dimensionless plot of screw characteristic curves in terms of

$$\frac{Q}{2\pi R_s^3 N \cos\phi} \quad \text{vs.} \quad \left(\frac{R_s}{2\pi\eta N \cos\phi}\right)\left(\frac{\Delta p}{x_1}\right)$$

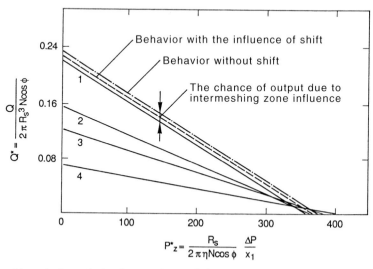

Figure 6.29   Dimensionless calculated screw characteristic curves [112] in terms of

$$\frac{Q}{2\pi R_s^3 N \cos\phi} \quad \text{vs.} \quad \frac{R_s}{2\pi\eta N \cos\phi}\frac{dp}{dx_1}$$

We now turn to non-Newtonian fluids. Here we work again with Eqs. 6.20 and 6.24, where we must introduce forms of the stresses σ_{12} and σ_{13} or σ_{12} and σ_{32}, respectively. These shear stresses are each related to shear rate through the relationship

$$\sigma_{ij} = \eta\frac{\partial v_i}{\partial x_j} \tag{6.29}$$

where η is the non-Newtonian viscosity. This viscosity depends on the invariants of the deformation rate tensor $\mathbf{d}$ through

$$\eta = \eta\left(tr\,d^2\right) \tag{6.30}$$

The work of Denson and Hwang [102] suggests velocity field of $v_1(x_2,x_3)$, and Szydlowski and White [111] suggest $v_1(x_2)$ and $v_3(x_2)$. Including the indicated velocity gradients leads to

$$\eta = \eta\left[\left(\frac{\partial v_1}{\partial x_2}\right)^2 + \left(\frac{\partial v_1}{\partial x_3}\right)^2 + \left(\frac{\partial v_3}{\partial x_2}\right)^2\right] \tag{6.31a}$$

For a power law fluid this would be

$$\eta = K\left[\left(\frac{\partial v_1}{\partial x_2}\right)^2 + \left(\frac{\partial v_1}{\partial x_3}\right)^2 + \left(\frac{\partial v_3}{\partial x_2}\right)^2\right]^{(n-1)/2} \tag{6.31b}$$

The first consideration of non-Newtonian flow in co-rotating twin screw extruder was in 1989 papers by Wang et al. [73, 130]. The velocity field was taken to have the form

$$v = v_1(x_2)e_1 + 0e_2 + v_3(x_2)e_3 \tag{6.32}$$

The viscosity function would depend on two velocity gradients

$$\eta = \eta\left[\left(\frac{\partial v_1}{\partial x_2}\right)^2 + \left(\frac{\partial v_3}{\partial x_2}\right)^2\right] \tag{6.33}$$

For a power law fluid

$$\eta = K\left[\left(\frac{\partial v_1}{\partial x_2}\right)^2 + \left(\frac{\partial v_3}{\partial x_2}\right)^2\right]^{(n-1)/2} \tag{6.34}$$

These authors substitute Eq. 6.34 into Eq. 6.29 to determine the shear stresses σ_{12} and σ_{32}. The shear stresses were introduced into Eq. 6.24, which were solved by the FAN method of Eq. 6.26. Screw characteristic curves were determined in the form of plots

$$\frac{Q}{2\pi R_S^3 N \cos\phi} \quad \text{vs.} \quad \frac{R_S^{n+1}}{K(2\pi R_S N \cos\phi)^n}\frac{dp}{dx_1} \tag{6.35}$$

with n as a running parameter. These plots show that screw pumping deteriorates with decreasing n. This is a well-known result for single screw extruders [114].

Rather rigorous non-Newtonian fluid analysis of flow in a self-wiping screw channel were given in 1989 papers by Lai-Fook et al. [115] and Wang and White [113]. Lai-Fook et al. [115] take the velocity field as

$$v = v_1(x_2,x_3)e_1 + v_2(x_2,x_3)e_2 + v_3(x_2,x_3)e_3 \tag{6.36}$$

and a viscosity function, which depends on four shear and two elongational velocity gradients. It should be understood that $\partial v_1/\partial x_2$ and $\partial v_3/\partial x_2$ are the dominant velocity gradients. The Equations of motion were formulated in the 1, 2, and 3 directions. This allows a full calculation of the helical velocity profile with v_1, v_2, and v_3 components in the channel cross section. Both Lai-Fook et al. [115] and Wang and White [113] present finite element (FEM) simulations. Lai-Fook et al. compute screw characteristic curves in the form of

$$\frac{Q}{2\pi R_s A_c N \cos\phi} \qquad vs. \qquad \frac{H^{n+1}}{K(2\pi R_s N \cos\phi)^n}\frac{dp}{dx_1} \qquad (6.37)$$

with n as a running parameter. Here again A_c is the channel cross section. Decreasing n is predicted by Lai-Fook et al. [115] to give poorer pumping capacity. This is also found by Wang and White [113], who report their computations in terms of the variables of Eq. 6.35.

Wang and White [113] have contrasted their computations for screw characteristic curves to the FAN method based on Eqs. 6.24, 6.28, and 6.32 to 6.34 with the FEM method described above (Fig. 6.30). The FEM calculations of both Denson and Hwang [102] and themselves are included and agreement is good.

The influence of the intermeshing region for non-Newtonian fluid including the full curvature of the problem has been considered in papers by Kalyon et al. [116] and Lai-Fook et al. [117]. Different non-Newtonian shear viscosity functions were used by the two groups,

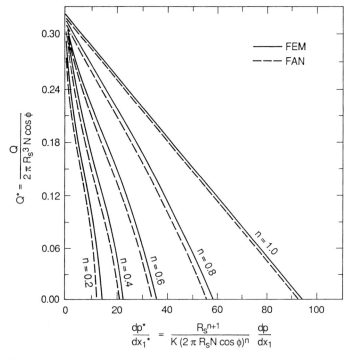

Figure 6.30   Dimensionless screw characteristic curves for power law fluids in self-wiping screw channels in terms of

$$\frac{Q}{2\pi R^3 N \cos\phi} \qquad vs. \qquad \frac{R_s^{n+1}}{K(2\pi R_s N \cos\phi)^n}\frac{dp}{dx_1}$$

FAN and FEM calculations are compared after Wang and White [113] both of whom use finite element mesh and velocity vectors calculated by Lai-Fook et al. (Fig. 6.31). They calculate screw characteristic curves, including the intermesh, and show it leads to larger pressure drops as was predicted earlier for Newtonian fluids [111]. Better agreement was found with experiments on a 50-mm Baker Perkins twin screw extruder.

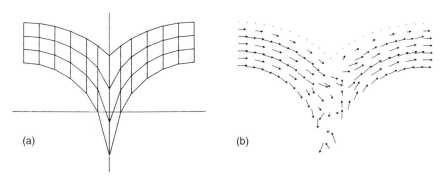

(a)   (b)

Figure 6.31   Non-Newtonian intermeshing region calculations of Lai Fook et al. [117]:   (a) finite element mesh;   (b) velocity vectors computed for intermeshing region

6.2.4.2.2   Kneading Disk Blocks

The first efforts at understanding the flow in kneading disk blocks is found in the doctorate dissertation of Werner [72]. His considerations of flow are kinematic rather than fluid mechanical in character. The first simulation of flow was by Szydlowski et al. [108] in 1987. These authors' formulation of the problem was based on hydrodynamics lubrication theory. They used Eq. 6.24 as their equations of motion. Newtonian fluid behavior was presumed, allowing them to reduce these equations to Eq. 6.25 and to

$$0 = -\frac{\partial p}{\partial x_1} + \eta \frac{\partial^2 v_1}{\partial x_2^{\,2}} \qquad\qquad \text{(6.38a, b)}$$

$$0 = -\frac{\partial p}{\partial x_3} + \eta \frac{\partial^2 v_3}{\partial x_2^{\,2}}$$

These equations were solved numerically using the FAN method of Tadmor et al. [112] as modified by Szydlowski et al. [108, 111] for moving boundaries. Considering the screw axis to be in the 1 direction and the 3 direction to be in the circumferential direction, the fluxes of Eq. 6.26 are

$$q_1 = -\frac{H^3}{12\eta}\frac{\partial p}{\partial x_1} \qquad\qquad \text{(6.39a, b)}$$

$$q_3 = \frac{1}{2}U_3 H - \frac{H^3}{12\eta}\frac{\partial p}{\partial x_3}$$

Eq. 6.39 was substituted into a flux balance, such as Eq. 6.27,

$$q_1(x_1 + \Delta x_1, x_3)\Delta x_3 + q_3(x_1, x_3 + \Delta x_3)\Delta x_1$$

$$-q_1(x_1, x_3)\Delta x_3 - q_3(x_1, x_3)\Delta x_1 = 0 \tag{6.40}$$

and the pressure fields determined. Meshes used on kneading disk blocks are shown in Fig. 6.32 and computed pressure fields are presented in Fig. 6.33. These pressure fields allow the calculation of pressure drop-flow rate relationships for the kneading disk block. Typical results are shown in Fig. 6.34.

Szydlowski et al. [109] in 1988 sought to model the influence on the flow by the oscillating local volume caused by the rotation of the kneading disk blocks. This volume oscillation acts as a peristaltic pump acting at the high frequencies determined by the screw rotation rates.

The problem of computing screw characteristic curves relating pressure drop and flow rate for non-Newtonian fluids was first considered by Szydlowski and White [118] in 1988 and is later considered by Wang et al. [74] in a 1989 paper. Here it is necessary to solve the lubrication equations using Eq. 6.34 for the shear viscosity rather than use a constant coefficient η. The equations of motion to be solved are

$$0 = -\frac{\partial p}{\partial x_1} + K \frac{\partial}{\partial x_2}\left[\left(\frac{\partial v_1}{\partial x_2}\right)^2 + \left(\frac{\partial v_3}{\partial x_2}\right)^2\right]^{(n-1)/2} \frac{\partial v_1}{\partial x_2} \tag{6.41a, b}$$

$$0 = -\frac{\partial p}{\partial x_3} + K \frac{\partial}{\partial x_2}\left[\left(\frac{\partial v_1}{\partial x_2}\right)^2 + \left(\frac{\partial v_3}{\partial x_2}\right)^2\right]^{(n-1)/2} \frac{\partial v_3}{\partial x_2}$$

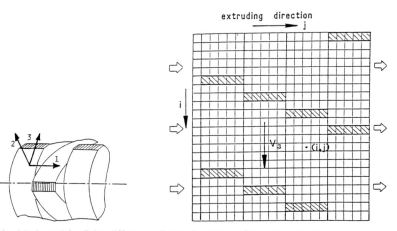

Figure 6.32   Mesh used for finite difference (FAN) simulation of kneading disk block

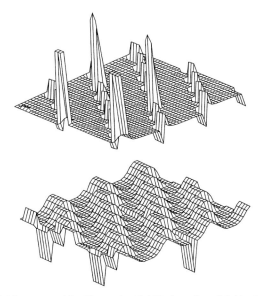

Figure 6.33   Pressure fields computed for Newtonian fluid in kneading disk block [109]

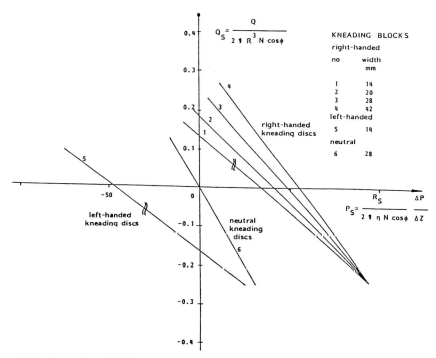

Figure 6.34   Screw characteristic curves

$$\frac{Q}{2\pi R_s^3 N \cos \phi} \qquad \text{vs.} \qquad \frac{R_s}{2\pi\eta N \cos \phi}\frac{dp}{dx_1}$$

for Newtonian fluids in kneading disk blocks [109]

These were solved using the FAN method of Eq. 6.40. Screw characteristic curves were computed. Typical results are shown in Fig. 6.35. As the fluid becomes increasingly non-Newtonian; i.e., as the power law n decreases below a value of 1.0, the pumping capacity becomes poorer.

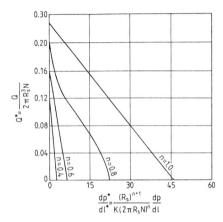

Figure 6.35   Screw characteristic curves

$$\frac{Q}{2\pi R^3 N \cos\phi} \quad \text{vs.} \quad \frac{R_s^{n+1}}{K(2\pi R_s N \cos\phi)^n}\frac{dp}{dx_1}$$

for power law non-Newtonian fluids in kneading disk blocks [74]

Finite element simulations of flow in kneading disk blocks were first discussed by Gotsis and Kalyon [110] and later in more detail beginning in 1992 by Yang and Manas-Zloczower [119], Lawal and Kalyon [120], and more recently Cheng and Manas-Zloczower [121]. Primary concern is with computing 3-dimensional flow fields. Some effort is made to calculate extrusion rate-pressure gradient relationships by Manas-Zloczower and her students Funatsu and his coworkers [122] have also made finite element calculations of flow in kneeding disc blocks.

6.2.4.3   Composite Modular Machine Behavior

The problem of simulation of a composite modular, intermeshing, co-rotating machine is first discussed in papers by White and Szydlowski [103] and Meijer and Elemans [70]. These efforts were extended in subsequent papers by Wang et al. [74], White et al. [75], Chen and White [104, 105], Potente et al. [107, 107a], and Bawiskar and White [79, 106]. The earliest efforts [70, 74, 75, 77, 103] were aimed at determining isothermal pressure and fill factor profiles in a twin screw extruder with a single feedport. This was the basis of Akro co Twin Screw-1. In this type of machine, the throughput is specified (by the feeders); thus, unlike single screw extruders, it is possible to calculate backward from the die system. The following procedure was used:

1. From the output of the twin screw extruder and the design of the die, the pressure at the end of the screw is calculated.
2. The screw characteristic curves are used to compute the backward pressure profile along the screw axis.
3. When the pressure falls to atmospheric, the screw channels are considered starved with the throughput Q being given by combining Eqs. 6.12 and 6.13:

$$Q_T = (2m-1)Q = (2m-1)\left[(1/2)\pi \, DN \cos\phi\right]A_c \Phi \tag{6.42}$$

where $2m-1$ is the number of effective thread starts for the self-wiping, co-rotating twin screw system, A_c is the cross section of a screw channel, N the screw rotation rate, and Φ the fill factor in the cross section.

4. Starvation is considered to continue until one comes to an element where the screw characteristic curve predicts a negative pressure gradient. Such negative pressure gradients require forward pressure flow, which is possible only with a filled screw/kneading disk block cross section. The screw channel becomes filled and the screw characteristic curves, whose calculation was discussed in the previous section, are used to calculate the pressure level at the beginning of the element.
5. The steps beginning with (2) are then repeated.

Nonisothermal modelling of flow of composite modular, co-rotating twin screw extruders was first given in papers by White and Chen [104, 105] and by Potente et al. [107a] in 1993/4. The energy equation for an incompressible fluid is of form [123]

$$\rho c\left[\frac{\partial T}{\partial t} + (v \bullet \nabla)T\right] = k\nabla^2 T + \sum_i \sum_j \sigma_{ij}\frac{\partial v_i}{\partial x_j} \tag{6.43}$$

If a velocity field of form

$$v = v_1(x_2)e_1 + 0e_2 + v_3(x_2)e_3 \tag{6.44}$$

is presumed, Eq. 6.43 becomes in the steady state

$$\rho c\left(v_1\frac{\partial T}{\partial x_1} + v_3\frac{\partial T}{\partial x_3}\right) = k\frac{\partial^2 T}{\partial x_2^2} + \sigma_{12}\frac{\partial v_1}{\partial x_2} + \sigma_{32}\frac{\partial v_3}{\partial x_2} \tag{6.45}$$

If we integrate Eq. 6.45 through the x_2 and x_3 directions we obtain

$$\rho c Q\frac{d\overline{\overline{T}}}{dx_1} = h_b W_b(T_b - \overline{\overline{T}}) + h_s W_s(T_s - \overline{\overline{T}}) + \sigma_{32}(H)W_b U_3 - Q\frac{dp}{dx_1} \tag{6.46}$$

where h_b and h_s are barrel and screw heat transfer coefficients and $\overline{\overline{T}}$ is the cup mixing temperature detined by

$$Q\overline{\overline{T}}(x_1) = \int_0^{W_b}\int_0^{H} v_1(x_2)T(x_2)dx_2 dx_3 \tag{6.47}$$

and W_b and W_s are barrel and screw perimeters.

The temperature change ΔT in a screw or kneading disk block element is obtained by integrating Eq. (6.46) along the length of an element.

In nonisothermal modeling of modular, composite twin screw extruders, it is necessary to use an iterative procedure. This is the basis of Akro co.-Twin Screw 2. One first presumes a temperature profile along the screw axis (e.g., the barrel temperature) and computes marching backward pressure and fill factor profiles. Such a computation following Chen and White [104, 105] proceeds backward to where one believes the melting occurs. This position is used as a starting point. Eq. 6.46 is then solved marching forward element by element along the screw profile for the cup mixing temperature profile $T(x_1)$ using the previously computed pressure and fill factor profile. In this manner, the cup mixing temperature is computed at the die. Generally, the die temperature obtained is different from the initial presumption. One now uses the temperature profile just calculated for the screw as the basis of a new backward calculation to redetermine the pressure and fill factor fields. When these are recalculated, we march forward again and compute a new axial temperature profile. This iterative procedure is continued until convergence is achieved.

Fig. 6.36 shows a calculation of a pressure profile, fill factor profile, and temperature profile for a particular modular screw design. This is based on Chen and White's Akro co-Twin Screw-2 [105]. A similar formulation and program called Sigma has been developed by Potente and his coworkers [107a] at the University of Paderborn.

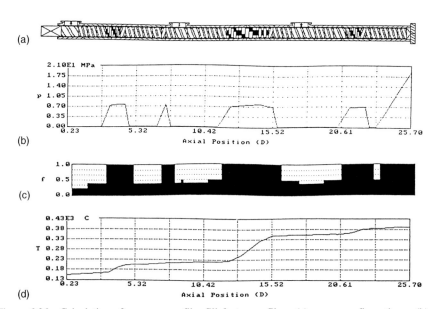

Figure 6.36   Calculation of pressure profile, fill factor profile: (a) screw configuration;   (b) pressure profile;   (c) fill factor profile;   (d) temperature profile

6.2.4.4   Melting and Composite Pumping Model

The melting of thermoplastics in a modular, co-rotating twin screw extruder has been analyzed by Bawiskar and White [78, 106]. For the case where the barrel temperature is higher than the melting temperature. A melting model based on a melt layer being initiated at the barrel wall was developed.

The model [78, 106] was based on Bawiskar and White [75, 78] observations that a melt layer formed along the surface of the heat barrel. This layer then grew toward the root of the screw. The mechanism of growth seemed to be viscous dissipation heating. They formulated this problem through an energy balance at the interface

$$-k_s \frac{\partial T_s}{\partial x_2}\bigg|_{x_2=H_m} + \rho v_{s2}\lambda = -k_m \frac{\partial T}{\partial x_2}\bigg|_{x_2=H_m} \qquad (6.48)$$

The energy balance on the melt layer is of the same form as Eq. 6.46. Specifically,

$$0 = h_b\left(T_b - T_m\right) + k_m \frac{\partial T_m}{\partial x_2}\bigg|_{x_2=H_m} + \left[\sigma_{12}\big|_b U_1 + \sigma_{32}\big|_b U_3\right] - q_1 \frac{\partial p}{\partial x_1} - q_3 \frac{\partial p}{\partial x_3} \qquad (6.49)$$

where heat convection terms are neglected because in the melt layer $\partial T / \partial x_j \left(j = 1,3\right)$ should be small. From these two Eq.s one may determine the variation in the melt layer thickness $H_m(x_1)$ along the screw axis. Typical calculations of melting profiles in modular, co-rotating twin screw extruders are shown in Fig. 6.37.

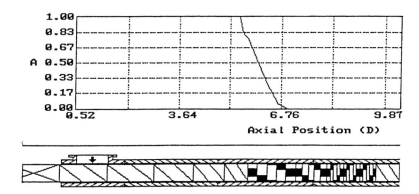

Figure 6.37   Computed melting profiles in a modular, co-rotating twin screw extruder

Bawiskar and White [106] have incorporated the melting model described above into the Akro co Twin Screw program as a Version 3. This makes it possible to calculate melting profiles as well as pressure and temperature profiles (plus torque, power consumption, residence time, etc.). Subsequent studies by Qian, Gogos and Todd [124a] have presented arguments that melting occurs by mechanical plastic deformation healing of pellets. More recent experiments in our laboratories [124b] indicate this is a primary mechanism when the barrel temperature is below the crystalline melt temperature in polyolefins but not at higher barrel temperatures. Melting may also begin at the screw surface. Potente and his coworkers [124] at the University of Paderborn have developed a different melting model and introduced it into their Sigma program.

6.2.4.5   Mixing

Mixing in modular, co-rotating twin screw extruders generally occurs in the kneading disk block regions. In this section we concentrate on modeling of flow in this region. The earliest efforts of this type involved the 1976 Dr-Ing dissertation of Werner [72], where he describes kinematics-driven mixing mechanisms. The first fluid mechanical modeling of the kneading disk block section was in 1987 by Szydlowski et al. [108], whose calculations of screw characteristic curves for kneading disk blocks were described in Section 6.2.4.2. These authors used Newtonian hydrodynamic lubrication theory based on fluxes defined by Eq. 6.39 and the flux balance of Eq. 6.40. They computed pressure and flux fields in kneading disk blocks. To reduce their computations to numbers, they integrated the flux field to determine circumferential for Q_c around the screws, longitudinal flow Q_L along the screw axis, Q_d, the flow over the kneading disk block tips, and $Q_{b'}$ the backward flow between the tips. They then defined the ratio factors:

$$f_c = \frac{Q_c}{Q_c + Q_L + Q_b + Q_d} \tag{6.50a}$$

$$f_L = \frac{Q_L}{Q_c + Q_L + Q_b + Q_d} \tag{6.50b}$$

$$f_b = \frac{Q_b}{Q_c + Q_L + Q_b + Q_d} \tag{6.50c}$$

$$f_d = \frac{Q_d}{Q_c + Q_L + Q_b + Q_d} \tag{6.50d}$$

Typical values of f_c, f_L, f_b, and f_d are computed for a power law fluid by Wang et al. [74] Fig. 6.38.

Szydlowski and White [109] in 1988 use an unsteady hydrodynamic lubrication form to model flow in a kneading disk block and predict peristaltic pumping motions of process fluid both forward and backward (see Fig. 6.39). These transient flows are associated by local variations in available spatial volume as a function of time caused by rotation of the modular screws.

The analyses described above compute only 2-dimensional fluxes in the machine and transverse directions and do not compute local velocity fields in the thickness direction. More powerful numerical tools, such as finite element methods, are required for this purpose. The first effort simulating velocity fields through the thickness direction of kneading disk blocks was by Gotsis and Kalyon [110] in 1989. More extensive results were given by Gotsis et al. [125] the following year. Extensive calculations were subsequently given in a 1992 paper by Yang and Manas-Zloczower [119]. Fig. 6.40a shows their calculations for the circumferential velocity profile in the plane perpendicular to the direction of flow in a 90° kneading disk block for a power law fluid ($n = 0.59$). Fig. 6.40b shows isovels

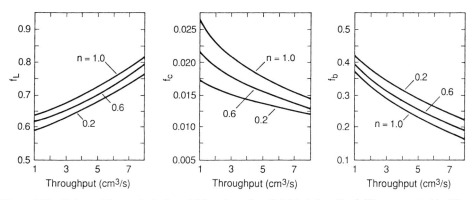

Figure 6.38   Values of factors f_c, f_L, f_b, and f_d for a kneading disk block (see Eq. 6.48) as computed by Wang et al. [74]

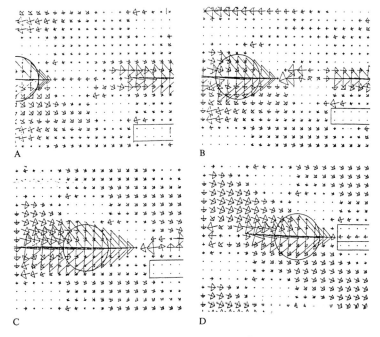

Figure 6.39   Peristaltic pumping motions of fluid in a kneading disk block as found by Szydlowski and White [111]

for this axial velocity field. It can be seen that the axial velocity field has negative as well as positive regions, as suggested by the work of Szydlowski et al. [108].

Lawal and Kalyon [120] have sought to use the concepts of chaos theory to investigate mixing in kneading disk blocks. Nonisothermal calculations for kneading disk blocks were first given by White and Chen [104, 105], who determined increases cup-mixing temperature along the screw axis for both screw and kneading blocks using Eq. 6.46.

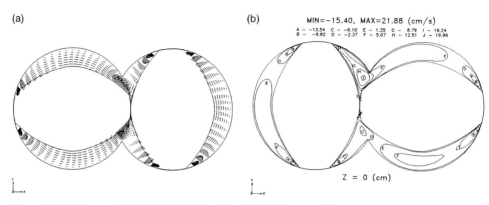

Figure 6.40   (a) v_2_v_3 velocity field in 90° kneading disk block perpendicular to direction of flow perpendicular to screw axis computed by Yang and Manas-Zloczower [119] FEM for power law fluid ($n = 0.59$) considering barrel fixed   (b) v_1 velocity field in axial direction of flow = contours showing position of equal velocity after Yang and Manas-Zloczower [119]

The formation of phase morphology during blending processes in a twin screw extruder has been modeled by Shi and Utracki [126], Huneault, Shi, and Utracki [127], and more recently by Delamare and Vergnes [128]. These models consider isolated droplets being drawn out into filaments and the filaments breaking by capillarity. Coalescence is empirically introduced. There is an implied limitation of this formulation to dilute systems and to the later stages of blending.

Lee and White [129] consider the problem of blending of pellets with a low-viscosity liquid or a lowmolecular weight solid that rapidly melts. The low-viscosity liquid wets the metal and acts as a lubricant to both polymer pellets and the metal of the screw and barrel. It is shown that this greatly delays the melting of the pellets, because of reduced viscous heating in the melt matrix which in turn retards the blending process. The predictions were compared to the same authors' earlier experiments [100].

6.2.4.6   Reactive Extrusion

The reactive extrusion process in a modular, co-rotating twin screw extruder has been modeled by various researchers in recent years. P.J. Kim and White [130] simulated the hydrolysis of ethylene-vinyl acetate copolymer in a modular, co-rotating twin screw extruder in a 1994 paper. Subsequently, Kye and White [131] modeled the polymerization of caprolactam and by B.J. Kim and White [132] for thermal/peroxide degradation of polypropylene and for maleating polypropylene. Basically, in each of these papers kinetic calculations are combined with the Akro-co Twin Screw 2 program. It is also important, especially in the case of polymerization, to include rheological property variations as a function of axial position along the twin screw extruder. Similar efforts at modeling polymer modification have been reported by Berzin and Vergnes [133] and more recently for the polymerization of caprolactone by Poulesquez et al. [133a].

6.2.5   Applications

Modular, co-rotating twin screw extruders are the most widely used of continuous mixing machines. They are applied to (1) compounding of particulates into thermoplastics, (2) blending of polymers (thermoplastics into thermoplastics, rubber into thermoplastics), (3) reactive extrusion, (4) devolatilization, and (5) food processing. In most, if not all of these areas, modular, co-rotating machines possess a dominant position. Modular machines are commercially available with screw diameters as small as 20 mm to as large as 300 mm and more.

6.3   Intermeshing Counter-Rotating Twin Screw Extruders

6.3.1   Technology

The concept of the intermeshing, counter-rotating twin screw pump and its behavior as a positive displacement pump seems very old. It appears fully developed in an 1874 American patent application by S.L. Wiegand [134] for sheeting baking dough. It is one of a large number of intermeshing, counter-rotating, positive-displacement pump designs found in the patent literature in the late 19th and early 20th centuries [134–142]. (Many of these machines were lobe and gear pumps [141, 142].) The patents promised steady flows independent of pressure head (Typical machines are contained in Fig. 6.41). The engineering literature of the 1920s also discusses intermeshing, counter-rotating twin screw pumps [143, 144]. Many companies seem to have commercialized twin screw and multiple screw pumps in the 1920s; among these was *Maschinenfabrik Paul Leistritz* of whom we will say more later. These pumps were used to handle viscous fluids such as lubricating oils and petroleum oils.

A 1925 patent application of Montelius [138] raises the issue of the advantages of screw pumps with more than two screws and screw pumps in which the different screws do not have same number of thread starts. According to Montelius, if one has a twin screw pump, the best design is for one screw to have g thread starts and the other to have $g + 1$ or $g - 1$ thread starts. If one has a three-screw pump in which two screws have g thread starts, the third screw, which would interact with the other, should have $2g - 2$ thread starts. More generally, if one has n screws with g thread starts, and N screws with G thread starts, they are optimized when

$$ng = NG + nN \tag{6.51}$$

Thus, if one screw operates with two external screws ($n = 2$, $N = 1$), it is best if the value of $2g$ for the external screws is $G + 2$. If G is 2, g should be 2.

In the early 1930s a joint program involving the *I.G. Farbenindustrie* in Frankfurt and *Maschinenfabrik Paul Leistritz* in Nuremburg was organized for the purpose of developing continuous kneaders for the ceramic and rubber industries and probably most importantly for handling of the *I.G. Farbenindustrie's* needs with coal-oil dispersions. Separate designs known as 'Knetpumpe' (kneading pumps) were developed by Kiesskalt et al. [145] of the

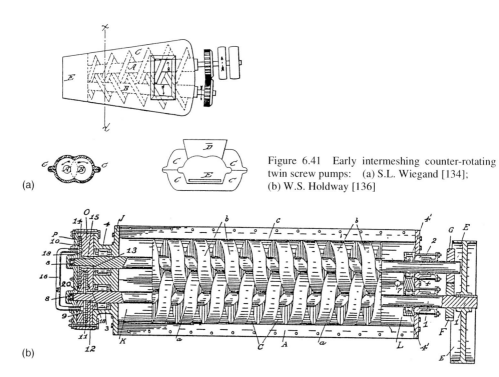

Figure 6.41   Early intermeshing counter-rotating twin screw pumps: (a) S.L. Wiegand [134]; (b) W.S. Holdway [136]

I.G. Farbenindustrie and Leistritz and Burghauser [146] of *Maschinenfabrik Paul Leistritz* (Fig. 6.42). These machines act to draw the material between the screws where it is masticated by the calendering action of the screws. Kiesskalt and his coworkers [147, 148] gave special attention to continuous mastication to produce ash-free coal-oil dispersions. The *I.G. Farbenindustrie* built not only twin screw kneading pumps but also multiple screw machines [149]. Subsequently, Kiesskalt [150–152] described his kneading pump in a series published papers. Kneading pumps were also described in a 1951 paper by Riess and Erdmenger [22] and more fully in a 1972 monograph by Herrmann [11].

Intermeshing, counter-rotating twin screw extruders came to be widely used for profile extrusion. In the early 1940s, *Maschinenfabrik Paul Leistritz* built twin screw extruders with screw diameters as large as 400 mm to extrude polyvinyl chloride profiles. In a 1944 patent application Steinmann and Heyne [153] of *Dynamit Nobel* developed an intermeshing twin screw extruder with conical screws.

In the years following the war, there was great interest, notably in Germany, Austria, and Italy, in developing an intermeshing, counter-rotating twin screw for extrusion of polyvinyl chloride profiles [154–159]. *Leistritz* withdrew from the manufacture of twin screw extruders, but many new companies entered the industry. These included *Schloemann AG* or Dusseldorf, who licensed a twin screw extruder design to Pasquetti [157], which they marketed as the Bitruder. This machine is discussed by Schaerer [154], Schutz [156], Prause [158], and Zielonowski [159], among others. Its screw design is shown in Fig. 6.43a.

Other machinery companies [154, 156, 158, 159], including *Nouvell Mapre SA* of Diekirch, Luxembourg (see Fig. 6.43b), *Gerhard Kestermann Zahnräder and Maschinen-*

fabrik of Bad Oeynhausen, West Germany, *Firma Wilhelm Anger*, OHG of Linz, Austria, and *Allgemeine Maschinenbau GmbH* (AGM), began to manufacture intermeshing, counter-rotating twin screw extruders. These machines usually had differing screw designs. AGM pioneered the manufacture of conical twin screw machines (see Fig. 6.43c).

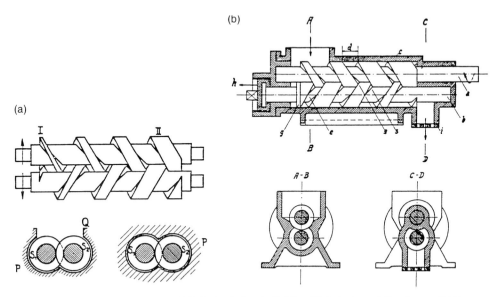

Figure 6.42   Kneading pumps: (a) I.G. Farbenindustrie [149];   (b) Leistritz and Burghauser [146]

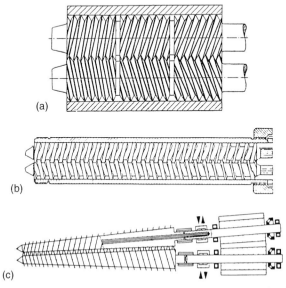

Figure 6.43   Intermeshing, counter-rotating twin screw designs: (a) Pasquetti twin screw design [157], which is the basis of the Bitruder first manufactured by Schloemann AG;   (b) screw design of Mapre; (c) conical screw design of AGM

The 1940s and 1950s also saw the development of intermeshing, counter-rotating continuous mixers. Such a machine was originally developed by Wilhelm Ellermann (see, eg., [11]) of *Krupp* in their Grusonwerke. This machine contained intermeshing screw and rotor sections machined only a shaft. A patent was applied for in 1941 [160]. The machine was marketed as the 'Knetwolf.' In the postwar period, Ellermann came to western Germany, where he contacted the firm *Josef Eck und Sohne Maschinenfabrik* in Dusseldorf. A patent was filed by Ellermann and *Eck* [161] in 1951 for a similar machine, which was marketed in the 1950s.

The concept of modular, intermeshing, counter-rotating multiscrew machine for the purpose of mixing/homogenizing as well as pumping is contained in a 1949 U.S. patent application by M.B. Sennet [162] of the *Delaval Steam Turbine Company*. A similar concept may be found in a German patent application of Hack [163], which was filed the same year.

In the late 1960s, H. Tenner [164–166] of *Maschinenfabrik Paul Leistritz* developed a modular, intermeshing, counter-rotating twin screw extruder for the purpose of compounding thermoplastics. The modular elements used in the Leistritz design, which is designated GG, are shown in Fig. 6.44. These are thin- and thick-flighted screws and special mixing elements. This Leistritz GG machine was further discussed in papers by Thiele and his coworkers [167–169] and by Lim and White [170].

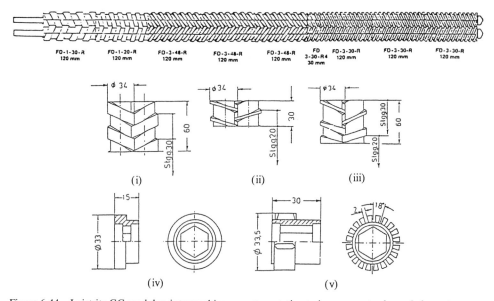

Figure 6.44   Leistritz GG modular, intermeshing, counter-rotating twin screw extruder and elements

A second modular, intermeshing, counter-rotating twin screw machine intended for compounding has been developed by *Japan Steel Works*. It is described in papers by Sakai and Hashimoto [171, 172]. This machine uses thin-flighted screw elements and kneading disk blocks as shown in Fig. 6.45.

A new modular, intermeshing, counter-rotating twin screw extruder has been devised by

W. C. Thiele [173] of *American Leistritz* in the 1990s. It has been described by the inventor as a counterflight machine and it involves multilobal cross sections with more shallow interscrew penetration (see Fig. 6.46).

A)   B)   C)

Figure 6.45   Japan Steel Works modular, intermeshing, counter-rotating twin screw extruder elements

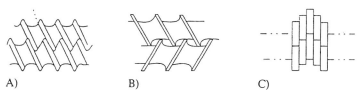

Counterflight Counterrotation

Screw length	Screw speed	Feed ports	Vent ports	Dispersive mixing	Distributive mixing	Self-wiping	Rapid clean-out	Output quality	Output rate
32-56	60-600	1-4	1-4	Capture in lobal event stresses	Vanes, gears, channels, narrow lobals	No	Yes	High	High
L/D	RPM								

Figure 6.46   Newer American Leistritz intermeshing, counter-rotating twin screw extruder designs

6.3.2   Experimental

6.3.2.1   Flow Visualization

Flow visualization studies of intermeshing, counter-rotating twin screw extruders were reported by Jewmenow and Kim [174], Janssen and Smith [175–177], and Sakai [69], all in the years 1973/8. Jewmenow and Kim [174] reported investigations in which a window was placed in a twin screw extruder and aluminum flake markers were placed in a poly-isobutylene process fluid. Photographs were taken through the windows and were used to determine streamlines. Sakai [69] made similar studies but used a transparent plastic barrel. Janssen and Smith [175–177] reported flow visualization studies on an intermeshing, counter-rotating twin screw with a polymethyl methacrylate barrel using an aqueous polyvinyl pyrrolidone solution.

6.3.2.2   Residence Time Distributions

Measurements of residence time distributions in intermeshing, counter-rotating twin screw extruders have been reported variously by Sakai [69, 171], Janssen, et al., [178], Rauwendaal [179], Wolf, Holen, and White [180], Potente and Schultheis [181], B.J. Kim and White [182], and Shon, Chang, and White [183]. The reader should be warned that these studies are for machines of different design, some intended for profile extrusion [178, 180] and others modular machines intended for compounding [179, 181, 182].

The experimental results of Sakai [69] are summarized in Fig. 6.47. These include results on single screw extruders, intermeshing, counter-rotating twin screw extruders, self-wiping, co-rotating twin screw extruders, and nonintermeshing (presumably counter-rotating) twin screw machine. The intermeshed, counter-rotating machine has the narrowest distribution of residence times, which Sakai associates with the C-chambers of the positive displacement pump action of the machine.

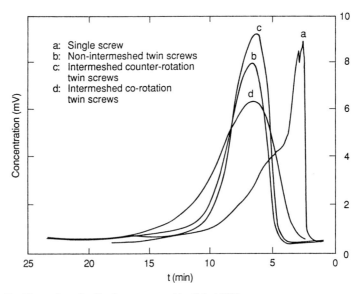

Figure 6.47   Residence time distributions measured by Sakai [69]

The studies of Janssen et al. [178] and Wolf et al. [180] are with the experimental technique of radioactive tracers to determine residence time distributions. The significance of their results is not well developed. Rauwendaal [179] reports experiments comparing a *Leistritz* intermeshing, counter-rotating twin screw extruder with the elements of Fig. 6.43 with a *Werner and Pfleiderer* modular, intermeshing, co-rotating twin screw extruder. He finds the intermeshing, counter-rotating twin screw extruder has a narrower residence time distribution than the co-rotating machine.

B. J. Kim and White [182] report residence times and residence time distributions for the intermeshing, counter-rotating and self-wiping, co-rotating modes of a *Japan Steel Works* twin screw extruder using equivalent screw configurations. The intermeshing, counter-rotating twin screw machine for all configurations has longer residence times and broader residence time distributions. This seems to be associated with the thin flights of the JSW twin screw extruder, which result in large backward leakage flows.

Shon et al. [183] compared a Leistritz GG modular, intermeshing, counter-rotating twin screw extruder with a modular, self-wiping, co-rotating twin screw extruder, a Kobelco continuous mixer, and a Buss Kneader. The intermeshing, counter-rotating twin screw machine gives the narrowest residence time distribution. This was especially the case when all mixing elements were removed (see Fig. 6.48).

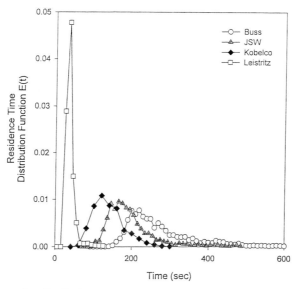

Figure 6.48   Residence time distributions measured by Shon et al. [183]: Leistritz refers to a modular, counter-rotating twin screw extruder; JSW refers to a modular, self-wiping, co-rotating twin screw extruder

6.3.2.3   Pumping Characteristics

The earliest published study of the pumping characteristics of modular, intermeshing, counter-rotating twin screw extruders seems to be that of Doboczky [184, 185] of Leistritz. This author compared the output of a polyvinyl chloride profile extruder with the positive displacement pump theory of Section 2.2.2.2 and Eq. 2.3.

$$Q = 2iV_cN \qquad (6.52)$$

where i is the number of thread starts, V_c is a volume of C chamber, Q is the flow rate, and N is the screw speed. Doboczky found the ratio of the observed to the theoretical output in a flood-fed machine to vary from about 50 to 90%, with 70% being the mean value. Doboczky also argued that the output of an intermeshing, counter-rotating twin screw extruder is about three times that for a single screw extruder of a similar size and screw speed.

A second study was reported in 1971 by Menges and Klenk [186, 187] using a *Schloc mann AG* Pasquetti Bitruder. They found that the output was proportional to the screw speed. The ratio of the experimental output to Eq. 6.52 was in the range of 37 to 41%.

A third study was published in 1976 by Janssen et al. [188] also using a Pasquetti-type machine with a 47.7-mm-diameter screw. The material investigated was polypropylene. The specific output Q/N was independent of pressure over a wide range of melt temperatures and screw speeds. Janssen et al. [188] removed the barrel of their machine and found that the machine was starved.

6.3.2.4   Melting

It was found by Cho and White [189] in studying the equivalent *Japan Steel Works* intermeshing, modular, counter-rotating and self-wiping, co-rotating screw configurations that melting always occurred earlier in the former machine. This is also implicit in the work of Lim and White [170], who contrasted the intermeshing, counter-rotating Leistritz GG machine and a *Werner and Pfleiderer* modular, co-rotating twin screw extruder in blending experiments. More recently Wilczynski and White [189a] have made basic studies of melting in intermeshing counter-rotating Ewin screw extruders. They find that melting is initiated by calendering stresses between the screws and by high barrel temperature induced melting.

6.3.2.5   Mixing

There have been few papers involving mixing in intermeshing, counter-rotating twin screw extruder. In a 1981 paper comparing a Leistritz GG modular, counter-rotating twin screw extruder and a modular, self-wiping, co-rotating twin screw extruder, Rauwendaal [179] concluded that the counter-rotating machine was superior in dispersing finely divided small particles, but was inferior in distributive mixing.

In a 1987 paper comparing the Japan Steel Works modular, intermeshing, counter-rotating machine with a modular, self-wiping, co-rotating machine, Sakai, Hashimoto, and Kobayoshi [172] concluded that the intermeshing, counter-rotating twin screw machine did a better job of dispersing small particles, but damaged glass fibers to a greater extent during mixing.

In 1994, Lim and White [170] reported a study of the blending of polyethylene and polyamide-6 in a Leistritz GG modular, intermeshing, counter-rotating twin screw extruder and described the phase morphology for different screw designs as a function of position along the screw. Comparisons were made to self-wiping, co-rotating twin screw machines. It was found for screw elements that counter-rotating machines mixed much better than co-rotating machines and that thin-flighted screws mixed much better than thick-flighted, fully intermeshed screws (see Fig. 6.49a, b). When special mixing elements were introduced, the phase reduction occurred more rapidly (Fig. 6.49c).

Shon and White [190] have compared the mixing of glass fibers into polypropylene in various types of continuous compounding machines. The most severe damage to glass fibers occurred with the Leistritz GG intermeshing, counter-rotating twin screw extruder.

6.3.3   Modeling

6.3.3.1   General

The literature on flow modeling in intermeshing, counter-rotating twin screw extruders is considerably different from that of other screw extrusion systems. This is because the basic mechanism of flow is the elements acting as a positive displacement pump and forming closed chambers, which forward the fluid being processed. In the case of single screw extruders, co-rotating twin screw extruders, the screw channels are open along their length.

In the intermeshing, counter-rotating twin screw machine they are closed. With open screw channels, it is convenient to embed the coordinate system in the root of the screw channel and determine velocity fields and flow rates in terms of it. This has been done for the single screw extruder in Section 2.4, the self-wiping, co-rotating twin screw extruder in Sections 2.6 and 6.2.4, and the tangential, counter-rotating twin screw extruder in Section 6.4. The literature for the intermeshing, counter-rotating twin screw extruder is traditionally based on flow in and leakage out of C-chambers.

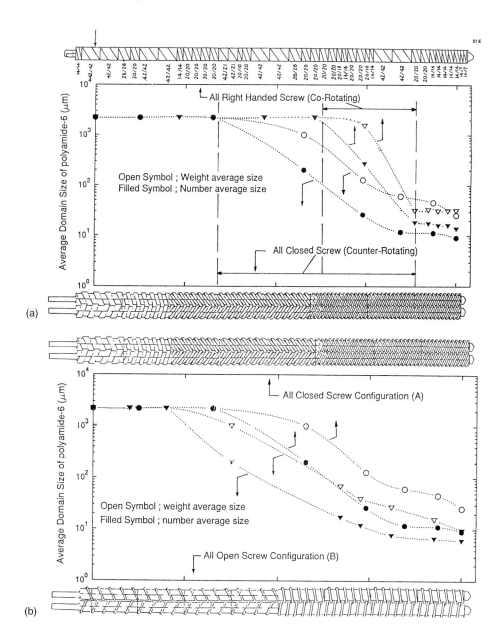

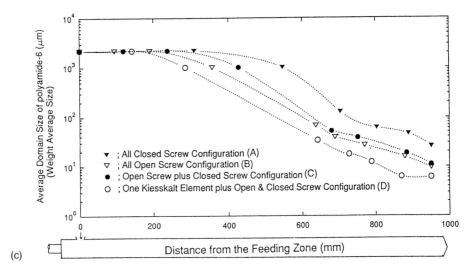

Figure 6.49   Comparison of reduction of size of polyamide-6 in a polyethylene polyamide-6 75/25 blend [170]:   (a) fully intermeshing, counter-rotating twin screw machine vs. self-wiping, co-rotating twin screw machine;   (b) fully intermeshed, counter-rotating machine vs. thin-flight intermeshed, counter-rotating machine;   (c) fully intermeshing, counter-rotating machine with different screw configuration containing mixing elements

The positive displacement pump C-chamber mechanism has long been well understood, as may be seen from the following quotations from the patent literature. Wiegand [134] in an 1874 patent application states that

> The dough enters the spiral or helical spaces between the screw threads and were a single screw only used would rotate with the screw excepting so far as it is detained by the friction of the portion in contact with the case; but using a screw made right and left, with the threads of each screw interlocking in the spaces of the other screw, the rotation of the dough is prevented, and the dough is forced lengthwise by the action of the screw towards the slot E in the end of the case.

Holdaway [136] in a 1915 patent application writes,

> The perimeters of said screws a and b contact with the interior of said intersecting cylinders thus forming a plurality of fluid chambers by which when the screws are operated, a given quantity of fluid is advanced from the inlet chamber to the outlet chamber and is forced through the outlet pipe.

Montelius [140] in a 1929 application states simply that

> These two screws act together as a piston which on the rotation of the screws advances the fluid, liquid or gas, present within the casing, in the direction along the axis of the screws.

The first published scientific paper that discusses analytically flow mechanisms in intermeshing, counter-rotating twin screw pumps seems to be that by Kiesskalt in 1927 [143]. He wrote the output Q as

$$Q = NV_c - Q_{leak} \tag{6.53}$$

where N is screw rotation into, V_c is C-chamber volume, and Q_{leak} is leakage flow. This was later elaborated on in the book of Schenkel [191], who wrote

$$Q = iNV_c \tag{6.54}$$

where i represents the number of screws and V_c is the chamber volume per screw. Doboczky [184, 185] in 1965 wrote for a twin screw machine

$$Q = 2mNV_c \tag{6.55}$$

where m is the number of thread starts.

Representations of V_c in greater detail are given in papers of 1966–71 by Menges and Klenk [186] and Klenk [187] and in 1975 by Janssen et al. [192]. The C-chamber volume may be written

$$V = V_i - V_2 - V_3 \tag{6.56}$$

Here V_1 is the volume of the cross section of one barrel half times the pitch S, i.e.,

$$V_1 = S\left[\pi R^2 - \frac{\alpha}{2} R^2 + \sqrt{\left(RH - \frac{H^2}{4} \right)\left(R - \frac{H}{2} \right)} \right] \tag{6.57a}$$

where α is the overlap angle, R the barrel radius, and H the channel depth. α is given by

$$\alpha - 2\tan^{-1}\frac{\sqrt{DH - H^2}}{D - H} \tag{6.57b}$$

where D is the screw diameter. V_2 is the volume of the screw root

$$V_2 = \pi S(R - H)^2 \tag{6.58}$$

and V_3 is the volume of the screw root

$$V_3 = \frac{S}{\sin\phi} \int_{R-H}^{R} b(r) 2\pi r \, dr \tag{6.59a}$$

where $b(r)$ is the width of the flight in the axial direction. If $b(r)$ is a constant value b, then

$$V_3 = \frac{\pi b S}{\sin \phi} \left[R^2 - (R - H)^2 \right]$$ (6.59b)

6.3.3.2   Leaking C-Chamber Models

Doboczky [184, 185] of *Leistritz* was the first to give serious attention to the problem of the imperfection of the C-chamber model outlined in the previous section. He concluded that there were leakage flows of various types that needed to be considered. He wrote the throughput an intermeshing, counter-rotating twin screw extruder as

$$Q = Q_c - Q_{CL} - Q_{PL} - Q_{FL}$$ (6.60)

Here Q_c is the C-chamber flow rate, Q_{CL} is the calendering leakage between the screws (called by Doboczky, 'Flankenschleppstrom'), Q_{PL} is the pressure leakage between the screws around the flights (called 'Durchstrom zwischen den Flanken'), and Q_{FL} is the leakage over the screw flights (called 'Leckstrom uber die Schneckenstege'). The characteristics of these flows is further explained in papers by Klenk [186, 187, 193] Janssen et al. [177, 188, 192, 194], and White and Adewale [195].

   A survey of the literature reveals that it is generally considered that calendering leakage is the most important of the leakage flows. The second in importance is the pressure leakage between the flanks of the screws. Third is the flight leakage over the screw flights, i.e.,

$$Q_{CL} > Q_{PL} > Q_{FL}$$ (6.61)

The simulation of flow in the calendering gap generally follows the Ardichvilli [196]–Gaskell [197] analysis of flow in a calender. It is based on the solution

$$0 = -\frac{\partial p}{\partial x_1} + \frac{\partial \sigma_{12}}{\partial x_2}$$ (6.62)

in the calender gap. Here 1 is the direction of flow and 2 the direction normal to the screw axes. The boundary conditions in a fixed stationary coordinate system are

$$U(-h) = \pi D N$$ (6.63a)

$$U(+h) = \pi D N$$ (6.63b)

The total flow is

$$Q_{cL} = \left[U(+h) + U(-h) \right] W h - \frac{2h^3}{3\eta} \frac{\partial p}{\partial x_1}$$ (6.64)

To calculate Q_{CL} we must know the pressure gradient. The pressure drop across the calender drop as a function of calender throughput can be obtained by solving Eq. 6.64 as a differential equation for $p(x_1)$. This has the form

$$\Delta p = \frac{3}{4}\pi\eta\frac{\sqrt{Dh}}{h_0^3}\left[\frac{3Q_{CL}}{4}\left(\frac{S}{2}-H\tan\psi\right)-\left[U(h)+U(-h)\right]h_0\right]$$

(6.65)

where $h(x_1)$ was represented as

$$h = h_0 + \frac{x^2}{D}$$

(6.66)

S is the screw pitch, and ψ the flight wall angle.

More recently Speur et al. [194] have reported a numerical simulation of flow in the calendering between screws using finite element techniques and the full Navier–Stokes Eq.s. A vortex is predicted at the entry to the calendering gap. Calculations show that the vortex intensity is determined by the value of the relative throughput and the shear thinning increased relative throughput and non-Newtonian character reduces the vortex. The finite element calculations for the magnitude of Q_{CL} agreed quite well with the lubrication theory analysis of Janssen et al.

6.3.3.3   FAN Analysis of Flow

A new simulation of flow using hydrodynamic lubrication theory and a modified Flow Analysis Network approach was been given by Hong and White [198, 199]. This approach is based on a cylindrical coordinate system with the z-coordinate axis coinciding with one of the screw axes and the radial direction leading from that axis to being perpendicular to the barrel. We presume the velocity field to be of the form

$$v = v_z(r)e_z + v_\theta(r)e_\theta + 0e_r$$

(6.67)

No velocity components in the r direction are considered and only shearing stresses in the r direction arise in a force balance. These have the form

$$0 = -\frac{\partial p}{\partial z} + \frac{1}{r}\frac{\partial}{\partial r}(r\sigma_{zr})$$

(6.68)

$$0 = -\frac{1}{r}\frac{\partial p}{\partial\theta} + \frac{1}{r^2}\frac{\partial}{\partial r}(r^2\sigma_{\theta r})$$

where σ_{zr} and $\sigma_{\theta r}$ are shear stresses in the z and θ directions. These shear stresses have the form

$$\sigma_{zr} = \eta\frac{\partial v_z}{\partial r}$$

(6.69a)

$$\sigma_{\theta r} = \eta r \frac{\partial}{\partial r}\left(\frac{v_\theta}{r}\right) \qquad (6.69b)$$

where η represents the shear viscosity. Hong and White present simulations for both Newtonian fluids [198], where η is a constant, and power law viscous non-Newtonian fluids [199], where the shear viscosity depends on both $\partial v_z/\partial r$ and $r\partial/\partial r(v_\theta/r)$.

As in the earlier models of flow in intermeshing twin screw machines, Hong and White divide their analysis into two parts, one considering flow between the screw and the barrel and the other considering the region between the screw. The boundary conditions between the screw and the barrel are the familiar ones based on a coordinate system rooted in the screw surface. There is no slip on the surface of the screw and the barrel translates relative to the screw, i.e.,

$$v_z\left(R_s\right) = v_\theta\left(R_s\right) = 0 \qquad (6.70)$$

$$v_z\left(R_B\right) = 0 \qquad v_\theta\left(R_B\right) = \pi D N$$

The equations are solved using flux balances of form

$$q_z(z + \Delta z, \theta) R_{Ch}\Delta\theta + q_\theta(z, \theta)\Delta z = 0 \; q_z(z+\Delta z, \theta)$$
$$-q_z(z, \theta) R_{Ch}\Delta\theta - q_\theta(z, \theta)\Delta z = 0 \qquad (6.71)$$

where R_{Ch} is a characteristic radius. Here q_z and q_θ are defined by

$$q_z = \frac{1}{R_{Ch}} \int_{R_s}^{R_B} r v_z\left(r\right) dr$$

$$q_\theta = \frac{1}{R_{Ch}} \int_{R_s}^{R_B} r v_\theta\left(r\right) dr \qquad (6.72)$$

In the region between the screws one must account for fluxes entering from circumferential motions. Thus, at the entrance to the region we have

$$q_z\left(z + \Delta z, \theta\right) R_{Ch}\Delta\theta + q_\theta\left(z, \theta + \Delta\theta\right)\Delta z$$
$$-q_z\left(z, \theta\right) R_{Ch}\Delta\theta - q_\theta^R\left(z, \theta\right)\Delta z - q_\theta^L\left(z, \theta\right)\Delta z = 0 \qquad (6.73a)$$

where q_θ^R and q_θ^L are fluxes from the two screws. In the central region between the screws, one must be concerned with the positive displacement effects caused by the moving flights. There are no longer entry q_θ^R and q_θ^L fluxes. Eq. 6.73a becomes

$$q_z(z+\Delta z,\theta)R_{Ch}\Delta\theta + q_\theta(z,\theta+\Delta\theta)\Delta z$$

$$-q_z(z,\theta)R_{Ch}\Delta\theta - q_\theta(z,\theta)\Delta z \qquad (6.73b)$$

$$-\left[U_\theta(R_\circ - R_i)_{(\theta+\Delta\theta,z)} - U_\theta(R_\circ - R_i)_{(\theta,z)}\right] = 0$$

At the exit to the interscrew region, there is a formulation similar to Eq.6.73a, where one considers the exiting of fluid to left- and right-handed screws.

Hong and White [198, 199] have determined screw characteristic curves for screws of different designs for both Newtonian and power law fluids. Typical results are shown in Fig. 6.50. These show that thick-flighted screws have very strong pumping ability, where thin-flighted screws resemble single screw extruders. Generally, the pumping characteristics for non-Newtonian shear thinning fluids are inferior to those for Newtonian fluids. Those are able to develop loss pressure because of greater backflows.

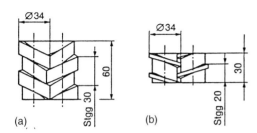

Figure 6.50 Screw characteristic curves for Leistritz GG modular elements predicted by Hong and White [199]:   (a) FD element;   (b) FF element

6.3.3.4   Three-Dimensional Solutions

Flow in an intermeshing, counter-rotating twin screw extruders has been modeled by Li and Manas-Zloczower [200] using a 3-dimensional finite element model, essentially a FIDAP fluid dynamics analysis package. They give their attention to the thick-flighted FD elements of the Leistritz GG machine. Presuming a velocity field, which is Newtonian at low shear rates and power law at high shear rates, they determine velocity fields in the C-chambers and seek to characterize the velocity field in the screw channels. They characterize the local velocity field with λ, the ratio of the absolute local rate of deformation to the vorticity, and determine its local variation. No attention is given to screw pumping characteristics.

6.3.3.5   Composite Pumping Model

In modular, intermeshing, counter-rotating twin screw extruders that are used for compounding one must consider the interaction of different screw elements on the flow. This problem was first considered by White and Szydlowski [103] and was worked out in detail by Wang et al. [73] for a modular, self-wiping, co-rotating twin screw extruder. Hong and White [198, 199] have applied this approach to intermeshing, counter-rotating twin screw extruders. The flow rate is presumed to be the same in all of the screw elements and the die. The die pressure is used to compute the pressure in front of the screws. This causes this region to be fully filled with polymer melt or other process fluid. The screw characteristics calculated by these authors [198, 199] (see Fig. 6.50) are then used to determine the pressure gradient backword along the modular screw sections. When the pressure falls to zero, starvation develops and the screw channels are only partially full. The starvation will continue backward until an element is reached that is not capable of pumping forward the specified flow rate Q. This gives rise to a negative pressure gradient. The screw channel will then fill up again and the pressure will rise. Screw characteristic curves are then used again to calculate the backward pressure profile along the screw axis.

Hong and White [199] also describe the application of energy balances to flow in a modular, intermeshing, counter-rotating twin screw extruder. They combine forward calculation of temperature profiles with backward calculations of fluid mechanics in a manner similar to Chen and White [104, 105] for modular co-rotating machines. Typical predicted pressure and temperature profile for a Leistritz modular GG machine is shown in Fig. 6.51.

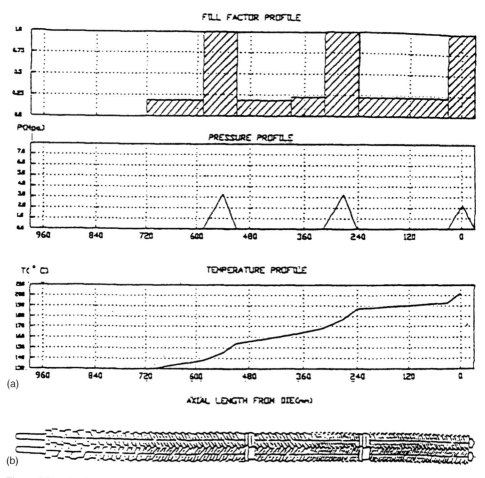

Figure 6.51   Predicted fill factor, pressure profile, and temperature along Leistritz GG modular, counter-rotating twin screw extruder by Hong and White

6.3.4   Screw Bending

A well-known phenomenon in intermeshing, counter-rotating twin screw extruders is the outward bending of screws, which contact and abrade the barrel. The phenomenon has been called the '8 o'clock–10 o'clock effect' because of the positions of the abrasion in the barrel. There has apparently been only one analytical study of the phenomenon. This is by White and Adewale [201]. These authors consider the influence of the high pressures developed by the melt in the small clearances of the calendering gap between the screws pushing the screws apart toward the barrel. Screw deflection is estimated by knowledge of the cross-section moment of inertia of the screws.

6.3.5   Applications

Intermeshing, counter-rotating twin screw extruders may, as described earlier, be divided into two classes: One of these is machines for profile extrusion; the second is machines for compounding. Profile extrusion is the major application. Here the extruders are flood fed and generally the screws have simple machined shafts. The profile extruders are used primarily for polyvinyl chloride and are used in applications such as the production of pipe and window frames.

In compounding applications, starve fed modular machines are primarily used. Here the major application has been to disperse additives, such as certain pigments and lubricants. Other applications where co-rotating twin screw extruder are used such as reactive extrusion are also possible. However, the machine characteristic of short residence times and high shear stresses must be considered. Short residence times and poor mixing can make the machine unfavorable for reactive extrusion [201a].

6.4   Tangential Counter-Rotating Twin Screw Extruders

6.4.1   Technology

Nonintermeshing, counter-rotating twin screw extruders have long been discussed in the patent literature [202, 203]. They first became of technological and industrial importance with the efforts of Fuller [204, 205] and the *Welding Engineers* company in the mid 1940s. The basic patent drawings are contained in Figs. 6.52 and 6.53. The machine was described in the engineering literature in papers published from 1949 to 1960 by various authors [206–209]. Fuller's machine was the first commercial modular twin screw extruder. The machine consisted of forward-pumping screws, nonpumping, solid barrier elements, and backward-pumping screws.

Various applications were developed for the tangential, counter-rotating twin screw extruder. These are described in patents of Street [210–213] and Skidmore [214–219] of *Welding Engineers*. The applications include compounding, devolatization, coagulation, and reactive extrusion.

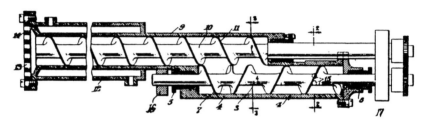

Figure 6.52   Basic tangential counter-rotating twin screw extruder of Fuller [204]

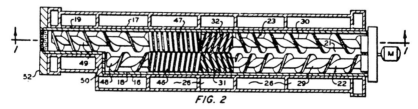

Figure 6.53   Modular, tangential, counter-rotating twin screw extruder of Fuller [205]

6.4.2   Experimental

6.4.2.1   Flow Visualization

There are many patents on tangential, counter-rotating twin screw extruders in the period of the 1950s to the 1970s. These are not, however, basic experimental studies. Flow visualization of a tangential, counter-rotating twin screw extruder begins with the papers of T. Sakai [69, 220]. The first paper [69] contrasts flow patterns in tangential, counter-rotating twin screw extruders with intermeshing, co-rotating and intermeshing, counter-rotating machines. Sakai used suspended pigments in a high-viscosity (~30-Pa) silicone oil in his studies. He found that flow in these is primarily along individual screws, but occasionally the marker strays over from one screw to the other. Sakai's second paper [220] investigates streamlines in different mixing sections of complex character.

A third flow visualization study using low-viscosity liquids was reported by Bigio and Penn [221] and subsequently Bigio et al. [222]. Their procedure is similar to that of Sakai in using a viscous liquid in a transparent plastic barrel. They proceed by following a neutralization reaction (potassium hydroxide and phenolphthalein) in a viscous liquid (corn syrup). This is found to occur much more rapidly, i.e., over shorter lengths in staggered as opposed to matched screw configurations; 50/50 staggered configurations were most effective.

Flow visualization studies showing the motions of a polyethylene melt in tangential, counter-rotating twin screw extruder in a barrel with glass windows have been reported by Min et al. [223]. Their results parallel those of Sakai [69].

The experimental studies reported above are for flood-fed machines. Nichols [224] was the first to consider the characteristics of starved and flood-fed regions in this machine. Under normal operating commercial tangential, counter-rotating twin screw extruders are starve fed with regions of fully filled character in front of the die and other locations in the machine. This behavior has been studied by Bang et al. [225]. They note that the twin screw extruders are filled in front of the die, as well as in front of backward-pumping elements and in the neighborhood of mixing elements. Generally, the extent of fill decreases with increasing screw speed.

Bigio and Wigginton [226] have reported flow visualization studies of starved flow. They give attention to interscrew fluid transfer. They find this does not occur when screw flights are matched but becomes increasingly important when the screw configurations are staggered relative to each other.

6.4.2.2   Screw Pumping Characteristics

The first experimental study of screw pumping in a tangential, counter-rotating twin screw extruder was by Kaplan and Tadmor [227] in 1974 using a 25-mm *Bausano* machine. Studies were made using matched screw flights. These were compared to an early model of flow in this geometry (Section 6.4.3.2). Inspection of their data suggests that when the die pressure goes to zero, the output of their machine approaches twice that of a single screw extruder. When the die pressure increases, the machine pumps relatively more poorly.

A second experimental study was published in 1983 by Nichols [228], who reported the behavior of two Newtonian oils in a series of *Welding Engineers* tangential, counter-rotating twin screw extruders. Screw characteristic curves Q vs. Δp were presented. These indicate that the pumping characteristics of these machines are significantly less than two equivalent single screw extruders. A most interesting result was that the pumping characteristics of the matched screw flight configurations were much superior to those of the staggered flight configurations (Fig. 6.54). Subsequently, Min et al. [223], using a flood-fed laboratory tangential twin screw machine that they had built, compared the pumping characteristics of matched- and staggered-flight twin screw machines. The experiments involved plasticating and melting as well as pumping. The matched-flight machines had much better pumping characteristics.

The pumping characteristics of individual screw element pairs in a tangential, counter-rotating twin screw extruder were determined by Bang et al. [225] at the University of Akron using a *Leistritz* 30/34 laboratory twin screw extruder. The machine was operated in a starved mode using a feeder with the element to be characterized placed at the end of the screw. The pressure at the die entrance was determined and was related to the length of fill L_f and the pressure gradient in the element through

$$p_T = \Delta p_{ends} + \left(\frac{dp}{dx} \right) L_f \qquad (6.74)$$

Here p_T is the die pressure determined with a pressure transducer, L_f a measured fill length, dp/dx the pressure gradient, and Δp_{ends} a pressure drop between the pressure transducer and the end of the screw. Δp_{ends} was determined experimentally at different screw rotation rates. Experimental screw characteristic curves are summarized in Fig. 6.55 for a polypropylene melt. These show results for forward- and backward-pumping screw elements of different design. For the backward-pumping elements, negative pressure gradients are necessary to achieve forward flow.

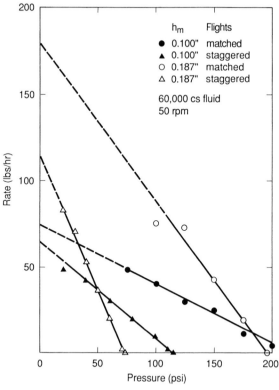

Figure 6.54   Nichols [228] studies of pressure losses in matched and staggered-screw, tangential, counter-rotating twin screw machines

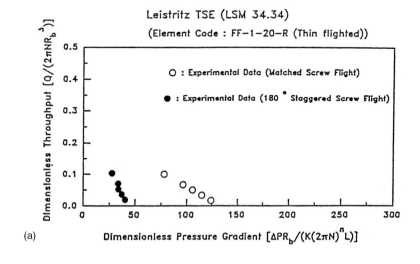

(a)

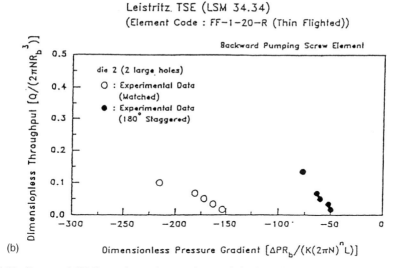

(b)

Figure 6.55   Bang et al. [225] experimental screw characteristics for Leistritz GG screw elements in tangential, counter-rotating twin screw mode for a polypropylene melt (a) Forward pumping (b) backward pumping

6.4.3   Modeling

6.4.3.1   General

Commercial tangential, counter-rotating twin screw extruders are modular machines. To understand the performance of these machines one must understand the flow both in individual screw element sections and in total modular machines.

6.4.3.2   Analytical Flow Models

The first effort at modeling flow pumping in tangential, counter-rotating twin screw extruders was by Kaplan and Tadmor [227] in 1974. This model was essentially analytical in character and considered flow in a matched flight machine. They recognized that the machine had a drag flow mechanism, unlike the intermeshing machine. While it is now generally considered outdated, the basic concepts of the model were sufficiently important to describe in Chapter 2 (Section 2.5.1). Kaplan and Tadmor consider the flow in the machine to be in two parts: (1) between screw and barrel and (2) between screws. The screw pumping characteristics were shown to be given by Eqs (2.29) to (2.30) and involve a forward drag flow and backward flows in both the screw channels and between the screws.

6.4.3.3   Quantitative Flow Models

Improved models of flow in this machine were published from the 1980s. In 1982, Nichols and Yao [229] described a modification of the Kaplan–Tadmor model involving an improved form of the Q_{leak} term, which gave a better fit of experimental data. A 1984 paper by Nichols (with the aid of J.T. Lindt) [230] gives a third modification of the Kaplan–Tadmor formulation. This again modifies the treatment of the interscrew flow term.

In 1988/89, a new generation of flow simulations for tangential, counter-rotating twin screw extrusion was initiated. In the papers by Nguyen and Lindt [231–233] finite element simulations were applied in cylindrical coordinates and used to compute velocity fields and Q–Δp screw characteristic curves for Newtonian fluids. In papers by M.H. Kim and White and their coworkers [234–237], flattened out screw and interscrew geometries were used with hydrodynamic lubrication theory to compute pressure fields, flux fields, and Q–Δp screw characteristic curves for both Newtonian and power law non-Newtonian fluids. More recently Bang and White [238–240] have simulated flow based on presuming a cylindrical coordinate system hydrodynamic lubrication theory approximation. This uses a coordinate system incorporated into one of the screw axes and a radial direction perpendicular to the axis in the direction of the barrel.

We develop the approach of Bang and White [238–240] below. The velocity field is taken to be of form

$$v = v_z(r)e_z + v_\theta(r)e_\theta + 0e_r \tag{6.75}$$

i.e., there is no radial velocity component and v_z and v_θ depend on the radial direction only. These equations of motion have the form

$$0 = \frac{\partial p}{\partial z} + \frac{1}{r}\frac{\partial}{\partial r}(r\sigma_{zr}) \tag{6.76a}$$

$$0 = \frac{1}{r}\frac{\partial p}{\partial \theta} + \frac{1}{r^2}\frac{\partial}{\partial r}(r^2\sigma_{\theta r}) \tag{6.76b}$$

where σ_{zr} and $\sigma_{\theta r}$ are shearing stresses in the z and θ directions. These are the same as Eqs. 6.68 for intermeshing, counter-rotating twin screw extruders. The shear stresses have the form

$$\sigma_{zr} = \eta\frac{\partial v_z}{\partial r} \tag{6.77a}$$

$$\sigma_{\theta r} = \eta r\frac{\partial}{\partial r}\left(\frac{v_\theta}{r}\right) \tag{6.77b}$$

Here η is the shear viscosity, which is considered constant for a Newtonian fluid and to vary with the shear rates $\partial v_z/\partial r$ and $r\partial/\partial r(v_\theta/r)$ for a non-Newtonian fluid. Analyses were presented for both Newtonian [238, 239] and power law non-Newtonian fluids [240].

The model used by Bang and White [238, 240], which builds on the earlier work of Kim and White [234–237], like Kaplan and Tadmor [227], divides the region to be analyzed into two sections: (I) between the screw and the barrel and (II) between the screws. The coordinate system was considered to be erected in the screw root. The boundary condition for the first region (I) between the screw and barrel are

$$v_z(R_s,z) = 0 \qquad v_z(R_o,z) = 0 \tag{6.78}$$

$$v_\theta(R_s,z) = 0 \qquad v_z(R_o,z) = 2\pi R_o N$$

where R_s is the radius of the screw and R_o the radius of the barrel. In the region between the screws

$$v_z(R_s,z) = 0 \qquad v_z(R_B,z) = 0 \tag{6.79}$$

$$v_\theta(R_s,z) = 0 \qquad v_\theta(R_B,z) = 0 \ \text{(screw root)}$$

$$= 2\pi(R_B - R_s)N \qquad \text{(screwflight)}$$

where R_B represents the radius required to reach the other screws. Bang and White [238–240] solved this problem for both Newtonian and non-Newtonian flow using a cylindrical coordinate balance on fluxes that is similar to Eq. 6.20. This has the form in the screw-barrel region of (compare Eq. 6.71).

$$q_z(z - \Delta z, \theta)R_{Ch}\Delta\theta + q_\theta(z,\theta - \Delta\theta)\Delta z$$

$$= q_z(z,\theta)R_{Ch}\Delta\theta + q_\theta(z,\theta)\Delta z \tag{6.80}$$

where

$$q_z = \frac{1}{R_{Ch}} \int_{R_s}^{R_B} rv_z dr$$

$$q_\theta = \frac{1}{R_{Ch}} \int_{R_s}^{R_B} v_z dr \tag{6.81}$$

are fluxes. R_{Ch} is a characteristic radius. At the entry position for the interscrew region the balance is of form

$$q_\theta^L(\theta - \Delta\theta^L, z)\Delta z + q_t^R(\theta - \Delta\theta^R, z)\Delta z$$

$$+ q_z(\theta, z - \Delta z)R_{Ch}\Delta\theta = q_\theta(\theta, z)\Delta z + q_z(\theta, z)R_{Ch}\Delta\theta \tag{6.82a}$$

q^L and q^R represent in-fluxes from the left- and right-handed screws. In the midst of the inter-screw region, we have instead

$$q(\theta - \Delta\theta, z)\Delta z + q_z(\theta, z - \Delta z)R_{Ch}\Delta\theta$$
$$= q_\theta(\theta, z)\Delta z + q_z(\theta, z)R_{Ch}\Delta\theta \qquad (6.82b)$$

There are no positive displacement terms as in Eq. 6.73b. At the exit to the interscrew region, there is a balance equation similar to Eq. (6.82a).

M.H. Kim and White [234–237] and Bang and White [238–240] have computed pressure fields and flux fields in this geometry. Typical flux fields are shown in Fig. 6.56. These compare flow fields in forward-pumping matched and staggered flight geometries. It may be seen that there are backward fluxes between the screws that are much greater for the staggered screws. This is because the interscrew clearances are generally larger and there are no restrictions as there are periodically with matched screw flight designs.

Calculations were also made by M.H. Kim and White [236] and by Bang and White [240] for backward-pumping screws in a tangential, counter-rotating twin screw extruders. The forward-pumping motion of the fluid is primarily between the screws by pressure flow. The flow in the screw channels is negative (see Fig. 6.57).

These authors have computed screw characteristic curves for Newtonian and power law fluids for both forward- and backward-pumping screws. Typical results are illustrated in Figs. 6.58 and 6.59, which show that the pumping capacity of tangential, counter-rotating machine is much less than two equivalent single screw extruders. This is

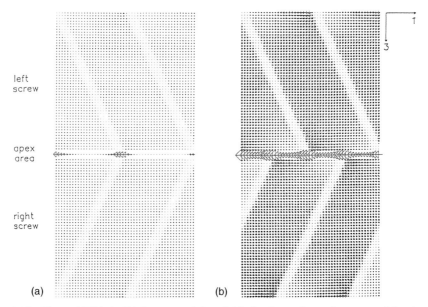

Figure 6.56   Kim and White's [236] computed flux fields for forward-pumping, tangential twin screw geometry: (a) matched;   (b) staggered

associated with backward leakage between the screws and reduced drag flow. The matched screws flights give much better pumping than staggered screw flights. Introducing non-Newtonian viscosity (shear thinning) through using a power law model reduces pumping capacity.

Bang et al. [225] have experimentally verified several of the screw characteristic curves described above.

Figure 6.57   Kim and White's [236] computed flow fields for backward-pumping, tangential twin screw geometry (a) matched screw flights (b) staggered screw flights

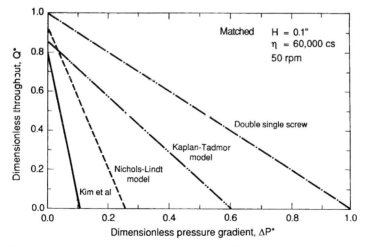

Figure 6.58   Theoretical screws characteristic curve for tangential, counter-rotating twin screw extruders – Newtonian Fluids

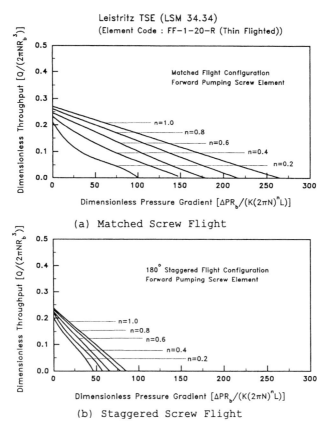

Figure 6.59   Theoretical screw characteristic curves for tangential counter-rotating twin screw extruders – power law fluids after Bang and White [239]

6.4.3.4   Composite Modular Pumping Modeling

The characteristics of modular tangential, counter-rotating twin screw extruders was first considered by White and Szydlowski [103]. More recently the behavior or modular machines of this type have been investigated both theoretically and experimentally by Bang et al. [225, 240]. The logic is very much that described in Section 6.2.4.3 for modular, co-rotating twin screw extruders.

The basic approach is to presume that the extrusion rate is known from the metered starved feeding. The die characteristics determine the pressure at the end of the screw. We then march backward from the die, element by element, toward the feed ports. The screw characteristic curves allow the calculation of the pressure profile. When the pressure falls to zero, starvation develops. Starvation continues until one reaches a nonpumping element, where the machine fills up again and becomes pressurized. We then march backward again using screw characteristic curves and starvation generally develops. Typical fill factor profiles are shown in Fig. 6.60.

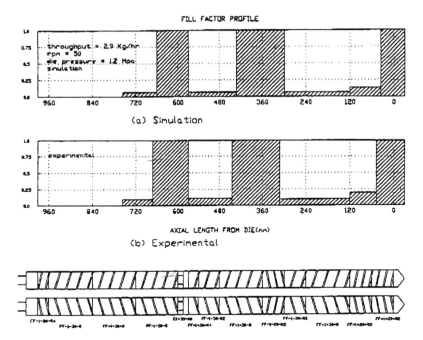

Figure 6.60   Typical experiment fill factors profile compering theory and experiment

6.4.4   Applications

These machines are used primarily for devolatilization, coagulation, reactive extrusion and complex continuous processing technologies in the polymerization industry. These are outlined in patents by Street [210–213] and Skidmore [214–219]. They tend not to be used for compounding.

6.5   Continuous Mixers

6.5.1   Technology

One generally considers continuous mixing machines with rotor sections resembling internal mixers as distinct from twin screw extruders. The history of these machines, at least in their early stages, has been more closely related to batch or internal mixers.

The first continuous mixer was proposed by Paul Pfleiderer [241] one of the principals of the new firm *Werner und Pfleiderer GmbH* soon after their founding in 1879. Pfleiderer [241, 242] had previously developed the 'sigma blade' batch mixer, which was used for baking dough [243]. This had been commercialized as the "Universal" Knet und Misch-Maschine. Pfleiderer's continuous mixer is shown in Fig. 6.61. It consisted of two counter-rotating shafts, each with screws and sigma blade sections machined along its length.

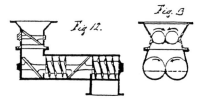

Figure 6.61   Pfleiderer's 1881 continuous mixer

A machine with some characteristics of a continuous mixers is contained in a 1924 German patent by Ahnhudt [244]. It involves two intermeshing, counter-rotating screws that lead to a pair of effectively intermeshing conical mixing rotors. This is shown in Fig. 6.62.

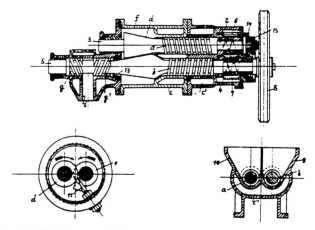

Figure 6.62   Ahnhudt's 1924 Continuous Mixer

In 1941, *Friederich Krupp and Co.* commercialized an intermeshing continuous mixer, which it called the "Knetwolf" [11] (Fig. 6.63). It was to be used for mastication of synthetic rubber. The patent was issued anonymously [245], but appears to be due to Wilhelm Ellermann [11]. The machine has an intermeshing, twin screw, counter-rotating, forward-pumping screw sections at the feed and the exit. Between these are rotor sections with nearly axial flights with the same curvature, giving slight forward-pumping capability. In the postwar period, W. Ellermann [246] developed a second intermeshing, counter-rotating mixer. This machine was commercially developed by *Josef Eck und Sohn Maschinenfabrik* of Dusseldorf [11] (see Fig. 6.64). It differs from the Knetwolf in the exit twin screw section being tangential rather than intermeshing and the flights in the rotor sections are backward pumping. A similar self-wiping, co-rotating machine was developed by Ellermann [55] in 1953 (Fig. 6.9b). The latter machine was commercially developed by *Krauss-Maffei* and is still manufactured by *Japan Steel Works*.

The most important of the continuous mixers was developed by *Farrel Corporation*. It is described in patents by E.H. Ahlefeld, A.J. Baldwin, P. Hold, W.A. Rapetzki, and H.R. Scharer [247, 248] and in later papers by Hold [249], Scharer [250], and Valsamis and

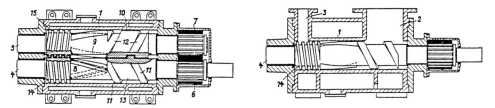

Figure 6.63   Ellermann's 1941 Krupp Knetwolf [245]

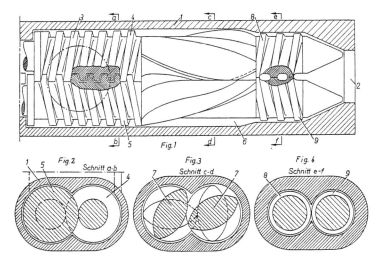

Figure 6.64   Ellermann's 1951 Eck mixtruder [246]

Canedo [251]. The machine is shown in Fig. 6.65. It involves a screw system that feeds nonintermeshing counter-rotating rotors. It was marketed by *Farrel Corp.* as the Farrel Continuous Mixer. Various rotor designs have been developed by Farrel through the years. Some of these rotors are shown in Fig. 6.66. The original intent of the *Farrel Corp.* was that the new machine would represent a continuous version of their well-established Banbury mixer and be used for continuous compounding of carbon black into rubber. However, the machine has been more widely used for compounding thermoplastics. *Farrel* in subsequent years licensed *Kobe Steel* of Kobe, Japan and *Pomini (Pomini-Farrel)* in Castellanza, Italy. These licenses have expired but both companies continue to manufacture these machines.

A second series of patents were filed from 1968 to 1974 by J.T. Matsuoka [252–257] of *Intercole Automation/Stewart Bolling,* based in Cleveland, Ohio. These patents describe a series of machines basically similar to those in the Ahlefeld et al. patents. The first patent is for a machine with two rotors of different design, one with two flights and a second with four flights (Fig. 6.67). The later patents of Matsuoka and Cantarutti deal with interfacing a single screw extruder with the mixing section of the continuous mixer, as shown in Fig. 6.68. The single screw extruder is arranged in a transverse manner and is built to directly interact with the continuous mixer.

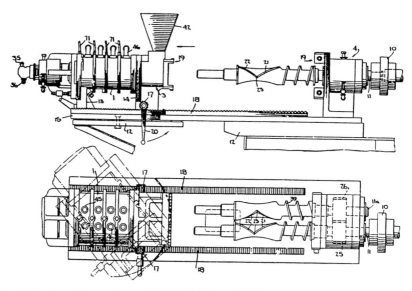

Figure 6.65   Farrel continuous mixer of 1962 Ahlefeld et al. [247]

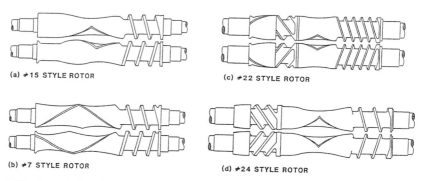

(a) ≠15 STYLE ROTOR

(c) ≠22 STYLE ROTOR

(b) ≠7 STYLE ROTOR

(d) ≠24 STYLE ROTOR

Figure 6.66   Rotor designs for Farrel continuous mixer [251]

Okada, Taniguchi, and Kamimori [258] of *Japan Steel Works* in a 1971 patent application describe a machine similar to the Farrel Continuous Mixer, but with a special barrel design and kneading region at the exit. The mixer rotor design is different and both the barrel and rotor decrease in diameter in the final section (Fig. 6.69). A machine of this type is described in a 1988 paper by Sakai [259], which deals with the grafting of maleic anhydride and styrene monomer onto ethylene-vinyl acetate copolymer.

In a 1980 patent application, Inoue et al. [260] of Kobe Steel described a new design for a continuous mixer, which is shown in Fig. 6.70. It consists of three sections: (1) forward-pumping screws that are slightly intermeshing, (2) a mixing rotor section with forward- and backward-pumping flights, and (3) a second mixing section that also has forward and backward pumping. There is a throttle between the two mixing rotor sections. The product stream exits downward in the latter part of the second rotor.

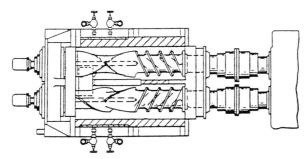

Figure 6.67   Matsuoka patent application of 1968 of a continuous mixer with two different rotors [252]

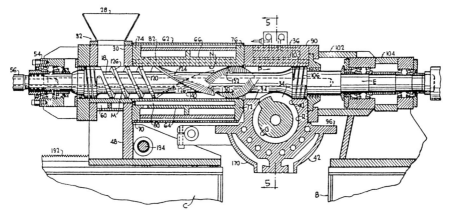

Figure 6.68   Matsuoka patent application of 1973 of a continuous mixer integrated into a transverse single screw extruder [256]

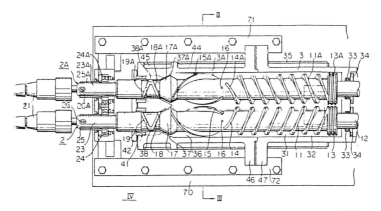

Figure 6.69   Okada, Taniguchi, and Kamimori (Japan Steel Works) continuous mixer [258]

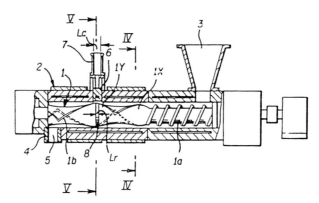

Figure 6.70   Inoue et al. (Kobe Steel) continuous mixer with two different rotor sections; basis of Kobe Steel L-series continuous mixer [260]

Farrel engineers have filed in more recent years patents related to improvements in the design and operation of continuous mixers. These introduce gear pumps to control the polymer stream exiting the machine and to control systems [261, 262].

6.5.2   Experimental

There have been few experimental studies of flow and mixing in a Farrel Continuous Mixer or similar continuous mixers. The first paper presenting any experimental data on this type of machine was published by Valsamis and Canedo [251] in 1989. They show a solidified carcass of a polyethylene/polystyrene blend removed from the machine plus cross sections of this carcass. They note that mean residence times are of order 10 to 25 s. An experimental residence time distribution plot was presented and was contrasted to dimensionless experimental results for a modular, co-rotating twin screw extruder.

More recently, Galle and White [263, 264] have presented an experimental study of flow, material distribution, residence time distribution, and blending in a Kobe Steel Nex T continuous mixer, which is a machine of similar but different detailed design from the Farrel Continuous Mixer. They found that the residence time distribution was strongly dependent on throughput and would approach the very low residence time values quoted by Valsamis and Canedo only at very high throughputs. The rotor used by these authors had a screw section, a forward-pumping rotor, a backward-pumping rotor, and a neutral rotor. The two later rotor sections were usually filled with melt and the screw section contained pellets. Melting occurred in the forward-pumping rotor section, which is also where most blending action took place. For high feed rates the beginning of fusion will occur in the screw section. Increasing the screw speed increases the rate of melting and gives a longer melt homogenization region.

Chang and White [264a] have described maleation of polypropylene in a Kobe Steel Nex T continuous mixer and compared it with results from various twin screw extruders.

6.5.3   Modeling

There have been two efforts at simulating flow in a Farrel Continuous Mixer. M.H. Kim and White [265] describe a Cartesian coordinate non-Newtonian hydrodynamic lubrication theory model of a fully filled Farrel Continuous Mixer. Bang and White [266] more recently have used a cylindrical coordinate non-Newtonian hydrodynamic lubrication theory model for a starved machine.

The simulation of Bang and White [266] for the Farrel Continuous Mixer is based on solving Eq. (6.81) through Eq. (6.88) in a power law fluid form. They determine the pressure and flux fields within the rotor region. Fluid is dragged over the barrel by the motions of the rotor flights and flows over the barrel. The direction of drag must be balanced by pressure gradients to maintain a constant flow rate. This leads to positive pressure gradients in forward-pumping sections and negative pressure gradients in backward- and neutral-pumping sections. Bang and White [266] presume a rotor consisting of a screw section, followed by a forward-pumping rotor section, a backward-pumping section, and then a neutral section. They calculate screw characteristic pumping curves for each of the rotor sections, as shown in Fig. 6.71. Such a design leads to the pressure profiles shown in Fig. 6.72.

Bang and White [266] consider the starved-flow characteristics of a Farrel Continuous Mixer. Using a specified pressure at the gate at the exit of the machine and the screw pumping characteristics curves of Fig. 6.72, they compute the pressure profile along the machine axis. When the pressure falls to zero, starvation is predicted to occur. They predict from their simulation that the level of starvation increases with increasing rotor speed and with decreasing pressure at the gate.

More recently Ishikawa et al. [267] have described a finite element simulation of flow in this type of machine, in particular for a Kobe Steel design. Velocity profiles, pressure fields and temperature distributions are calculated.

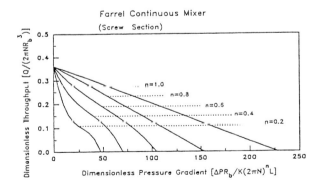

(a) Screw Element Section

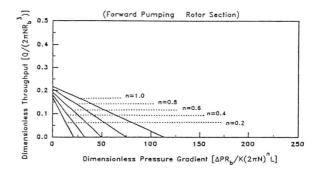

(b) Forward Pumping Rotor Section

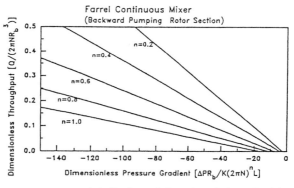

(c) Backward Pumping Rotor Section

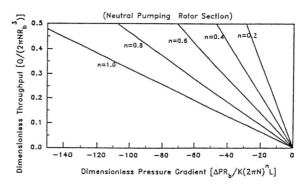

(d) Neutral Pumping Rotor Section

Figure 6.71   Screw pumping characteristic curves for the different section of a Farrel continuous mixer according to Bang and White [266]

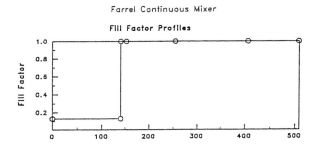

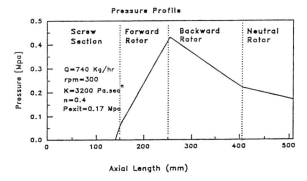

Figure 6.72   Predicted pressure profile for flow in a Farrel continuous mixer according to Bang and White [266]

6.5.4   Applications

These machines are used primarily for compounding thermoplastics such as polyolefins or styrenies with mineral filters such as calcium carbonate and talc. In the polymerization industry, they are used to compound antioxidants into newly polymerized thermoplastics. They may also be used for grafting monomers such as maleic anhydride onto thermoplastics.

References

1. Wunsche, A., German Patent 131, 392 (1901)
2. Easton, R.N., British Patent (filed 25 September 1916) 109, 663 (1917)
3. Easton, R.N., U.S. Patent (filed 2 June 1920) 1, 468, 379 (1923)
4. Colombo, R., Italian Patent (filed 6 February 1939) 370, 578 (1939)
5. Colombo, R., British Patent (filed 2 May 1946) 629, 109 (1949)
6. Colombo, R., U.S. Patent (filed 7 August 1947) 2, 563, 396 (1951)
7. Colombo, R., Canadian Patent (filed 27 December 1949) 517, 911 (1955)
8. Greenwood, S.H., *Rubber World* (1953) 129, p. 73
9. White, J.L., *Twin Screw Extrusion: Technology and Principles* (1990) Carl Hanser Verlag, Munich
10. Erdmenger, R., *Chem. Ing. Tech.* (1964) 36, p. 175
11. Herrmann, H., *Schneckenmaschinen in der Verfahrenstechnik* (1972) Springer, Berlin
12. Erdmenger, R., *Schneckenmaschinen für die Hochviskose-Verfahrenstechnik* (1978) Bayer AG, Leverkusen

13. Meskat, W. German Patent (filed 17 October 1943) 852, 203 (1952)
14. Meskat, W., Erdmenger, R., German Patent (filed 7 July 1944) 862, 668 (1953)
15. Meskat,W., Erdmenger, R., German Patent (filed 28 July 1944) 872, 732 (1953)
16. Erdmenger, R., German Patent (filed 24 September 1949) 815, 641 (1951)
17. Erdmenger, R., German Patent (filed 29 September 1949) 813, 154 (1951)
18. Telle, O., Erdmenger, R., German Patent (filed 22 July 1953) 945, 086 (1956)
19. Erdmenger, R., German Patent (filed 28 July 1953) 940, 109 (1956)
20. Nelson, W.K., U.S. Patent (filed 4 June 1931) 1, 868, 671 (1932)
21. Erdmenger, R., U.S. Patent (filed 20 September 1950) 2, 670, 188 (1954)
22. Riess, K., Erdmenger, R., *VDI Zeitschr.* (1951) 93, p. 633
23. Meskat, W., Pawlowski, J., German Patent (filed 10 December 1950) 949, 162 (1956)
24. Riess, K., Meskat,W., *Chem. Ing. Tech.* (1951) 23, p. 205
25. Riess, K., *Chem. Ing. Tech.* (1955) 27, p. 457
26. Erdmenger, R., U.S. Patent (filed 17 August 1959) 3, 122, 356 (1964)
27. Erdmenger, R., *Chem. Ing. Tech.* (1962) 34, p. 751
28. Fritsch, R., Fahr, G., Kunststoffe (1959) 49, p. 543
29. Herrmann, H., *Kunststoffe und Gummi* (1964) 3, p. 217
30. Herrmann, H., *Chem. Ing. Tech.* (1966) 38, p. 25
31. Boden, H., Ocker, H., Pfaff, G., Wotz, W., U.S. Patent (filed 13 November 1964) 3, 305, 894 (1967)
32. Fritsch, R., U.S. Patent (filed 24 February 1966) 3, 456, 317 (1969)
33. Fritsch, R., Kuhner, H.H.O., U.S. Patent (filed 23 February 1964) 3, 392, 962 (1968)
34. Ocker, H., U.S. Patent (filed 15 November 1968) 3, 525, 124 (1970)
35. Koch, H., U.S. Patent (filed 20. November 1968) 3, 608, 868 (1971)
36. Ocker, H., U.S. Patent (filed 18. November 1970) 3, 682, 086 (1972)
37. Ocker, H., U.S. Patent (filed 7 September 1971) 3, 764, 114 (1973)
38. Herrmann, H., Ocker, H., U.S. Patent (filed 7 February 1972) 3, 749, 375 (1973)
39. Loomans, B.A., Brennan, A.K., Jr., U.S. Patent (filed 21 March 1962) 3, 195, 868 (1965)
40. Loomans, B.A., Brennan, A.K., Jr., U.S. Patent (filed 21 March 1962) 3, 198, 491 (1965)
41. Brennan, A.K., U.S. Patent (filed 14 November 1969) 3, 618, 902 (1971)
42. Anonymous, German Patent (filed 24 July 1943) 895, 058 (1953)
43. Illing, G., Zahradnik, F., U.S. Patent (filed 21 July 1964) 3, 371, 055 (1968)
44. Illing, G., U.S. Patent (filed 23 October 1964) 3, 536, 680 (1970)
45. Illing, G., *Mod. Plastics (Ava)* (1959) 72
46. Wheeler, D.A., Irving, H.F., Todd, D.B., U.S. Patent (filed 30 October 1969) 3, 630, 689 (1971)
47. Erdmenger, R., Ullrich, M., Germandonk, R., Pedain, J., Quiring, B., Wingler, F., U.S. Patent (filed 21 January 1971) 3, 725, 340 (1973)
48. Ullrich, M., Meisert, E., Eitel, A., U.S. Patent (filed 17 July 1974) 3, 963, 679 (1976)
49. Erdmenger, R., U.S. Patent (filed 26 March 1963) 3, 254, 367 (1966)
50. Loomans, B.A., U.S. Patent (filed 6 March 1973) 3, 714, 350 (1975)
51. Colombo, R., U.S. Patent (filed 27 December 1961), 114, 171 (1963)
52. Todd, D.B., U.S. Patent (filed 27 July 1977) 4, 136, 968 (1979)
53. Kraffe de Laubarede, L., U.S. Patent (filed 10 July 1950) 2, 631, 016 (1953)
54. Kraffe de Laubarede, L., British Patent (filed 22 April 1953) 738, 784 (1955)
55. Ellermann,W., German Patent (filed 31July 1953) 935, 634 (1955); U.S. Patent (filed 12 October, 1953) 2, 693, 348 (1954)
56. Proksch,W., *Kunststoffe und Gummi* (1964) 3, p. 426
57. White, J.L., in *Polymer Mixing: Technology and Engineering* edited by White, J.L., Coran, A.Y., Moet. A., Carl Hanser Verlag, Munich (2001)
58. Colombo, R., U.S. Patent (filed 15 January 1964) 3, 252, 182 (1966)
59. Rathjen, C., Ullrich, M., European Patent Application (filed 10 February1981) 049, 835 (1982)
60. Burbank, F., Brauer, F., Anderson, P., *SPE Antec Tech. Papers* (1991) 37, p. 149
61. Grillo, J., Anderson, P., Papazoglu, E., *SPE Antec Tech. Papers* (1992) 38, p. 20
62. Ahlefeld, E.H., Baldwin, A.J., Hold, P., Rapetzki,W.A., Scharer, H.R., U.S. Patent (filed 15 May 1962) 3, 154, 808 (1964)
63. Valsamis, L.C., Canedo, E., *Int. Polym. Process* (1994) 9, p. 225

64. Zimmermann, H.G., U.S. Patent 3, 170, 566 (1965)
65. Booy, M.L., *Polym. Eng. Sci.* (1978) 18, p. 978
66. Todd, D.B., Irving, H.F., *Chem. Eng. Proc.* (1969) 65 (9) p. 85
67. Todd, D.B., *Polym. Eng. Sci.* (1925) 15, p. 437
68. Armstroff, O., Zettler, H.D., *Kunststofftechnik* (1973) 12, p. 240
69. Sakai, T., *Gosei Jushi* (1978) 24, p. 7
70. Meijer, H.E.H., Elemans P.H. M., *Polym. Eng. Sci.* (1988) 28, p. 275
71. Kim, P.J., White, J.L., *Int. Polym. Process* (1994) 9, p. 108
72. Werner, H., Dr.-Ing. Dissertation, University of Munich (1976)
73. Nichols, K.L., Jayaraman, K., Grulke, E.G., Papers presented at the 4th annual Polymer Processing Society Meeting, Orlando, May (1988)
74. Wang, Y., White, J.L., Szydlowski, W., *Int. Polym. Process* (1989) 4, p. 262
75. White, J.L., Montes, S., Kim, J.K., *Kautschuk Gummi Kunststoffe* (1990) 43, p. 20
76. Bawiskar, S., White, J.L., *Int. Polym. Process* (1995) 10, p. 105, Paper presented at the 10th annual Polymer Processing Society meeting Akron, March (1994)
77. Potente, H., Melisch, U., *Int. Polym. Process* (1996) 11. p. 101
78. Todd, D.B., *Int. Polym. Process* (1993) 8, p. 113
79. Bawiskar, S., White, J.L., *Polym. Eng. Sci.* (1998) 38, p. 727
80. Curry, J., *SPE Antec Tech. Papers* (1995) 41, p. 92
81. Hornsby, P.R., *Plastics Compounding* (1983) (Sept./Oct.) 65
82. Bur, A.J., Gallant, F.N., *Polym. Eng. Sci.* (1991) 31, p. 1365
83. Kye, H., White, J.L., *J. Appl. Polym. Sci.* (1994) 52, p. 1249
83a Shon, K.J., Chang, D. and White, J.L. *Int. Polym. Process* (1999) 14, p. 35
84. Todd, D.B., *SPE Antec Tech. Papers* (1989) 35, p. 168; *Int. Polym. Process* (1991) 6, p. 143
85. Gogos, C.G., Esseghir, M., Yu, D.W., Todd, D.B., *SPE Antec Tech. Papers* (1994) 40, p. 270
86. Wang, N.H., Sakai,T., Hashimoto, N., *Int. Polym. Process* (1998) 13, p. 27
87. Todd, D.B., *SPE Antec Tech. Papers* (1988) 34, p. 58
87a. White, J.L., Kim, E.K., Keum, J.M., Jung, H., *Polym. Eng. Sci.* (2001) 41, p. 1448
88. Bigio, D., Erwin, L., *SPE Antec Tech. Papers* (1985) 31, p. 45
89. Ess, J.W., Hornsby, P.R., *Polym. Testing* (1986) 6, p. 205
90. Ess, J.W., Hornsby, P.R., *Plastics Rubber Process* (1987) 8, p. 147
91. Kalyon, D.M., Sangani, H.N., *Polym. Eng. Sci.* (1989) 29, p. 1018
92. Sinton, S.N., Crowley, J.C., Lo, G.A., Kalyon, D.M., Jakob, C., *SPE Antec Tech. Papers* (1990) 36, p. 116
93. Plochocki, A.P., Dagli, S.S., Mack, M.H., *Kunststoffe* (1988) 78, p. 3
94. Lim, S., White, J.L., *SPE Antec Tech. Papers* (1992) 38, p. 2682 *Int. Polym. Process* (1993) 8, p. 119
95. Sundaraj, U., Macosko, C.W., Rolando, R.J., Chan, H.T., *Polym. Eng. Sci.* (1992) 32, pp. 2682, 1814
96. Bordereau, V., Shi, Z.H., Utracki, L.A., Sammut, P., Carrega, M., *Polym. Eng. Sci.* 32, (1992) 32, p. 1846
97. Lim, S., White, J.L., *Polym. Eng. Sci.* (1994) 34, p. 221
98. Kim, J.K., Lee, S.C., Park, H.K., *Int. Polym. Process* (1995) 10, p. 19
99. Cho, J.W., White, J.L., *SPE Antec Tech. Papers* (1995) 41, p. 321; *Int. Polym. Process* (1996) 11, p. 21
100. Lee, S.H., White, J.L., *SPE Antec Tech. Papers* (1996) 42, p. 316; *Int. Polym. Process* (1997) 12, p. 316
101. Herrmann, H., Burkhardt, U., In *The 5th Leobener Kunststoffkolloquium Doppelschnelken-Extruder* (1978) Lorenz Verlag, Vienna, p. 11
102. Denson, C.D., Hwang, B.K., *Polym. Eng. Sci.* (1980) 20, p. 965
103. White, J.L., Szydlowski, W., *Adv. Polym. Technol.* (1987) 7, p. 419
104. Chen, Z., White, J.L., *SPE Antec Tech. Papers* (1992) 38, p. 1332; White, J.L., Chen, Z., *Polym. Eng. Sci.* (1994) 34, p. 229
105. Chen, Z., White, J.L., *SPE Antec Tech. Papers* (1993) 39, p. 3401; *Int. Polym. Process* (1994) 9, p. 310
106. Bawiskar, S., White, J.L., *Int. Polym. Process* (1997) 12, p. 331
107. Potente, H., Ansahl, J., Wittemaier, R., *Int. Polym. Process* (1990) 5, p. 208
107a Potente, H., Ansahl , J., Klarholz, B., *Int. Polym. Process* (1994) 9, p. 11
108. Szydlowski, W., Brzoskowski, R., White, J.L., *Int. Polym. Process* (1987) 1, p. 207
109. Szydlowski, W., White, J.L., *Int. Polym. Process* (1988) 2, p. 142
110. Gotsis, A.D., Kalyon, D.M., *SPE Antec Tech. Papers* (1989) 35, p. 44

111. Szydlowski,W., White, J.L., *Adv. Polym. Technol.* (1987) 7, p. 177
112. Tadmor, Z., Broyer, E., Gutfinger, C., *Polym. Eng. Sci.* (1974) 14, p. 660
113. Wang, Y., White, J.L., *J. Non-Newt. Fluid Mech.* (1989) 32, p. 19
114. Griffith, R.M., *IEC Fund* (1962) 1, p. 180
115. Lai-Fook, R.A., Senouci, A., Smith, A.C., Isherwood, D.P., *Polym. Eng. Sci* (1989) 29, p. 433
116. Kalyon, D.M., Gotsis, A.D., Yilmazer,U., Gogos, C.G., Sangani, H., Aral, B., Tsenoglu, C., *Adv. Polym. Technol.* (1988) 8, p. 337
117. Lai-Fook, R.A., Li, Y., Smith, A.C., *Polym. Eng. Sci.* (1991) 31, p. 1157
118. Szydlowski,W., White, J.L., *J. Non-Newt. Fluid Mech.* (1988) 28, p. 29
119. Yang, H.H., Manas-Zloczower, I., *Polym. Eng. Sci.* (1992) 32, p. 1411
120. Lawal, A., Kalyon, D.M., *Polym. Eng. Sci.* (1995) 35, p. 1325
121. Cheng, H., Manas-Zloczower, I., *Polym. Eng. Sci.* (1997) 37, p. 1087
122. Ishikawa, T., Kihara, S. and Funatsu, K., *Polym. Eng. Sci.* (2000) 40, p. 357
123. Bird, R.B., Stewart, W.E., Lightfoot, E.N., *Transport Phenomena* (1960) Wiley, New York
124. Potente, H., Melisch, U., Fleche, J., *SPE Antec Tech. Papers* (1996) 42, p. 334
124a. Qian, B., Gogos, C. and Todd, D.B., *Paper presented at PPS-17 Polymer Processing Society Annual Meeting* Montreal 2001
124b. Jung, H. and White, J.L., *SPE Antec Tech Papers* (2002) 48 (in press)
125. Gotis, A.D., Ji, Z., Kalyon, D.M., *SPE Antec Tech. Papers* (1990) 36, p. 139
126. Shi, Z.H., Utracki, L.A., *Polym. Eng. Sci.* (1992) 32, p. 1834
127. Huneault, M.A., Shi, Z.H., Utracki, L.A., *Polym. Eng. Sci.* (1995) 35, p. 115
128. Delamare, L., Vergnes, B., *Polym. Eng. Sci* (1996) 36, p. 1685
129. Lee, S.H., White, J.L., *Int. Polym. Process* (1998) 13, p. 247
130. Kim, P.J., White, J.L., *Int. Polym. Process* (1994) 9, p. 33
131. Kye. H., White, J.L. *Int. Polym. Process* (1996) 11, p. 129
132. Kim, B.J., White, J.L., *Polym. Eng. Sci.* (1997) 37, p. 576
133. Berzin, F., Vergnes.B., *Int. Polym. Process* (1998) 13, p. 13
133a. Poulesquen, A., Vergnes, B., Cassagnau, Ph., Giminez, J., Michel, A., *Int. Polym. Process* (2001) 16, p. 31
134. Wiegand, S.L., U.S. Patent (filed 28 April 1874) 155, 602 (1874)
135. Werner, K., German Patent (filed 12 July 1912) 281, 104 (1912)
136. Holdaway. W.S., U.S. Patent (filed 2 June 1915) 1, 218, 602 (1917)
137. Soc. Anonym. Est., Olier, A., British Patent (filed 8 April 1922) 180, p. 638 (1922)
138. Montelius, C.O.J., U.S. Patent (filed 20 March 1925) 1, 698, 802 (1929)
139. Leistritz, P., Burghauser, F., German Patent (filed 24 April 1926) 453, 727 (1927)
140. Montelius, C.O.J., U.S. Patent (filed 1 March 1929) 1, 965, 557 (1934)
141. Schmidt, R., U.S. Patent (filed 6 June 1930) 1, 846, 692 (1932)
142. Ungar, G.A., U.S. Patent (filed 11 June 1930) 1, 846, 700 (1930)
143. Kiesskalt, S., *Z. VDI* (1927) 71, p. 453
144. Montelius, C., *Teknisk Tidskrift-Mekanik* (1933) 61 (6), p. 9
145. Kiesskalt, S., Tampke, H., Winnacker, K., Weingaertner, E., German Patent (filed 26 July 1925) 652, 990 (1937)
146. Leistritz, P., Burghauser, F., German Patent (filed 1 December 1935) 682, 787 (1939)
147. Kiesskalt, S., Tampke, H., Winnacker, K., Weingaertner, E., German Patent (filed 25 May 1935) 676, 045 (1939)
148. Kiesskalt, S., Tampke, H., Winnacker, K., Weingaertner, E., German Patent (filed 14 June 1935) 690, 829 (1940)
149. I.G. Farbenindustrie, Italian Patent (filed 6 April 1939) 373, 183 (1939)
150. Kiesskalt, S., *VDI Z.* (1942) 86, p. 752
151. Kiesskalt, S., *VDI Z.* (1950) 92, p. 551
152. Kiesskalt, S., *Kunststoffe* (1951) 41, p. 414
153. Steinmann, H., Heyne, F., German Patent (filed 1 April 1944) 846, 012 (1952)
154. Schaerer, A.J., *Kunststoffe* (1954) 44, p. 105
155. Baigent, K., *Trans. Plastics Inst.* (1956) p. 134
156. Schutz, F.G., *SPE J.* (1962) p. 1147
157. Pasquetti, C., British Patent (filed 7 June 1950) 677, 945 (1952)

158. Prause, J.J., *Plastics Technology* (1968) (March), p. 52
159. Zielonowski, W., *Kunststoffe* (1978) 58, p. 394
160. Anonymous, German Patent (filed 31 January 1941) 750, 509 (1945)
161. Ellermann, W., German Patent (filed 4 July 1951) 879, 164 (1953)
162. Sennet, M.B., U.S. Patent (filed 14 September 1949) 2, 581, 451 (1952)
163. Hack, E., German Patent (filed 2 August 1949) 815, 103 (1951)
164. Tenner, H., *Kunststoffberater* (1976) 6
165. Tenner, H., *Kunststoff J.* (1987) 12, p. 102
166. Tenner, H., *Plastuerarbeiter* (1989) 40, 11
167. Thiele, W.C., *SPE Antec Tech. Papers* (1983) 29, p. 127
168. Thiele, W.C., Petrozelli, W., Loring, D., *SPE Antec Tech. Papers* (1990) 36, p. 120
169. Thiele, W.C., Petrozelli, W., Martin, C., *SPE Antec Tech. Papers* (1991) 37, p. 1849
170. Lim, S.H., White, J.L., *Int. Polym. Process* (1994) 9, p. 33
171. Sakai, T., Hashimoto, N., *SPE Antec Tech. Papers* (1986) 32, p. 360
172. Sakai, T., Hashimoto, N., Kobayoshi, N., *SPE Antec Tech. Papers* (1987) 33, p. 146
173. Thiele, W.C., In *Plastics Compounding: Equipment and Processing* Todd, D.B. (Ed.) (1998) Carl Hanser Verlag, Munich
174. Jewmenow, S.D., Kim, W.S., *Plaste Kautschuk* (1973) 20, p. 356
175. Janssen, L.P.B.M., Smith, J.M., *Proceedings of the congress on Polymer Rheology and Plastics Processing* (1975) PRI/BSR Loughborough p.160, Sept.
176. Janssen, L.P.B.M., *Plastics and Rubber: Processing* (1976) p. 90
177. Janssen, L.P.B.M. *Twin Screw Extrusion* (1978)Elsevier, Amsterdam
178. Janssen, L.P.B.M., Hollander, R.N., Spoor, M.W., Smith, J.M., *AIChE J.* (1979) 25, p. 345
179. Rauwendaal, C., *SPE Antec Tech. Papers* (1981) 38, p. 618
180. Wolf, D., Holen, W., White, D.H., *Polym. Eng. Sci.* (1986) 26, p. 640
181. Potente, H., Schultheis, S.M., *Int. Polym. Process* (1989) 4, p. 247
182. Kim, B.J., White, J.L., *Int. Polym. Process* (1995) 10, p. 213
183. Shon, K.J., Chang, D.H., White, J.L., *Int. Polym. Process* (1999) 14, p. 44
184. Doboczky, Z., *Plastverarbeiter* (1965) 16, p. 57
185. Doboczky, Z., *Plastverarbeiter* (1965) 16, p. 395
186. Menges, G., Klenk, K.P., *Plastverarbeiter* (1966) 17, p. 791
187. Klenk, K.P., *Plastverarbeiter* (1971) 22, p. 33 *Plastverarbeiter* (1971) 22, p. 105
188. Janssen, L.P.B.M., Pelgrom, J.L., Smith, J.M., *Kunststoffe* (1976) 66, p. 724
189. Cho, J.W., White, J.L., *Int. Polym. Process* (1996) 11, p. 21
189a. Wilczynski, K. and White, J.L., *Int. Polym. Process* (2002) 16, p. 257
190. Shon, K.J., White, J.L., *Polym. Eng. Sci.* (1999) 39, p. 7757
191. Schenkel, G., *Kunststoff Extrudertechnik* (1963) Carl Hanser Verlag, Munich
192. Janssen, L,P.B.M., Mulders, L.P.H.R.M., Smith, J.M., *Plastics Polym.* (1975) June, p. 93
193. Klenk, K.P., *Plastverarbeiter* (1971) 22, p. 189
194. Speur, J.A., Mavridas, H., Vlachopoulus, J., Janssen, L.P.B.M., *Adv. Polym. Tech.* (1987) 7, p. 30
195. White, J.L., Adewale, A.O., *Int. Polym. Process* (1993) 8, p. 210
196. Ardichvilli, G., *Kautschuk* (1938) 14, p. 23
197. Gaskell, R.E., *J. Appl. Mech.* (1950) 17, p. 334
198. Hong, M.H., White J.L., *Int. Polym. Process* (1998) 13, p. 342
199. Hong, M.H., White, J.L., *Int. Polym. Process* (1999) 14, p. 36
200. Li, T., Manas-Zloczower, I., *Polym. Eng. Sci.* (1994) 34, p. 551
201. White, J.L., Adewale, A.O., *Int. Polym. Process* (1995) 10, p. 15
201a. Lee, B.H. and White J.L., *Int. Polym. Process* (2001) 16, p. 172
202. Ahnhudt, H., German Patent (filed 19 May 1923) 397, 961 (1924)
203. Loomis, E.G., U.S. Patent (filed 11 October 1929) 1, 990, 555 (1935)
204. Fuller, L.J., U.S. Patent (filed 10 October 1943) 2, 441, 222 (1948)
205. Fuller, L.J., U.S. Patent (filed 15 May 1945) 2, 615, 199 (1952)
206. Anonymous, *Mod. Plastics* (1949) June, p. 87
207. Street, L.F., *Rubber World* (1950) 123, p. 58
208. Street, L.F., *Rubber and Plastics Age* (1960) p. 1519

209. Schultz, F.C., *SPE J.* (1962) 18, p. 1147
210. Street, L.F., U.S. Patent (filed 16 August 1952) 2, 733, 051 (1956)
211. Street, L.F., U.S. Patent (filed 23 February 1954) 3, 068, 514 (1962)
212. Street, L.F., U.S.Patent (filed 15 August 1957) 3, 085, 288 (1963)
213. Street, L.F., U.S.Patent (filed 31 March 1960) 3, 078, 511 (1963)
214. Skidmore, R.H., U.S. Patent (filed 28 December 1959) 3, 082, 816 (1963)
215. Skidmore, R.H., U.S. Patent (filed 8 April 1970) 3, 742, 093 (1973)
216. Skidmore, R.H., U.S. Patent (filed 27 September 1973) 3, 917, 507 (1975)
217. Skidmore, R.H., U.S. Patent (filed 13 December 1974) 3, 993, 292 (1976)
218. Skidmore, R.H., U.S. Patent (filed 29 August 1978) 4, 110, 843 (1978)
219. Skidmore, R.H., U.S. Patent (filed 13 February 1978) 4, 148, 991 (1979)
220. Sakai, T., *Kobunshi Ronbunshu* (1984) 38, p. 279
221. Bigio, D., Penn, D., *SPE Antec Tech. Papers* (1988) 34, p. 59
222. Bigio, D., Conner, J., Vashihat, A., *SPE Antec Tech. Papers* (1989) 35, p. 133
223. Min, K., Kim, M.H., White, J.L., *Int. Polym. Process* (1988) 3, p. 165
224. Nichols, R.J., Personal communication, June 1988
225. Bang, D.S., Hong, M.H., White, J.L., *Polym. Eng. Sci.* (1998) 38, p. 485
226. Bigio, D., Wigginton, M., *SPE Antec Tech. Papers* (1990) 36, p. 1905
227. Kaplan, K., Tadmor, Z., *Polym. Eng. Sci.* (1974) 14, p. 58
228. Nichols, R.J., *SPE Antec Tech. Papers* (1983) 29, p. 130
229. Nichols, R.J., Yao, J., *SPE Antec Tech. Papers* (1982) 28, p. 416
230. Nichols, R.J., *SPE Antec Tech. Papers* (1984) 30, p. 6
231. Nguyen, K., Lindt, J.T., *SPE Antec Tech. Papers* (1988) 34, p. 93
232. Nguyen, K., Lindt, J.T., *SPE Antec Tech. Papers* (1989) 35, p. 145
233. Nguyen, K., Lindt, J.T., *Polym. Eng. Sci.* (1989) 29, p. 709
234. Kim, M.H., Szydlowski, W., White, J.L., Min, K., *Adv. Polym. Technol.* (1989) 9, p. 87
235. Kim, M.H., White, J.L., *SPE Antec Tech. Papers* (1989) 35, p. 49
236. Kim, M.H., White, J.L., *Int. Polym. Process* (1990) 5, p. 201
237. Kim, M.H., White, J.L., *J. Non-Newt. Fluid Mech.* (1990) 37, p. 37
238. Bang, D.S., White, J.L., *SPE Antec Tech. Papers* (1993) 39, p. 2763
239. Bang, D.S., White, J.L., *Int. Polym. Process* (1996) 11, p. 109
240. Bang, D.S., White, J.L., *SPE Antec Tech. Papers* (1997); 43. *Int. Polym. Process* (1997) 12, p. 278
241. Pfleiderer, P., German Patent (filed 10 June 1881) 18, 797 (1882)
242. Pfleiderer, P., German Patent (filed 4 April 1879) 10, 164, (1880)
243. White, J.L. *Rubber Processing: Technology, Materials and Principles* (1995) Hanser Munich
244. Ahnhudt, H., German Patent (filed 29 May 1923) 397, 961 (1924)
245. Anonymous, German Patent (filed 31 January 1941) 750, 509 (1945)
246. Ellermann, W., German Patent (filed 4 July 1951) 879, 164 (1953)
247. Ahlefeld, E.H., Baldwin, A.J., Hold, P., Rapetzki, W.A., Scharer, H.R., U.S. Patent (filed 15 May 1962) 3, 154, 808 (1964)
248. Ahlefeld, E.H., Baldwin, A.J., Hold, P., Rapetzki, W.A., Scharer, H.R., U.S. Patent (filed 24 July 1964) 3, 239, 878 (1966)
249. Hold, P., *Adv. Polym. Technol.* (1964) 4, p. 281
250. Scharer, H.R., *Adv. Polym. Technol.* (1985) 5, p. 65
251. Valsamis, L.N., Canedo, E.L., *Int. Polym. Process* (1989) 4, p. 247
252. Matsuoka, J.T., U.S. Patent (filed 5 September 1968) 3, 565, 403 (1971)
253. Matsuoka, J.T., Cantarutti, A., U.S. Patent (filed 19 July 1969) 3, 700, 374 (1972)
254. Matsuoka, J.T., Cantarutti, A., U.S. Patent (filed 4 January 1971) 3, 723, 039 (1972)
255. Matsuoka, J.T., U.S. Patent (filed 17 February 1972) 3, 764, 118 (1973)
256. Matsuoka, J.T., U.S. Patent (filed 7 May 1973) 3, 829, 067 (1973)
257. Matsuoka, J.T., Cantarutti, A., U.S. Patent (filed 12 September 1974) 3, 923, 291 (1975)
258. Okada, T., Taniguchi, K., Kamimori, K., U.S. Patent (filed 17 October 1971) 3, 802, 670 (1974)
259. Sakai, T., *SPE Antec Tech. Papers* (1988) 34, p. 1853
260. Inoue, K., Ogawa, K., Fukii, T., Asai, T., Hashizume, S., U.S. Patent (filed 27 March 1980) 4, 332, 481 (1982)

261. Scharer, H. R., D'Amato, S.A., Hold, P., Hobner, M., U.S. Patent (filed 5 November 1976) 4, 310, 251 (1982)
262. Valenzky, D., Markhardt, G.T., U.S. Patent (filed 22 November 1985) 4, 707, 139 (1987)
263. Galle, F., White, J.L., *SPE Antec Tech. Papers* (1997) 43, p. 271
264. Galle, F., White, J.L., *Int. Polym. Process* (1999) 14, p. 241
264a. Chang, D. and White J.L., *Polym. Eng. Sci.* (in press)
265. Kim, M.H., White, J.L., *Int. Polym. Process* (1992) 7, p. 15
266. Bang, D.S., White, J.L., *Polym. Eng. Sci.* (1997) 37, p. 1210
267. Ishikawa, T., Kihara, S., Funatsu, K., Amaiwa, T. and Yano K., *Polym. Eng. Sci.* (2000) 40, 365

Subject Index

Name-Author Index